Giardia lamblia in Water Supplies— Detection, Occurrence, and Removal

AN AWWA TECHNICAL RESOURCE BOOK

 American Water Works Association

ISBN 0-89867-331-3

Table of Contents

Foreword v

INTRODUCTION

Review of Giardiasis in Water Supply
  Giardia *Task Group, Research Committee, Rocky Mountain Section, AWWA* 3

Giardia lamblia and Water Supply
  Shun Dar Lin 20

METHODS FOR DETECTING *GIARDIA*

Giardia Methodology for Water Supply Analysis
  Frank W. Schaefer III and Eugene W. Rice 31

A Reference Method for Detecting *Giardia* Cysts in Water
  Walter Jakubowski 36

Detection and Identification of *Giardia* Cysts Using Immunofluorescence
and Phase Contrast Microscopy
  Judith A. Sauch 42

Giardia Methods Workshop
  Walter Jakubowski 49

Determination of the Use of Solid Particle Samplers for *Giardia* Cysts in Natural Waters
  William S. Brewer 61

GIARDIASIS OUTBREAKS AND *GIARDIA* OCCURRENCE

Waterborne Disease Outbreaks—1946–1980: A Thirty-Five-Year Perspective
  Edwin C. Lippy and Steven C. Waltrip 67

A Waterborne Outbreak of Giardiasis in Camas, Wash.
  J.C. Kirner, J.D. Littler, and L.A. Angelo 75

Tracing a Giardiasis Outbreak at Berlin, New Hampshire
  Edwin C. Lippy 81

Giardiasis in Washington State
  Floyd Frost, Lucy Harter, Byron Plan, Karen Fukutaki, and Bob Holman 90

Engineering Defects Associated With Colorado Giardiasis Outbreaks June 1980–June 1982
  Richard J. Karlin and Richard S. Hopkins 93

Causes of a Waterborne Giardiasis Outbreak
  Thomas E. Braidech and Richard J. Karlin 97

Occurrence of *Giardia* in Connecticut Water Supplies and Watershed Animals
  Henry Adams and Arthur Bruce 100

Cross Transmission of *Giardia*
  R.B. Davies, K. Kukutaki, and C.P. Hibler 103

TREATMENT TECHNIQUES FOR *GIARDIA* REMOVAL

Removal of *Giardia lamblia* Cysts by Flocculation and Filtration
Jogier Engeset and Foppe B. DeWalle 109

Alternative Filtration Methods for Removal of *Giardia* Cysts and Cyst Models
Gary S. Logsdon, James M. Symons, Robert L. Hoye Jr., and Michael M. Arozarena 117

Surrogate Indicators for Assessing Removal of *Giardia* Cysts
David W. Hendricks, Mohammed Y. Al-Ani, William D. Bellamy, Charles P. Hibler,
and John M. McElroy 125

Evaluating Sedimentation and Various Filter Media for Removal of *Giardia* Cysts
Gary S. Logsdon, V. Carol Thurman, Edward S. Frindt, and John G. Stoecker 136

Control of *Giardia* Cysts by Filtration: The Laboratory's Role
Gary S. Logsdon, David W. Hendricks, Gordon R. Pyper, Charles P. Hibler, and Robert Sjogren 142

Removing *Giardia* Cysts With Slow Sand Filtration
William D. Bellamy, Gary P. Silverman, David W. Hendricks, and Gary S. Logsdon 160

Slow Sand Filter and Package Treatment Plant Evaluation, Operating Costs and Removal
of Bacteria, *Giardia*, and Trihalomethanes
Gordon R. Pyper 169

Removal of *Giardia lamblia* Cysts by Drinking Water Treatment Plants
Foppe B. DeWalle, Jogier Engeset, and William Lawrence 180

Filtration of *Giardia* Cysts and Other Substances Volume 1: Diatomaceous Earth Filtration
Kelly P. Lange, William D. Bellamy, and David W. Hendricks 186

Filtration of *Giardia* Cysts and Other Substances Volume 2: Slow Sand Filtration
William D. Bellamy, Gary P. Silverman, and David W. Hendricks 191

Filtration of *Giardia* Cysts and Other Substances Volume 3: Rapid Rate Filtration
Mohammed Al-Ani, John M. McElroy, Charles P. Hibler, and David W. Hendricks 204

Pilot Testing and Predesign of Two Water Treatment Processes for Removal
of *Giardia lamblia* in Palisade, Colorado
James L. Ris, Ivan A. Cooper, and W. Russell Goddard 214

Effect of Halogens on *Giardia* Cyst Viability
Ernest A. Meyer 239

Index 241

Foreword

Giardia lamblia in Water Supplies is a compilation of selected articles from recent issues of *Journal AWWA*, AWWA conference proceedings, and USEPA project summaries. A technical resource handbook, this AWWA publication features discussions of the detection of *Giardia lamblia* cysts in water, their occurrence in water supplies, and techniques for their removal. The possible contamination of drinking water by *Giardia* cysts has been a major concern of researchers during the last decade, and recent giardiasis outbreaks have brought it to the forefront of the attention of many water utilities. Although timely, this compilation is not all-inclusive. The reader is encouraged to examine references found within the individual articles and to review related literature through use of Waternet, the AWWA Technical Library's computerized bibliographic data base. As an additional aid to the reader, the index at the back of this handbook will prove useful in accessing the considerable information available in the various articles.

The opinions expressed within the publication are those of the individual authors and do not constitute a statement of AWWA policy.

AWWA Technical Resource Series
Analyzing Organics in Drinking Water
Giardia lamblia in Water Supplies

AWWA Management Resource Series
Managing Water Rates and Finances
Computer-Based Automation in Water Systems
Energy and Water Use Forecasting
Challenges in Water Utility Management
Water Conservation Strategies

Introduction

REVIEW OF GIARDIASIS IN WATER SUPPLY

Report by <u>Giardia</u> Task Group
of the
Research Committee for the Rocky Mountain Section
American Water Works Association
September 1982

DISCLAIMER

This report has been developed by the Giardia Task Group at the request of the Research Committee of the Rocky Mountain Section of the American Water Works Association. It is offered for the purpose of informing members of the conclusions drawn by the Task Group. The report has been reviewed by the Rocky Mountain Section and has been approved for publication. It is intended that it be used for information purposes only, and is not to be construed as proclaiming any policy or position on the part of the Rocky Mountain Section or of the American Water Works Association.

ACKNOWLEDGEMENT

The Research Committee of the Rocky Mountain Section of the American Water Works Association extends its appreciation to Richard P. Arber, who served as chairman of the Giardia Task Group and to the following who offered their expertise in serving as members:

Dr. Robert L. Champlin, University of Wyoming
Tom Braidech, USEPA
Dr. David W. Hendricks, Colorado State University
William D. Bellamy, Colorado State University
Rick Karlin, Colorado Department of Health
Dick Margetts, Project 7, Montrose, Colorado
George Budd, Richard P. Arber Associates

Thanks are also extended to the Rocky Mountain Section for its support in this endeavor.

INTRODUCTION

In recent years, outbreaks of giardiasis as a result of infection by a pathway that has been presumed to be waterborne has resulted in considerable publicity. These outbreaks have been attributed to the flagellated, protozoan type of pathogen, <u>Giardia lamblia</u>. Because <u>Giardia lamblia</u> can infect several animal hosts, outbreaks can occur in water supplies which have no upstream human activity. Beavers have been of particular concern as animal hosts because of the potential for habitation in streams which are tributary to water supplies.

The protozoa has two forms: a free-living or trophozoite form and an encapsulated cyst form. Although both forms can be excreted by host organisms, the encapsulated cyst form is more resistant and tends to be the predominant form outside of the host. In addition, the trophozoite is not resistant to the initial stages of digestion and therefore the cyst is thought to be the likely infectious form. Cysts undergo an excystation process when they reach the small intestine and transform to the active trophozoite stage which reproduces and infects the host. A wide range of symptoms have been noted which include severe diarrhea, weight loss, and sometimes retarded growth in children.

A significant number of documented outbreaks of giardiasis have occurred in the Rocky Mountain region, resulting in considerable concern to the public as well as to professionals of the water industry. As a consequence, the Research Committee of the Rocky Mountain Section of AWWA determined that a significant need existed for ascertaining the state-of-the-art and research needs relative to giardiasis. As a consequence, a Giardia Task Group was developed to perform this evaluation. This report summarizes the findings of that task group.

OCCURRENCE

Although reported as a possible waterborne pathogen as early as 1946 in Tokyo, <u>Giardia lamblia</u> was first documented as the causative agent of the intestinal disease, giardiasis, in the United States in 1966. Since that time, there have been numerous outbreaks of giardiasis throughout the United States, principally in cool mountainous areas of the country.

Outbreaks of the disease have been reported from Colorado, Utah, New Hampshire, New York, Washington, California, Montana, Idaho, Vermont, Tennessee and Wyoming. Eleven giardiasis outbreaks have been documented in Colorado. Seven of these occurred during the period between spring of 1980 and spring of 1982 when a USEPA/CDC (Center for Disease Control) sponsored waterborne disease surveillance program was carried out by the Colorado Department of Health.

These latter seven outbreaks may not indicate an increase in the prevalence of <u>Giardia lamblia</u> in Colorado, but instead the higher likelihood of identifying waterborne disease when a specifically designed surveillance program is in place. Conversely, increased use of Colorado watersheds is undoubtedly a factor.

The Rocky Mountain region is considered a high risk area for giardiasis outbreaks for a number of reasons. Readily available surface sources of water in this area has led to reliance on this easily contaminated supply as the principal source of drinking water. <u>Giardia lamblia</u> cysts are associated with surface water and have only been found in well supplies that have been contaminated with sewage (2 cases).

The apparent "high quality" (low turbidity) of such sources has resulted in the practice of using minimal or no treatment. Resistance of the cysts to chlorination at routine dosages and the difficulty of removing particles in its size range without full treatment allows the passage of cyst through treatment plants even though coliform and turbidity standards are being met.

t Giardia lamblia cysts can remain viable for up to two
Thus, the cold environment of mountain streams is
al of the cysts and subsequent passage into community
ns.

eptible to Giardia lamblia infection. Beavers, because of
re significant sources of the organism and have been
tbreaks. Domestic dogs are commonly infected and it
they, along with humans, may be the source of wildlife
human activity in a watershed does not seem to be a
contamination.

e Colorado Department of Health at Dr. Charles Hibler's
niversity have indicated widespread presence of Giardia
waters throughout the state. Cysts determination were
e with Standard Methods, modified by use of one-micron
en-micron filters specified. This modification was based
on the results of field tests comparing cyst recovery on different filters. Although not every surface source has been sampled, adequate information is available to indicate that all surface water sources are at risk.

HEALTH EFFECTS

Acute giardiasis is manifested by many symptoms; the most common of which is diarrhea. Other usual symptoms that accompany the disease are, in descending order of occurrence: weakness, weight loss, abdominal cramps, nausea, greasy stool, abdominal distention, flatulence, vomiting, belching, and fever. The diarrhea is thought to be caused by low-grade inflammation of the intestine brought about by the attachment of the parasites to the mucosa. Most of the remaining frequent symptoms appear to be secondary to the diarrhea, and either caused by it or the accompanying water loss in severe cases.

Malabsorption can occur with heavy infestations of Giardia. This phenomenon is brought about by either the trophozoites becoming so numerous that intestinal villi are covered, thereby limiting the area available to absorb nutrients, or the trophozoites causing an increase in cell turnover leading to an accumulation of mucus and cytoplasm covering the intestine. Malabsorption can lead to vitamin deficiencies, interference with protein absorption, increased fat content of the feces and, in children, a failure to thrive.

There are also reports of the parasite causing low levels of all forms of gamma globulin (hypogammaglobulinemia) and diseases of the bile tract, liver, pancreas, and large bowel. In addition, Giardia has been implicated with nodular lymphoid hyperplasia, mesenteric adenitis and rectal bleeding.

Several studies of chronic giardiasis are recorded in the literature where the patients carried the parasite for up to five years. In these cases, constipation was common alternating with episodes of diarrhea. Also associated with this syndrome were nasal allergy, asthma and an intolerance to milk.

There are four drugs available for use in treating giardiasis. These are Atabrine, Flagyl, Furoxone and Humatin. Each of these drugs is effective in curing the disease and all except Humatin have some deleterious side effects.

Atabrine (quinacrine) is effective in 80 to 90 percent of the cases. Its side effects, which are not common at the recommended dosage, are nausea, vomiting, dizziness, mild diarrhea, colic, headache, urticario, central nervous system stimulation, severe skin rash, and/or yellow discoloration of the urine, skin or sclerae.

Flagyl (metronidazole), not approved for this use by the Food and Drug Administration, is effective in 70 to 80 percent of the cases and has been shown to cause tumors in mice and rats. Frequent side effects include nausea, headache, dry mouth, and metallic taste. Occasionally, vomiting, diarrhea, vertigo, paresthesia, skin rash, and dark discoloration of the urine occur.

Furoxone (fuazolidone) is the drug of choice when treating small children because it can be obtained in liquid form. Side effects include vomiting, fever and nausea. Like Flagyl, it too is a suspected carcinogen.

Humatin (paromomycin) has been proposed for use in cases where the patient is pregnant. The main use of this drug has been as an amebicide. It is poorly absorbed by the small intestine and thus has minimal toxic effect on the patient and fetus. It has been shown that the side effects of the other drugs have caused spontaneous abortions and possible teratogenicity. The usual side effects of Humatin are nausea, increased gastrointestinal motility, and diarrhea. Occasionally the patient may develop rash, headache, vertigo, abdominal pain and vomiting. It is apparent that more research needs to be done to confirm the safety of this drug on pregnant patients.

There is little evidence to support a theory of immunity with respect to giardiasis. It does seem, though, from the small numbers of people affected during an outbreak, that certain people are either immune to Giardia or become asymptomatic carriers. This may mean that there are many more people who are infected with the parasite than previously suspected. In additon, there is some speculation that people who are repeatedly exposed suffer fewer effects than those who are exposed for the first time.

With all of the symptoms and side effects, nowhere in the literature has it yet to be shown that giardiasis has not been associated with any mortalities or any lasting adverse physiological effects. However, it is still regarded as a significant problem and its effect on public attitudes to the potability of water should not be minimized.

GIARDIA LAMBLIA SAMPLING AND ANALYSIS TECHNIQUES

The literature has generally addressed the subject of Giardia lamblia sampling and analysis as having three interrelated steps: (1) sampling a sufficient volume of water to assure cyst collection, (2) analyzing the sample to determine if cysts are present, and (3) determining the viability of cysts that are detected.

Sampling

A number of techniques have been used in attempts to sample water supplies and distribution systems for Giardia cysts. Grab samples, membrane filters, sand filters, diatomaceous earth filters and cartridge filters have all been used with varying degrees of success. The cartridge filter has proven to be the best technique for field samples, which require filtration of large quantities of water to concentrate cysts to a detectable level. Work by USEPA has indicated significant variations in the percent of cyst recovery achievable with cartridge filters. This suggests that although this approach is suitable for qualitative assessments, it does not render reliable quantitative results. USEPA has evaluated different cartridge filters and was decided that the 7 um nominal porosity Orlon filter was the best. This type of filter gave positive results for Giardia cysts at three water supplies in Colorado and also detected Giardia cysts in the raw and finished water of the water treatment plant at Berlin, New Hampshire during an outbreak in 1977. This sampling system is described in the 15th edition of Standard Methods.

In 1979 the Colorado Department of Health, Water Quality Control Division began extensive sampling for Giardia. They used essentially the same sampling arrangement as the USEPA with the exception of the cartridge filter. They found that 1 um porosity cotton yarn-wound cartridge was the filter element that gave the best results.

In laboratory research, smaller quantities of water are involved and membrane filters have provided the best results. Polycarbonate membrane filters have been used with the most success, rendering quantitative as well as qualitative results. Cartridge filters are qualitative in nature and consequently not suitable for the quantitative needs of laboratory research.

Analysis

Counting cysts is the only available technique for analyzing Giardia samples. To date, there is no suitable in vitro culturing technique such as those used for bacterial analyses. Counting techniques include the hemocytometer, particle counter, and microscopic counting (cover slip technique). The hemocytometer has been used successfully in research and the cover slip technique has been used for ambient sample analysis as well as research.

The hemocytometer technique starts by concentrating the sample. The sample can be a grab sample or the washwater from a filter used to concentrate a sample. After concentrating by settling, centrifuging or a combination of both, an aliquot of the concentrated sample is introduced into the hemocytometer and counted (staining may be necessary).

Particle counting is only applicable when the cysts are the predominant particle in the 5 to 10 um size range. This is necessary since there is no positive indentification of the particles counted.

The cover slip technique has been used by numerous laboratories for the identification of Giardia cysts. The first step in the procedure is to concentrate the sample by settling and/or centrifuging. Next, the cysts are stained with Lugol's Iodine; then they are resuspended with a zinc sulfate solution in a centrifuge tube. The mixture is then centrifuged with a cover slip in contact with the surface of the liquid in the tube. The cysts float to the top and adhere to the cover slip. The cover slip is removed and placed on a slide and the cysts counted.

The accuracy of the hemocytometer and the cover slip method was investigated by the USEPA. Their results indicated that the hemocytometer method will detect from 2 to 9 times as many cysts as the cover slip method. However, the results also indicate that the cover slip method gave more reproducible results. Comparison of these techniques at Colorado State University resulted in almost the opposite conclusion, i.e., that the cover slip method gave results 9 times higher than those of the hemocytometer. Additional research will be required to reconcile the differences in these observations.

Viability

Determining Giardia cyst viability is important when conducting disinfection studies since cysts can be identified even though they have been inactivated by a disinfectant. The techniques that have been used are: eosin dye exclusion, excystation and infectivity.

Infectivity studies with human subjects are impractical for monitoring or even for occasional testing; consequently, the subject will not be discussed here. It does, however, appear that this type of analysis may be required to correlate excystation to viability.

Cyst stain (eosin dye) exclusion has been used as a test procedure to demonstrate cyst viability; however, staining is probably not an accurate method for determining viability. The results of some of the past disinfection research have been questioned because of the inadequacy of this approach.

Giardia excystation has been attempted since 1927. Basically the process involves placing the cyst in an environment which simulates the human digestive system. The development of motile trophozoites from the suspension of cysts is then observed. This method is believed to be the most accurate of those presently available.

TREATMENT METHODS

General

Research on the effectiveness of treatment processes for the removal of Giardia lamblia is a difficult undertaking due to the problems associated with laboratory culture techniques. Because of these difficulties, specific research on treatment techniques for Giardia lamblia removal from municipal water supplies has evolved slowly. Some of the earlier conclusions have been based on laboratory research performed with E. histolytica cysts (an amoeba associated with amoebic dysentary) and Giardia muris cysts (a form of Giardia not infectious to humans). Much of the knowledge which has developed has evolved from field evaluations of treatment inadequacies at outbreak sites.

Some of the older research efforts on the effectiveness of chlorination have been called into question because of misleading results which may have arisen from the use of cyst stain techniques for determining cyst viability. Recent and current research efforts have undertaken direct evaluations with Giardia lamblia and the more accurate excystation technique for estimating cyst viability is now being used. As a consequence, a better understanding of treatment process effectiveness is starting to develop.

Based on the present understanding of treatment processes, the consensus is that treatment systems should consist of an effective particulate removal process followed by an effective disinfection process. Evaluations of outbreaks indicate that they have been associated with one or more of the following when treating surface waters or polluted groundwaters:

- o Lack of or failure of the filtration process.

- o Lack of or failure of the disinfection process.

- o Lack of or failure of a chemical coagulation process.

It should be noted that these processes are required even when the source waters are low in turbidity and have low levels of coliform bacteria. This derives from the fact that the presence of _Giardia_ cysts in water sources is not correlated with either of these water quality parameters.

Particulate Removal Processes

The particulate removal processes which have been investigated include the conventional chemical coagulation, flocculation and rapid sand filtration process, diatomaceous earth filtration, and slow sand filtration. Research to date indicates that each of these approaches is capable of effective removal of cysts when properly operated. Selection of processes will depend on the circumstances surrounding each situation and will be governed by relative costs and judgments by the appropriate decision makers.

Cartridge filtration has also been suggested as a potential method for cyst removal. The effectiveness of this system in concentrating cysts suggests that it may be an appropriate cyst removal system. However, little research has been performed on this process. The use of cartridge filters is most likely to be limited to individual homes or small systems. It should be noted that one recent outbreak has been attributed to a system that used an improperly maintained cartridge filter in conjunction with a chlorinator that was turned off.

Chemical Coagulation, Flocculation and Filtration

A review of past outbreaks indicates an association between the lack of adequate coagulation and filtration processes and the occurrence of _Giardia_ outbreaks. In most cases, outbreaks have occurred where coagulation, flocculation and filtration processes have been absent and chlorination was relied upon as the sole treatment process. In other cases, outbreaks have been attributed to inadequate or nonexistent chemical coagulation prior to disinfection and to media loss from the filters. In many cases these deficiencies have also coincided with poor or non-existent disinfection and polluted wells which were not adequately treated.

In the discussion that follows, no distinction is made between coagulation, flocculation and filtration process with or without a sedimentation process. Research to date does not permit an adequate assessment of the differences between the two approaches.

The most comprehensive research effort that has been reported to date is the work by Logsdon and his co-workers at USEPA. Because of difficulties in culturing _Giardia lamblia_, this work was performed using more convenient surrogate parameters. These consisted of _Giardia muris_ cysts, a protozoa similar to _Giardia lamblia_, and 9 um radioactive particles. Both surrogates are within the 7 to 12 um size range of _Giardia lamblia_ cysts.

The results from this investigation indicated that a properly operated chemical coagulation, flocculation and filtration process is capable of achieving a high degree of removal of the surrogate parameters. Conversely, inadequate coagulant dosages resulted in significant reductions in the level of removal.

The removal of _Giardia muris_ cysts was evaluated over an entire filter cycle. A significant improvement in removal capability occurred during the course of a "curing" period which follows a filter backwash. In this case, the "curing" period lasted for 30 minutes, after which a high degree of removal was maintained until a breakthrough occurred as a result of high head loss at the end of the filter run. In addition, sudden increases in hydraulic loading during a run were seen to cause upsets which resulted in momentary cyst breakthroughs.

A significant observation in this work was that satisfactory cyst removals required treatment that was capable of attaining turbidity levels which were much less than the MCL of 1.0 ntu (0.3 ntu or less with the lower turbidities required where the raw water turbidities were low). Removal of cysts was seen to roughly parallel changes in turbidity, and slight increases in turbidity were associated with phenomena that resulted in large changes in the concentration of cysts in the filtered water.

DeWalle and co-workers have recently reported on filtration studies which directly evaluated the removal of Giardia lamblia cysts. Their results confirmed the need for an effective coagulation process ahead of filtration; showing greater than 99 percent cyst removal by filtration at an effective coagulant dose and less than 15 percent cyst removal in an uncoagulated water. This work also indicated a relationship between the pH of the coagulation process and the effectiveness of filtration. The most effective pH region was found to be from 6.4 to 7.0 in the water that they examined.

In addition to the information from direct research of Giardia removal, the need for effective coagulation and flocculation processes is suggested by existing filtration theory. To be removed by a filter, a suspended particle must contact the media particles and "stick" to these particles once it has made contact. Although removal of suspended particles by physical straining of particles from the liquid stream is limited, transport mechanisms which typically account for particle contacts are enhanced by increasing the particle size, a function which coagulation and flocculation perform. In addition, coagulants perform the function of increasing the attraction forces between particle surfaces, thereby making them more likely to "stick" upon contact. Therefore, without proper coagulation, the effectiveness of the filtration process will be reduced.

Based on review of the literature discussed herein, the following items should be considered in the design and operation of conventional filtration plants:

o   Effective coagulation and flocculation processes should be included as a part of a conventional filtration system. Consideration should be given to a good mixing system that achieves rapid dispersal of the coagulant, addition of the proper coagulant dose, and adjustment of the pH of coagulation to optimize the process. Flocculation or mixing basin design should be adequate to achieve particle agglomeration. (These needs may be different for processes that eliminate sedimentation and those that incorporate sedimentation.) Inlet and outlet designs between the coagulant addition point and the filters should be such that floc break-up due to turbulence is minimized.

o   Although removal of Giardia cysts roughly parallels reductions in turbidity, it is likely that a finished water turbidity significantly less than the existing MCL of 1.0 ntu will have to be produced. The need for low finished water turbidities is greatest for those cases where low raw water turbidities are experienced. Attainment of the AWWA goal of 0.1 ntu has been suggested by some.

o   The work of Logsdon and co-workers suggests that significant breakthrough of cysts is associated with small changes in turbidity. If this is the case, then meeting a set turbidity level may not be sufficient to prevent breakthrough. Operation must also be directed towards limiting changes in turbidity in individual filters. As a consequence, continuous monitoring of the turbidity exiting each filter in a plant should be considered.

o   Because of the filter "curing" phenomenon, a filter-to-waste capability should be considered. The filter-to-waste period should be established based on observed changes in turbidity at the beginning of a filter run. Past practice has been to use a period that is typically in the range of 2 to 20 minutes (Water Treatment Plant Design, 1969), however, more research is required before a specific recommendation can be made. It should be noted that Logsdon et. al. observed improved removals over a period of the initial 30 minutes of a filter run.

o   Care should be taken not to operate a filter beyond breakthrough. Even slight increases in turbidity at the end of a filtration cycle may be associated with large increases in the breakthrough of Giardia cysts. Consideration should be given to backwashing a filter when an increase of turbidity is observed as opposed to waiting for a selected turbidity level to be reached.

o   Filters should be designed and operated to reduce hydraulic surges that have been shown to cause short-term breakthrough. Surges may result from changes in the selected flowrate, from unstable flow control valves that "hunt," or in the case of filters that are stopped and re-started from day to day. In addition to hydraulic controls to avoid problems, consideration should be given to the addition of polymer as a filter aid to minimize floc shear due to hydraulic changes.

o   Filter beds should be maintained in good condition and a concerted effort should be made to eliminate potential cross-connections. Effective filter wash procedures should be used to prevent filter cracking and mud ball formation. Water depth over the filter and filter cleaning should be adequate to prevent the onset of negative head conditions.

Diatomaceous Earth Filters

Diatomaceous earth filters were developed for the Army in the 1940s as a means for removing amoebic E. histolytica cysts, an etiological agent associated with amoebic dysentary.

Because these cysts are in the same size range as Giardia cysts, diatomaceous earth filtration would also be expected to be effective in the removal of Giardia cysts and their use for this purpose has been investigated.

Diatomaceous earth filters consist of a septum which provides the support for the filter media and a coating of diatomaceous earth which forms the media. A wide range of gradations is available for the media with a size in the range of 10 to 25 um being typical for media in water treatment practice. This is substantially smaller than the media in sand filters, and unlike conventional coagulation/flocculation/filtration processes, straining is the primary mechanism for particle removal by diatomaceous earth filters.

Typically the filter is contained within a pressure vessel and flow is induced through the filter as a result of either a pressure on the inlet side or a vacuum on the discharge side, depending on the location of the pump. It should be noted that a vacuum type of filter affords the option of containing the filter in a vessel that is open to the atmosphere.

Because of problems associated with dissolved gas coming out of solution at the lower pressures in vacuum type filters, pressure filters have typically been used in municipal water treatment practice. However, recent advances in septum material may minimize this concern and make vacuum type filters more attractive.

Operation consists of three phases:

o   <u>Precoat</u> - Prior to the initiation of a filter run, the septum must be coated with a layer of diatomaceous earth to form the filter media. This is accomplished by continuously recycling a suspension of media through the filter until a clear recycle stream develops. In the early work on the removal of <u>E. histolytica</u> by diatomaceous earth, it was noted that the quality of this recycle water had an impact on breakthrough of the filter and it was recommended that finished water be used for this purpose.

Logsdon and co-workers have indicated that the quantity of precoat has a significant impact on the effectiveness for the removal of <u>Giardia</u> cysts. For the water used in their investigation, improvements in removal effectiveness were seen at precoat application rates up to 1.0 Kg/m^2 (0.2 lb/ft^2). Beyond that rate, little improvement was noted.

o   <u>Filtration</u> - During the filtration phase of operation, water is continuously passed through the layer of diatomaceous earth which is held in place by the resulting force. It is important to maintain the flow of water throughout a filter run so that the media will be maintained in place. Designs should incorporate features that prevent an automatic restart after a loss of flow. If a continuous process is not to be used, then the cake should be removed at the end of each run. As an alternative, consideration could be given to a small recycle pump that could provide a flow sufficient to hold the cake on the septum between filter runs.

Throughout a filter run, it is advantageous to continue the addition of diatomaceous earth through the addition of a body feed which prevents the accumulation of all of the filtered material at the surface of the precoat layer. As a consequence, a more permeable filter media is maintained and head loss builds up at a slower rate. In addition, Logsdon and co-workers noted a decrease in the degree of cyst removal for runs which included no body feed. Consequently, body feed is recommended. However, because this is essentially a physical process, momentary loss of body feed should not result in a sudden decrease in treatment efficiency.

The dosage of body feed depends on the quantity of solids present in the raw water. Typically, dosages are selected based on a ratio of body feed concentration to turbidity.

In their work, Logsdon and co-workers evaluated the correlation between removals of turbidity and cysts. They observed that elevated finished water turbidity could be coincidental with a high degree of removal of cysts and concluded that turbidity could not be used as a basis for terminating a filter run. Typically, the decision to backwash is based on the buildup of head loss. No convenient operational parameter is associated with effective treatment and it is necessary to maintain good operating procedures in order to consistently achieve a high degree of cyst removal.

Although diatomaceous earth filters are effective in the removal of cyst-size particles in a wide range of media gradations, consideration should also be given to the requirements for the removal of bacteria and viruses when selecting media. Research indicates the need for use of a finer grade of media for effective removal of these biological agents. As an alternative, use of alum and polymer as a precoat for the media might be considered, especially in the case of virus removal. Polymer has also proved effective as a coagulant when added with the body feed ahead of the filter. Problems have occurred when alum has been added in this manner because the gelatinous floc has plugged the media.

o   Backwash - At the end of the filter run, the diatomaceous earth is removed from the septum by an air-bumping process, sluicing with water, or by reversing the direction of water flow.

The spent diatomaceous earth is then washed with the backwash water. An effective backwash is required to adequately clean the septum, just as is the case for granular media filtration.

Slow Sand Filtration

Slow sand filters differ from rapid sand filters in that a deeper bed and a substantially lower hydraulic loading rate are used. In some cases, it has been found to be advantageous to pretreat the raw water by settling. However, pretreatment by coagulation is frequently not required. Nonetheless, consideration might be given to providing capability for chemical coagulation and sedimentation for use in cases where seasonally high turbidities are encountered.

The process relies on the development of a Schmutzdecke in the upper layer of the filter. This layer contains a large number of microorganisms that perform a significant role in the removal process.

In the past, this process has proved to be effective in the removal of bacteria and viruses, and research is presently being conducted at Colorado State University to evaluate the effectiveness for the removal of Giardia. Work to date suggests that it is an effective process for cyst removal under conditions of proper operation. High levels of cyst removal are maintained at turbidities up to 2.0 ntu, suggesting that removal is less sensitive to turbidity when compared to rapid sand filters.

The filters are periodically cleaned by scraping when the accumulated head loss becomes excessive. Because of a decline in filter performance, filter-to-waste should be practiced for one or more days after cleaning. In addition, several weeks may be required to adequately establish a biological population throughout a filter during start-up and filter-to-waste may be advisable over this period of time.

Of particular concern with respect to application of the process in this region are the effects of cold weather on performance. Deterioration of filter performance in cold water and freezing problems have been noted at some installations.

Disinfection Processes

Hoff has suggested that early research on the effectiveness of chlorination has substantially overstated the resistance of Giardia cysts to chlorine. Much of the early work was based on the use of staining techniques that have subsequently been found to overstate the viability of cysts. Present work, based on the use of an excystation approach that is thought to give a more realistic estimate of cyst viability, yields a more optimistic view. However, it still appears that higher chlorine levels or longer contact times will be needed for the destruction of Giardia cysts as compared to the levels required for the destruction of bacteria. This suggests that coliform bacteria are not an effective monitor for Giardia disinfection. In addition, it suggests that consideration should be given to maximizing chlorine contact through a water treatment plant by using plug flow type contacting basins to minimize the effects of short-circuiting.

Using the excystation approach, Jarroll and co-workers investigated the effect of chlorine on <u>Giardia</u> cysts in a range of free chlorine residual from 1 to 8 mg/l in chlorine demand free water. Water temperature, pH, chlorine residual and contact time were found to have a significant effect. A decrease in water temperature and an increase in pH in the range of 6 to 8 decreased the disinfection effectiveness. It was observed that a residual of 4 mg/l over a contact period of 60 minutes was required to kill all cysts at each pH with lower dosages found to be effective as pH decreased. It should be noted that a residual of 4 mg/l is not normally used in water treatment practice. However, a residual of this magnitude might be considered under certain emergency conditions.

In additon to chlorine, ultraviolet irradiation has been examined. Rice and Hoff found cysts to be resistant to even high dosages of radiation, suggesting that higher than normal radiation levels may be required for <u>Giardia</u> cyst destruction.

Several emergency disinfectant preparations for small quantities of water were evaluated by Jarroll and co-workers. These included tincture of iodine, saturated iodine, bleach, a chlorine based preparation and two iodine based preparations. Effectiveness was evaluated for the normally recommended quantities and contact times of these preparations. All methods were effective at a temperature of 20°C, whereas only the chlorine based preparation and one of the iodine based preparations were effective under all conditions at 3°C. These results suggest that a degree of caution is appropriate when using these types of preparations. It was indicated that the effectiveness in cold water might be improved by the use of either higher dosages or longer contact times. In addition, it was pointed out that boiling of the water represented an effective alternative in cases where uncertainty exists.

To date, little work has been performed to evaluate the effectiveness of ozone, bromine compounds or chlorine dioxide. Additional work will be required to determine the effectiveness of these agents.

RESEARCH NEEDS

Following the review of the literature, the Task Group evaluated the state-of-the-art and identified major areas of research needs. Although some of the areas might be currently under investigation, until the work is published, evaluation of its adequacy cannot be made.

Below is a list of research needs identified by the RMAWWA Giardia Task Group.

1) Techniques for sampling and analyzing for <u>Giardia lamblia</u> cysts should be improved. Present techniques are time-consuming, costly and reproducibility is questionable. In addition, further work is needed to develop a more appropriate indicator organism.

2) "Indicator" parameters for the coagulation/flocculation/filtration process should be developed to aid the treatment plant operator in controlling the processes and determining treatment effectiveness. The potential for using turbidity as an operational control parameter needs further investigation. In addition, consideration should be given to the use of degree of turbidity reduction as a supplemental operational parameter.

3) Procedures for monitoring the effectiveness of diatomaceous earth filters and slow sand filters need to be further developed.

4) The effectiveness of various disinfectant chemicals should be determined. Relationships of cyst kill with dosage, pH, detention time and temperature should be further developed. Further research is needed to evaluate chlorine disinfection at lower residuals and longer retention times.

5) The extent of the occurrence of _Giardia_ in raw waters should be further evaluated.

6) A rational procedure for selection of polymers as coagulant and filter aids needs to be developed. This will help to establish a stronger floc that is less likely to result in cyst breakthroughs.

7) The effect of different filter media and different hydraulic loadings should be further evaluated. Process sensitivity and ease of operation should be considered.

8) The effect of transient hydraulic conditions during a filter cycle should be established. These include the effects of sudden decreases or increases in hydraulic load as well as the need for a filter-to-waste mode. In addition, more research is needed to provide a basis for establishing a filter-to-waste period.

9) The effectiveness of direction filtration (with coagulant addition) as it relates to _Giardia_ removal should be evaluated. In particular, the level of operator skill associated with coagulation for direct filtration approaches should be assessed.

10) The use of chemical coagulants in combination with diatomaceous earth filters should be assessed further. Both precoat and pretreatment methods should be evaluated.

11) The effectiveness of alternative coagulant chemicals for _Giardia_ cyst removal should be established.

12) Cross transmission of _Giardia_ between humans and other mammals should be verified. Some research has indicated that cross transmission between certain mammals and humans does not occur.

13) Problems associated with operation of slow sand filters in cold climates needs to be evaluated.

It is apparent from the literature that there is much to be learned about the effects, occurrence, analytical methods, and treatment of _Giardia lamblia_ in water supplies. It is important for the water industry to learn as much as possible about _Giardia_ so that we can continue to reliably serve the public with the highest quality drinking water possible.

REFERENCES

1. Anonymous "Nutritional Consequences of Giardiasis." <u>Nutrition Reviews,</u> 38:11 (Nov. 1980).

2. Barbour, A.G.; Nichols, C.R. and Fukushima, T. "An Outbreak of Giardiasis in a Group of Campers." <u>Am. J. of Trop. Med. & Hygiene,</u> 25:3 (1976), 384-389.

3. Bellamy, W.D. and Hendricks, D.W. <u>Removal of Giardia lamblia from Water Supplies,</u> unpublished First Year Progress Report No. 5836-82-1, Dept. Civil Eng., Colorado State Univ., (March 1982).

4. Bingham, A.K.; Jarroll, E.L.; Meyer, E.A. and Randulescu, S. <u>"Giardia sp."</u> Physical Factors of Excystation In Vitro and Excystation Versus Eosin Exclusion as Determinants of Viability." <u>Exp. Parasitology,</u> 47:2 (1979), 284-291.

5. Chang, S.L. and Kabler, P.W. "Detection of Cysts of <u>Entamoeba histolytica</u> in Tap Water by the Use of Membrane Filter." <u>Am. J. Hygiene,</u> 64 (1956), 170-180.

6. Chester, A.C. and MacMurray, F.G. "Giardiasis as a Chronic Disease." <u>Gastroenterology,</u> 80:5 (1981), 1123.

7. Colorado Department of Health, unpublished results.

8. Craun, G.F. "Waterborne Outbreaks in the United States 1971-1978." Proc. AWWA 1980 Annual Conf., Part I, (June 15-20, 1980).

9. Davies, R.B. and Hibler, C.P. "Animal/Reservoirs and Cross-Species Transmission of <u>Giardia.</u>" <u>Waterborne Transmission of Giardiasis.</u> Report EPA-600/9-79-001, (1979), 104-126.

10. DeWalle, F.B. <u>Removal of Giardia lamblia Cysts in Drinking Water Plants,</u> unpublished reports, First and Second Quarterly Report for EPA Grant CR 806127, Univ. of Wash., Department of Environmental Health, (1979).

11. DeWalle, F.B. <u>Removal of Giardia lamblia Cysts in Drinking Water Plants,</u> Project Summary for EPA Grant CR 806127, Univ. of Wash., Department of Environmental Health, (1982).

12. Drew, J.H. "Biliary Giardiasis and Pancreatitis." <u>Medical Journal of Australia</u> (April 4, 1981).

13. Engeset, J. and DeWalle, F.B. "Removal of <u>Giardia lamblia</u> Cysts by Flocculation and Filtration." Proc. AWWA 1979 Annual Conference (1979).

14. Erlandsen, S.L. "Scanning Electron Microscopy of Intestinal Giardiasis: Lesions of the Microvillous Border of Villus Epithelial Cells Produced by Trophozoites of <u>Giardia.</u>" Proc. of the Workshop on Advances in Biomedical Applications of the Scanning Electron Microscope, IIT Research Institute, Chicago, Illinois (April 1974).

15. Frost, F.; Plan, B. and Liehty, B. "<u>Giardia</u> Prevalence in Commercially Trapped Mammals." Journal of <u>Environmental Health,</u> 42:5 (March/April 1980), 245-249.

16. Goodman, M.J.; Pearson, K.W.; McGhie, D.; Dutt, S. and Deodhar, S.G. "Lampylobacter and Giardia lamblia Causing Exacerbation of Inflammatory Bowel Disease." The Lancet, (Dec. 6, 1980), 1247.

17. Hoff, J.C. "Disinfection Resistance of Giardia Cysts; Origins of Current Concepts and Research in Progress." Waterborne Transmission of Giardiasis. Report EPA-600/9-79-001, (1979), 231-239.

18. Feely, D.E. and Erlandsen, S.L. "In Vitro Analysis of Giardia Trophozoite Attachment."

19. Jakubowski, W. Detection of Giardia Cysts in Drinking Water: State of the Art, Report EPA-600/D-81-258, (1981).

20. Jakubowski, W. and Ericksen, T.H. "Methods for Detection of Giardia Cysts in Water Supplies." Waterborne Transmission of Giardiasis. Report EPA-600/9-79-001, (1979), 193-210.

21. Jakubowski, W.; Chang, S.; Ericksen, T.H.; Lippy, E.C. and Akin, E.W. "Large-Volume Sampling of Water Supplies for Microorganisms." J. Am. Water Works Assoc., 70:12 (1978), 702-706.

22. Jarroll, E.L.; Bingham, A.K. and Meyer, E.A. "Effect of Chlorine on Giardia lamblia Cyst Viability." Appl. and Envirn. Micro., 41:2 (1981), 483-487.

23. Jarroll, E.L.; Bingham, A.K. and Meyer, E.A. "Giardia Cyst Destruction: Effectiveness of Six Small-Quantity Water Disinfection Systems." Am J. Trop. Med. Hyg., 29:1 (Jan. 1981), 8-11.

24. Jarroll, E.L.; Bingham, A.K. and Meyer, E.A. "Inability of an Iodination Method to Destroy Completely Giardia Cysts in Cold Water." The Western Journal of Medicine, 132:6 (June 1980), 567-569.

25. Johnston, L. "An Investigation of Animal Hosts for Giardia lamblia in the Bridger Wilderness Area." US Forest Service, (Jan. 1979).

26. Johnston, T.S. "Diagnosis and Treatment of Five Parasites." Drug Intelligence and Clinical Pharmacy, 15:2 (Feb. 1981).

27. Juranek, D.D. In discussion of: "Water Filtration Techniques for Removal of Giardia Cysts and Cyst Models." Waterborne Transmission of Giardiasis. Report EPA-600/9-79-001, (1979), 256.

28. Lippy, E.C. "Waterborne Disease: Occurrence is on the Upswing." J. Am. Water Works Assoc., 73:1 (Jan. 1981).

29. Kirner, J.C.; Littler, J.D. and Angelo, L.A. "Waterborne Outbreak of Giardiasis in Camas, Washington." J. Am. Water Works Assoc., 70:1 (1978).

30. Kreutner, A.K.; Delbene, V.E. and Amstey, M.S. "Giardiasis in Pregnancy." Am. J. of Obstetrics and Gynecology, 140:8 (Aug. 1981), 895-901.

31. Kyronseppa, H. and Petterson, T. "Treatment of Giardiasis: Relative Efficacy of Metronidazole as Compared with Tinidazole." Scandinavian Journal of Infectious Diseases, 13:4 (1981).

32. Levin, M.A.; Fisher, J.R. and Cabelli, V.J. "Quantitative Large-Volume Sampling Technique." Applied Micro., 28:3 (1974), 515-517.

33. Lippy, E.C. "Tracing a Giardiasis Outbreak at Berlin, New Hampshire." J. Am. Water Works Assoc., 73:1/70:9 (1978), 512-520.

34. Lippy, E.C. "Waterborne Disease: Occurrence is on the Upswing." J. Am. Water Works Assoc., 73:1 (Jan. 1981).

35. Lippy, E.C. "Water Supply Problems Associated with a Waterborne Outbreak of Giardiasis." Waterborne Transmission of Giardiasis. Report EPA-600/9-79-001, (1979).

36. Logsdon, G.S.; Symons, J.M. and Hoye, R.L. "Water Filtration Techniques for Removal of Cysts and Cyst Models." Waterborne Transmission of Giardiasis. Report EPA-600/9-79-001, (1979), 240-256.

37. Logsdon, G.S.; Symons, J.M.; Hoye, R.L. and Arozarena, M.M. "Alternative Filtration for Removal of Giardia Cysts and Cyst Models." J. Am. Water Works Assoc., 73:2 (Feb. 1981), 111-118.

38. Luchtel, D.L.; Laurence, W.P. and DeWalle, F.B. "Electron Microscopy of Giardia lamblia Cysts." App. Envir. Micro., 40:4 (1980), 821-832.

39. Meyer, E.A. "The Propagation of Giardia Trophozoites In Vitro." Waterborne Transmission of Giardiasis. Report EPA-600/9-79-001, (1979), 211-216.

40. Meyer, E.A. and Jarroll, E.L. "Giardiasis." American Journal of Epidemiology, 3:1 (Jan. 1980).

41. Moore, G.T.; Cross, W.M.; McGuire, D.; Mollohan, D.S.; Gleason, N.N.; Healy, G.R. and Reeves, E.B. "Epidemic Giardiasis at a Ski Resort." New England J. of Med., 281 (1969), 402-407.

42. Poley, J.R. and Rosenfeld, S. "Malabsorption (MA) in Giardiasis: Presence of a Mucoid Pseudomembrane as seen by Scanning (SEM) and Transmission (TEM) Electron-Microscopy." Pediatric Research, 15:8 (1981), 1200.

43. Rey, C.; Escribano, J.C.; Foz, M.; Vidal, M.T. and Salvador, R. "Mesenteric Adenitis Secondary to Giardia lamblia." Digestive Diseases and Sciences, 25:12 (Dec. 1980), 968-971.

44. Roberts-Thomson, I.C.; Anders, R.F. and Mitchell, G.F. "Evaluation of the Humoral Immune Response to Giardia lamblia." Clinical and Experimental Pharmacology and Physiology, 8:4 (July/Aug. 1981).

45. Roberts-Thomson, I.C. and Anders, R.F. "Serum Antibodies in Adults with Giardiasis." Gastroenterology, 80:5 (1981), 1262.

46. Smith, P.D.; Horsbaugh, C.R. and Brown, W.R. "In Vitro Studies on Bile and Deconjugation and Lipolysis Inhibition by Giardia lamblia." Digestive Diseases and Sciences, 26:8 (Aug. 1981).

47. Rice, E.W.; Hoff, J.C. and Schaefer, F.W. "Inactivation of Giardia Cysts by Chlorine." Appl. and Envirn. Micro., 42:3 (1981), 546-547.

48. Rice, E.W. and Hoff, J.C. "Inactivation of Giardia lamblia Cysts by Ultraviolet Irradiation." Appl. and Envirn. Micro., 42:3 (1981), 546-547.

49. Rice, E.W. and Schaefer, F.W. "Improved In Vitro Excystation Procedure to Giardia lamblia Cysts." J. of Clinical Micro., 14:6 (1981), 709-710.

50. Shaw, P.K.; Brodsky, R.E.; Lyman, D.O.; Wood, B.T.; Hibler, C.P.; Healy, G.R.; McCleod, K.I.E.; Stahl, W. and Schultz, M.G. "A Community-Wide Outbreak of Giardiasis with Evidence of Transmission by a Municipal Water Supply." Annal Intr. Med., 87:4 (1977), 426-432.

51. Sherman, P. and Liebman, W.M. "Apparent Protein-Losing Enteropathy Associated with Giardiasis." Amer. Jour. of Diseases of Children, 134:9 (Sept. 1980).

52. Sobsey, M.D.; Wallis, C.; Henderson, M. and Melnick, J.L. "Concentration of Enteroviruses from Large Volumes of Water." Applied Micro., 26:4 (1973), 529-534.

53. Syrotynski, S. and Reamon, T.A. "Giardia as Related to Water." Proc. AWWA Annual Conf., New Orleans, LA (1976).

54. Water Treatment Plant Design, Prepared by ASCE, AWWA, CSSE, Publ. by AWWA, New York (1969).

55. Witherell, L.E. and Herbert, L.W. "An Epidemiological Investigation of Two Outbreaks of Giardiasis in Vermont During 1977." J. New Eng. Water Works Assoc., 91, (1977), 102-118.

56. Wright, R.A. "Giardia Infection from Water" (Letter), Annals of Int. Med., 8:2 (1975), 589-590.

57. Wright, R.A.; Spencer, H.C.; Brodsky, R.E. and Vernon, T.M. "Giardiasis in Colorado: An Epidemiological Study." Amer. Journ. of Epidemiology, 105:4 (1977).

Giardia lamblia and Water Supply

Shun Dar Lin

Giardiasis is now recognized as the most common identified waterborne intestinal disease in the United States. This article describes the morphology and life cycle of the causative agent, *Giardia lamblia*, the detection of *Giardia* and the effectiveness of water treatment processes in removing it, and the epidemiology and clinical aspects of giardiasis.

For most of the 300 years since its discovery, *Giardia lamblia* was considered nonpathogenic because many of the individuals who harbored it did not get sick. Only in the last two decades has this organism been recognized as the most common identified pathogenic intestinal protozoan in the United States and as the cause of flagellate diarrhea (giardiasis).

Often called "traveler's disease," giardiasis is an affliction not only of travelers but also of skiers, campers, hikers, hunters, and fisherman who drink untreated water from mountain streams or inadequately treated water in recreational areas. Although outbreaks of waterborne giardiasis usually occur in small communities where water receives little or no treatment, some outbreaks have been associated with municipal and semipublic supplies that met the US Environmental Protection Agency's coliform and turbidity standards. No federal or state standard has been established to protect the public from giardiasis, and research into the removal and control of *Giardia* cysts is handicapped by the lack of an effective method for detecting the cysts in water.

This article describes the morphology, life cycle, and detection of *Giardia lamblia*; the pathogenesis, transmission, diagnosis, treatment, prevention, and epidemiology of giardiasis; and the removal of *Giardia* cysts by water treatment practices.

Morphology

Giardia lamblia was first discovered in 1681 by Leeuwenhoek, who found the organism in his own stools.[1] The taxonomy of this species has been confused in the past, and even today it is often called *G. intestinalis* or *G. enterica*. It was named for Giard, who studied the parasite, and for Lambl, who first described it.[8] *Giardia lamblia* is a flagellated protozoan belonging to the class Zoomastigophorasida, order Diplomonado-

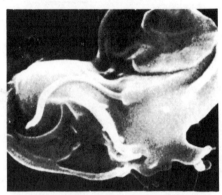

Giardia *trophozoites multiply by binary fission, the nuclei dividing first, followed by the locomotive apparatus and the sucking disc, and finally the cytoplasm.*

rida, and family Hexamitidae.[3] *Giardia lamblia* has a trophozoite form (Figure 1) and a cyst form (Figure 2).

Trophozoite. The trophozoite of *G. lamblia* is one of the most easily recognized intestinal protozoa.[1,3-8] Its body is dorsoventrally flattened and pear-shaped, is bilaterally symmetrical, and has two anterior, vesicular nuclei and four pairs of flagella, which are used for swimming. Each nucleus has a large endosome, which, combined with a large sucking disc (adhesive disc) on the ventral side, gives the appearance of a "monkey face." Using the sucking disc, it can attach to flat surfaces.

The dorsal side of the trophozoite is convex. Along the longitudinal median line there are two parallel, slender median rods or axostyles, which are not true axostyles.[1,3] A pair of large, curved, transverse, darkly staining median bodies lie behind the sucking disc. These median bodies are unique to *Giardia* and are composed of microtubules arranged either irregularly or in ribbons. It has been suggested that they may help support the posterior end of *Giardia* or be involved in its energy metabolism.

The length of the trophozoite ranges from 9 to 21 μm, the width from 5 to 15

μm, and the thickness from 2 to 4 μm. Movement is erratic, occurring as a swaying and dancing motion. Because of their appearance and movement, *G. lamblia* trophozoites are easily identifiable in fresh preparations.

Since the trophozoites usually live in the duodenum and jejunum of humans, they appear only in the feces. Both trophozoites and cysts can be found in diarrhea feces along with unusual amounts of mucus, but trophozoites do not appear consistently.

Cyst. In an unfavorable environment the parasite encysts, and it is the cysts that are usually found in stools. During encystation, the flagella are retracted internally and appear as four pairs of delicate curved rods. The thick, resistant cyst wall is distinctly visible, and all cell contents usually shrink away from the posterior end, leaving a clear space. Characteristically oval or ellipsoid and slightly asymmetric in shape (8-14 μm long, 7-10 μm wide),[5,6] cysts from different hosts or even from the same host exhibit variation in size and shape.

Tombes et al[9] used scanning electron and phase microscopy to study the surface of *Giardia* cysts collected from feces of mice, guinea pigs, dogs, mule deer, and humans. They observed that the surfaces were relatively smooth without any characteristic features and that cysts from different hosts appeared to be indistinguishable. However, the cyst size they reported (approximately 10 × 6 μm) was smaller than described elsewhere.

Young cysts have only two nuclei, but mature ones contain four, located at one end. Starch inclusion in young cysts gives them the name "blue bodies."[5]

The median rods, fibrils, flagella, and endosome usually stain conspicuously. The characteristics of cysts can be examined in a formalinized stool specimen, with iodine solution added to render them more conspicuous.[4] *Giardia* trophozoites and cysts can be identified with microscopy after fixation (Schaadinn's fixative) and staining (Heidenhain's hematoxylin).[3] Cysts can be concentrated either by zinc sulfate flotation or by formalin–ether sedimentation. Sugar solution for flotation is unsatisfactory.[3]

Rendtorff and Holt[10] reported that *G. lamblia* cysts can survive in tap water and remain infective for 16 days, the longest time period tested.

Life cycle

Giardia lamblia lives in the duodenum, jejunum, and upper ileum of humans (Figure 3), with the sucking disc of the trophozoite fitting over the surface of an epithelial cell.[1,6] In a severe infection the parasites cover nearly every cell. The trophozoites multiply in the human intestine by binary fission,[1,8] the nuclei dividing first, followed by the locomotive apparatus and the sucking disc, and finally the cytoplasm.

When conditions in the duodenum are unfavorable, encystment occurs, usually in the large intestine. During encystment, the cell divides into two, so that each newly formed cell has only two nuclei. The sucking disc and flagella, however, multiply so that each newly formed cell contains one sucking disc and four flagella.

It has been calculated that as many as 14 billion parasites could occur in a diarrhea stool, but in a moderate infection a stool may contain 300 million cysts.[1] However, *Giardia* does not appear consistently in the stools of all patients, as there are three patterns of excretion, i.e., high, low, and mixed.[11]

Infection of humans occurs by ingestion of cysts. The cysts pass through the stomach and into the duodenum, where division of the cytoplasm is immediately completed. Within 30 min of ingestion, a cyst hatches out two trophozoites. The parasites proceed to multiply and to colonize the duodenum. To avoid the high acidity of the duodenum, *Giardia* often tend to be located in the biliary tract and gall bladder.[8]

Transmission

Transmission of giardiasis from host to host may occur by ingestion of *G. lamblia* cysts in contaminated drinking water and foods or by infected persons who may have cysts on their body or clothes. Campers, hunters, and hikers who drink untreated water from mountain streams are at increased risk. An obvious source of human contamination could not be identified in many of the water supplies studied. It is suspected that some wild and domestic animals, such as beaver, rats, rabbits, and dogs, may play an important role in the transmission of giardiasis to humans. Also, the incidence of giardiasis is found to be high in homosexual males,[12] suggesting that fecal-oral transmission by sexual activity is possible.

Davies et al[13] recently reported that *Giardia* cysts obtained from humans, dogs, cats, and beaver were not host-specific. *Giardia* cysts from humans infected dogs, cats, and beaver; *Giardia* from beaver infected dogs; *Giardia* from dogs or cats were cross-transmitted. It can be assumed that cysts from dogs, cats, and beaver would infect humans.

Pathogenesis

Giardia lamblia has long been considered nonpathogenic or an opportunistic pathogen for humans because cysts are often found in large numbers in the stools of asymptomatic persons. The response to exposure to *Giardia lamblia* is quite variable, with some individuals more sensitive to its presence than others. Children are more frequenlty affected than adults, and there is considerable evidence to suggest that protective immunity can be acquired.[1] Yardly and Bayless[14] discussed factors that affect human susceptibility, including infective dosage, host, enteric bacterial infection, and malnutrition.

Rendtorff[15] conducted a series of studies with five volunteer prisoners and found that none of the five was infected after receiving a single *Giardia lamblia* cyst orally in a gelatin capsule. Of 22 men who received 25-cyst dosages, only 6 became infected. However, various dosages from 10 to 1 million cysts given to 40 other men produced 100-percent infection. The results of these experiments suggest that (1) a small number of cysts in water can readily produce an infection in a susceptible host, (2) the dosage is not related to the persistence of infection, (3) examination of a stool from a *Giardia*-positive person might have a negative result, and (4) there are possibly different parasite strains involved in infection by *Giardia* cysts.

Giardiasis is not fatal but can be extremely uncomfortable. *Giardia lamblia* is capable of producing discomfort by its toxic effects in allergic individuals, by its traumatic and irritative effects, and by its spoliative action, i.e., by diverting nutrients.[8] It is not, however, a tissue invader. The incubation period is about eight days, and the prepatent period, before *Giardia* can be detected in the stool, is 10-30 days.[7]

G. lamblia attaches itself by means of its sucking disc to the surface of the epithelial cells of the intestine and bowel.[8] A dense coating of *Giardia* trophozoites on the intestinal epithelium interferes with the absorption of fats and other nutrients. The parasite does not lyse host cells but appears to feed on mucous secretions.[1] This may cause a disturbance of intestinal function and irritation or low-grade inflammation of duodenal or jejunal mucosa. Consequently, the patient suffers acute or chronic diarrhea and steatorrhea. At various times during the infection the stools may be watery, semisolid, greasy, bulky, and foul-smelling, but they never contain blood. At Aspen, Colo., during an outbreak of giardiasis, about five

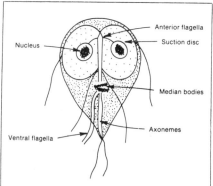

Figure 1. *Giardia lamblia* trophozoite

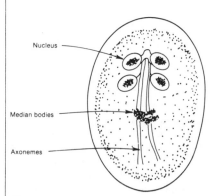

Figure 2. *Giardia lamblia* cyst

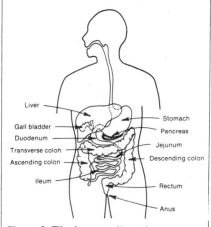

Figure 3. The human digestive system

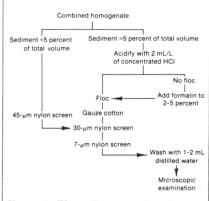

Figure 4. Flow diagram of sediment processing for detecting *Giardia* cysts in water

stools per day were observed for one to two months.

The symptoms range from mild diarrhea, malaise, weakness, fatigue, dehydration, weight loss, distension, flatulence, anorexia, cramplike abdominal pain, and epigastric tenderness to steatorrhea and malabsorption.[1-8] Some may be due to interference with fat absorption as well as to mechanical irritation of the bowel. They may last for long periods and may be intractable or hard to eliminate. The bile ducts and gall bladder may also be infected, causing mild catarrhal cholangitis, cholecystitis, jaundice, and colitis.[1,6]

Diagnosis

Chronic diarrhea as well as the symptoms mentioned previously can be clinically diagnosed. The cases can be categorized as follows:[8] (1) silent case: no symptoms; (2) intestinal: chronic enteritis and acute enterocolitis; (3) general: fever, anemia, and allergic manifestations; and (4) chronic cholecystopathy. The acute stage usually lasts only three to four days and is often not recognized at the time as being due to giardiasis, but sometimes may last for months.[16]

Kamath and Murugasu[17] evaluated the four laboratory diagnostic methods for detecting G. lamblia by studying 21 Malaysian children who had acute or chronic diarrhea or were suffering from malabsorption. These researchers found that mucosal impression smears and intestinal biopsies gave the highest number of positive tests—12 infections. Duodenal aspirates resulted in 10 positive tests, and stool examination was positive in only six cases. Because the biopsy method is tedious and time-consuming, Kamath and Murugasu recommended mucosal impression smears for G. lamblia as an initial step in routine intestinal biopsies. Giardia in the duodenum can also be detected by "Enterotest," which involves passing through the body a lead-weighted gelatin capsule on a string to which the parasites cling.[18]

Nevertheless, microscopic examination of the stool is widely used, and identification of distinctive Giardia cysts in formed stools or cysts and trophozoites in liquid stools is definitive. The drawbacks are that the excretion of the parasite may be intermittent and that the recovery methodology is imperfect. Stool samples should be collected properly on alternate days, using both direct smears and concentration with zinc sulfate or with the formalin–ether technique. Examination of the duodenal contents and intestinal biopsy are used only for confirmation when stool examinations are negative.[16]

Treatment of giardiasis

Usually, only the symptomatic patient requires treatment. However, it has been

TABLE 1
Outbreaks of waterborne giardiasis in the United States resulting from water system deficiencies

Year	Number of Outbreaks	Number of Cases	Reference
1965	1	123	26
1969	1	19	26
1970	1	34	26
1961–70	3	176	27
1971–72	3	112	28
1972	4	124	26
1973	4	73	26
1974	4	4 930	26
1971–74	12	5 127	29
1975	1	9	26
1976	3	629	26
1977	4	1 058	26
1971–78	24	12 004	30
1979	2		31

suggested that all members of the patient's family be treated simultaneously to avoid reinfection.[1]

Three drugs are available in the United States for treating giardiasis: quinacrine,* metronidazole,† and furazolidone.‡ The details of treatment are discussed elsewhere.[16] Quinacrine is widely used and effective in the treatment of giardiasis;[4,8,16] metronidazole is an alternative for children; furazolidone is an antibiotic and not widely used. Drugs found useful in other countries include acranil,[8] diiodohydroxyquin,[14] tinidazol, and nimorazole.[16] However, all of these drugs have potential problems.

Prevention

The precautions generally practiced for the prevention of amebiasis are applicable to giardial infection and are based on breaking up the chain of transmission. Water treatment, food sanitation, and personal hygiene are the most important measures for the prevention of giardiasis. Human feces should be disposed of in a sanitary way, water supplies should be protected against fecal contamination by proper treatment and good watershed management practices, and drinking water should be boiled when necessary.

Commonly practiced chlorination (0.5 mg residual chlorine/L) of water supplies is inadequate as the sole treatment for destruction of cysts. Rapid sand filtration preceded by coagulation removes nearly all cysts; properly operated slow sand filters or diatomaceous earth filters completely remove them. Cross-connections between public water supplies and private auxiliary supplies, and backflow connections in plumbing systems should be avoided. For small quantities of water, Jarroll et al[19] claimed that a saturated solution (double strength) of iodine effectively destroyed Giardia cysts at 20°C with a 20-min exposure.

The personal health and sanitary practices of people who prepare and serve foods in public eating places should be supervised. Hand washing after defecation and before preparing or eating food is important. Special attention should be given to the education of preschoolers about personal hygiene.

Epidemiology

Giardia lamblia occurs worldwide and is the etiologic agent in waterborne giardiasis outbreaks. In about 50 percent of the disease outbreaks, however, the etiology is unknown. Healy[20] concluded from a literature review that, on the basis of stool examinations, the prevalence of G. lamblia infection among children in the United States ranged from 4 to 22 percent, whereas the rate for adults was from 2 to 15 percent. This variation may be due to differences in sampling frequencies and test procedures. Giardia lamblia was found in 7.4 percent of 35 299 people tested in the United States.

Harter et al[21] recently conducted a survey of intestinal parasites in two counties of Washington state. They reported that the prevalence of G. lamblia among 518 one- to three-year-old children was 7.1 percent and was unrelated to sources of drinking water (surface or well), day-care center attendance, or parental occupation. Conditions identified as high-risk for Giardia infection included a history of drinking untreated surface water and the presence in the family of two or more siblings between the ages of three and ten. In 47 families with children that had acquired a primary infection, 10 (21.3 percent) other family members subsequently became infected with Giardia as secondary cases.

Fifty-two percent of 419 Indochinese immigrants from Vietnam, Laos, and Cambodia in San Diego County, Calif., had intestinal parasites. Most of the refugees studied had symptoms and hematologic findings suggestive of parasitism.[22] Eighteen percent were found

*Atabrine, Winthrop Laboratories, New York, N.Y.
†Flagyl, Pharmaceutical Specialties, Morgantown, N.C.
‡Furoxone, Eaton Laboratories, Landsdale, Pa.

to be infected by *G. lamblia*. The infection was found to be unrelated to gender but was more prevalent in children.[22] In Prachinburi Province, Thailand, *G. lamblia* was found in 54 (11.9 percent) of 453 Cambodian refugees residing in Ban Kaeng Holding Center.[23] Among Amerindian populations of Amazonia, Lawrence et al[24] found that 24 percent of the inhabitants of acculturating villages were infected with *G. lamblia*, whereas only 5 percent were infected in newly contacted villages. The lowest occurrence of *G. lamblia* (1.5 percent) was recorded in Labrador, Canada, among 401 people between the ages of 1 and 72.[25]

Outbreaks of waterborne giardiasis in the United States have been reported in AWWA committee reports and other literature.[26-46] Detailed reviews were made by Craun.[26,30] The number of outbreaks resulting from water system deficiencies and the number of cases of giardiasis are summarized in Table 1. Craun[30] indicated that 24 recognized outbreaks of waterborne giardiasis occurred during 1971-78. Ten outbreaks were caused by surface waters treated with chlorine only, and eight were caused by untreated surface water. Ineffective filtration of surface water was responsible for the three outbreaks with the largest number of cases. Two outbreaks resulted from untreated groundwater supplies; the cause of one was undetermined. Nine outbreaks occurred in Colorado, three in Utah, and two each in Montana, New Hampshire, and Vermont. Six cases occurred in other states. Giardiasis outbreaks continue to increase in the United States, and undoubtedly many more cases occur than are reported.

Examples of outbreaks. During the fall and winter of 1954-55, an outbreak of gastroenteritis, which was characterized by the exceptionally prolonged duration of many of the cases, occurred in Portland, Ore. The outbreak affected at least 50 000 people of both sexes and all ages. An epidemiologic survey concluded that *G. lamblia* was the only possible etiologic agent for such an outbreak.[32]

The first documented outbreak of waterborne giardiasis in the United States was probably the outbreak at Aspen, Colo., during the 1965-66 ski season. An epidemiologic survey showed that 11 percent of 1094 skiers were infected.[33] Gleason et al[34] found that 5 percent of the permanent population remained infected after the epidemic.

Until the 1974 outbreak at Rome, N.Y., waterborne giardiasis outbreaks generally involved small municipal systems or semipublic systems in mountainous areas of the United States, particularly in New England, the Pacific Northwest, and the Rocky Mountains.[26,29] The Rome outbreak (November 1974-June 1975) was the largest outbreak of waterborne giardiasis ever reported in the United States. Epidemiologic studies estimated that 4800-5300 people (10 percent of the population) were affected.[33] This was the first time that *G. lamblia* cysts were actually recovered from water and clarifier sediment samples.[26,35,36] The city of Rome used a surface source of good bacteriological quality. Chlorination with 0.8 mg combined residual chlorine/L was the only treatment. The watershed was sparsely populated, but the source may have been contaminated by human and animal wastes.

During late April and early May of 1976 a giardiasis outbreak occurred in Camas, Wash. Approximately 600 of the population of 6000 people showed clinical signs of infection.[37] The city used surface water and well water sources—two mountain creeks and seven deep wells. The well waters were not contaminated, but *G. lamblia* cysts were detected in raw surface water and in samples from the distribution system. The watersheds of the two surface water sources were well isolated, with no human habitation and limited human activity.[36,37] *Giardia* cysts were isolated from trapped beaver, and it is possible that beaver active in the watershed were carriers of the organism. The water treatment system included mixed-media pressure filters and chlorination. Prior to the outbreak, interruptions in chlorination often occurred and the pretreatment processes were ineffective. *Giardia lamblia* cysts passed through the filter because of media loss, media disruption, and inadequate pretreatment. During the outbreak, however, the finished water was found to meet turbidity and coliform count standards.[37]

A second outbreak of waterborne giardiasis resulting from inadequate filtration occurred in Berlin, N.H., during April and May 1977.[38-40] About 7000 of the city's population of 15 000 people were affected. The majority (76 percent) of the infections were asymptomatic and ran a self-limited course without treatment.[40] The Berlin water plant used the Upper Ammonoosuc River and the Androscoggin River as independent water supply sources. *Giardia lamblia* cysts were recovered from both raw water sources and from the treated water. The Ammonoosuc plant uses eight pressure filters and chlorination without coagulation-sedimentation. Numerous deficiencies in the condition and operation of the filters could have permitted cysts to pass through the filters. Water from the Androscoggin River was treated conventionally with chemical coagulation, upflow clarification, rapid sand filtration, and chlorination. An engineering evaluation concluded that contamination resulted from joint leakages and faulty construction of a common wall separating the filtered and unfiltered waters, which allowed the water to bypass the filters.[38,39] Humans or beaver could have been responsible for contaminating the source waters.[38,40]

An epidemiologic study indicated that 429 (58 percent) of 741 residents of Vail, Colo., reported diarrheal illness during February, March, and April 1978.[41] Within a two-day period in April, 37 cases of laboratory-confirmed giardiasis were reported. There were statistically significant differences between those who drank no water and those who drank one or more glasses each day, but no difference was noted on the basis of age, sex, or length of residence. None of the seven separate Vail Valley water districts could be implicated as the source of the outbreak.

A waterborne giardiasis outbreak related to chlorination occurred in Bradford, Pa., from September to December 1979.[31] Epidemiologic investigations indicated that 5.4 percent of the residents had giardiasislike illnesses, with 407 cases recorded. *Giardia* cysts were isolated from raw and treated waters and from the stool of a beaver trapped in one of the water supply reservoirs. The city water was treated only with chlorination, and high coliform counts and turbidity had been observed prior to the outbreak.

Since the increase in travel to Russia began in 1970, giardiasis has become a "travel disease" for people who visit the Soviet Union. Twenty-three of 80 members (29 percent) of the Olympic Boxing Team and 84 of 164 (51 percent) US scientists were infected with giardiasis after visiting the Soviet Union in 1970. *Giardia lamblia* cysts were detected in the stools of some of these travelers, and epidemiologic investigations confirmed that the giardiasis was caused by consumption of tap water in Leningrad.[26]

In October 1976 a high incidence of diarrhea was reported among an American tour group that had traveled to Madeira Island, Portugal.[44] Results of a mail survey indicated that 39 percent of the responding 859 travelers had had diarrhea; in 42 percent of these, the diarrhea lasted for longer than a week. Twenty-seven of 58 (47 percent) ill patients had *G. lamblia* cysts in their stools. Tap water and ice cream or raw vegetables consumed on the island were implicated. There was no evidence of continuing transmission of giardiasis to American tourists visiting Madeira 8 to 12 months after the outbreak.

Twenty-three of 54 campers who hiked and camped in the Uinta Mountains of Utah from Sept. 5 to 17, 1974, had diarrhea during and after the trip.[45] The stools of 22 out of 28 (79 percent) symptomatic campers contained *G. lamblia* cysts, whereas the stools of 4 out of 14 (29 percent) asymptomatic campers contained cysts. Although the remote stream used as a water source by the campers was implicated as the vehicle of transmission, *G. lamblia* cysts were not found in the stream water or in the intestines or feces of mammals living in the drainage area, possibly because of poor methods of cyst recovery. On the basis of data from 25 other groups of campers, Barbour et al[45] concluded that campers who use water from mountain streams risk acquiring giardiasis.

An endemic giardiasis outbreak occurred among travelers visiting Cattail Cove State Park, south of Lake Havasu City, Ariz., from June through September 1979.[31] Most of the cases were reported in California. Four of 74 (5.4 percent) cases were stool-positive for *Giardia* cysts. The shallow well water serving the recreation area was chlorinated and pumped to two storage facilities. Cross-connections and poor plumbing may have resulted in contamination of the drinking water by an irrigation system that sprayed sewage effluent.

Thus, outbreaks of giardiasis can be caused by lack of filtration, improper filter operations, inadequate chlorination, cross-connections, or drinking contaminated mountain stream waters. The AWWA Committee on Waterborne Disease recommended that complete reporting of such outbreaks be continued and that the data be used by regulatory agencies to establish standards and surveillance programs.[46]

Detection methods

Culture. *Giardia* (*G. duodenalis, G. muris,* and *G. agilis*) trophozoites from several hosts have been successfully cultured monoxenically or axenically on artificial media.[47-53] Most of the cultures used have been complex and have incorporated animal serum. Meyer and his co-workers[49-51] reported that *Giardia* trophozoites could be isolated and cultured in the presence of yeast. Andrews et al[54] described a method that used density-layer centrifugation and nylon fiber columns for purification of *G. muris* trophozoites from the intestine of the mouse. They claimed that centrifugation of trophozoites on metrizamide medium, (specific gravity 1.10), followed by incubation on a nylon fiber column at 37°C for 120 min, yielded up to 10 million viable, purified trophozoites from each infected mouse. Feely and Eriandsen[55] used Hanks balanced salt solution and low temperatures for isolation and purification of *Giardia* trophozoites from rat intestines.

No satisfactory method for culturing *G. lamblia* cysts in water samples is currently available. The difficulty lies in getting the trophozoites to excyst on laboratory media.

Clinical methods. Of the four methods for detecting *G. lamblia* cysts in patients,

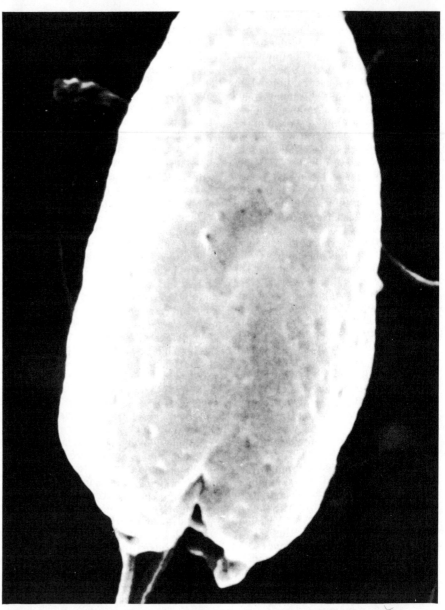

Within 30 min of ingestion, a Giardia *cyst hatches out two trophozoites, which multiply and colonize the small intestine. In a severe infection, the parasites cover nearly every surface cell of the duodenum, jejunum, and upper ileum.*

the mucosal impression smears technique is the best for routine purposes, but stool examination is the most widely used technique.

During examination of stools for intestinal protozoa, the direct saline wet-mount method is most commonly used, and cultivation is least used. There is no standard, recommended procedure for preparing, fixing, preserving, and staining stool samples to detect *G. lamblia*. Stool samples can be concentrated by zinc sulfate flotation or formalin-ether sedimentation; the former is the better method for recovery of *G. lamblia* cysts. According to the study by Levine,[3] sugar flotation is not recommended. Polyvinyl-alcohol fixative, Schaudinn's fixative, and formalin are used for fixing *Giardia* cysts. Samples can be preserved in formalin or Lugol's iodine solution. Gomori-Wheatley tri-

chrome stain, Lugol's solution, iodine, Noland's solution, and Kohn's Chlorazol Black E can be used for staining cysts. Merthiolate- iodine formalin, sodium acetate acid- formalin, and phenol-alcohol-formalin can be used for fixing and staining cysts.

It is difficult to assess the true prevalence of giardiasis in a population. The variability in the proportion of a population that is infected by *G. lamblia* (as recorded by various stool surveys) probably results not only from differences between various populations but also from other factors: (1) differences in stool examination techniques used, (2) the use of preservatives, (3) the number of stools examined, and (4) the inherent variability with which *G. lamblia* is excreted.[20] When a giardiasis outbreak is reported, the methods of stool examination should be detailed.

Recently developed methods, such as membrane filtration with syringe,[10] excystation,[56] electron microscopy,[57] and the fluorescent-antibody technique,[58] have improved the detection and knowledge of *G. lamblia*. The viability of *Giardia* cysts suspended by the excystation procedure was consistently lower than the viability of those determined by the eosin-exclusion techique. However, since some cysts that take up the dye are not viable, these results may be in question.

Detection in water. The detection of *G. lamblia* cysts in water is more difficult than detection by clinical methods because the concentrations of cysts in water are usually low. Jakubowski[59] reviewed the methodologies for detecting *Giardia* cysts in water. The excystation technique may subsequently lead the way to the development of culture methods, but currently microscopic examination is the method of choice. For detecting *G. lamblia* cysts in water, concentration and separation techniques are critical.

The method of detecting *G. lamblia* cysts in water that is included in the fifteenth edition of *Standard Methods*[60] was first developed by Jakubowski et al.[61] The development of the methodology, the so-called "EPA method," has been described in detail by Jakubowski.[59] The method includes concentration, separation, and microscopic examination of the cysts.

The water sample is filtered aseptically through a *Giardia*-sampling device, which consists of an inlet hose, a plastic filter holder with a 10-in.- (25-cm-) long yarn-wound orlon filter (7 μm porosity), an outlet hose, a water meter, and a limiting orifice flow-control valve with a flow rate of 3.15×10^{-5} m^3/s. The line pressure of 100 to 500 kPa is satisfactory, but a pump should be used if there is no water pressure. A sample volume of 500 gal (1900 L) should be filtered over an 18- to 24-h period;[60] a 100-gal (380-L) sample volume is suggested when concurrent epidemiologic or animal surveys are performed.

When the device is completely drained, the filter holder is opened and the filter cartridge is removed aseptically and placed in a labeled plastic bag, which is then sealed. The labeled bag is in turn placed in a second plastic bag, which is sealed, refrigerated, and transported to the laboratory for separation.

In the laboratory, the filter cartridge is aseptically removed and placed in a tray or pan; the orlon fibers are cut into four approximately equal lengths and removed from the stainless steel support. Each fiber portion is then blended at low speed for 10 s with 250 mL isotonic saline or distilled water and filtered through a coarse screen. All four fiber portions are then placed in a sealable

plastic bag that has one corner open, and the bag is squeezed to extract the residual fluid, which is combined with the homogenate (filtrate) in a 1- to 1.5-L graduated cylinder. The filtrate is allowed to settle for 15 min.

If the amount of sediment from the combined homogenate is less than 5 percent of the total volume, the homogenate is filtered through 47-mm-diameter, 45-, 30-, and 7-μm-porosity nylon screens. The 7-μm screen is transferred with forceps to the inside wall of a 50-mL beaker that contains 1 to 2 mL of isotonic saline.

The procedure followed after settling is outlined in Figure 4. Two drops of 7-μm-screen washing concentrate are mixed with one drop of Lugol's iodine solution on a clean microscope slide with a cover slip. The slide is examined under 10× and 43× objectives.

A step-by-step analysis of the procedure indicated that 42 percent of cysts were lost in the preparation of a homogenate with an orlon filter. There was a 58 to 71 percent loss of cysts when a 7-μm-mesh nylon screen process was used and a >90 percent loss if flocculation was used to clarify the homogenate. The overall efficiency of the EPA method is between 3 and 15 percent, with an average of 6.3 percent.[59]

A better detection and identification methodology for *G. lamblia* cysts in water is urgently needed. Improvements to the EPA method should include techniques for concentration, separation of the cysts from the concentrating material, identification of the cysts, evaluation of potential interfering organisms, and determination of viable cysts. The method should be cost-effective, less time-consuming, and relatively uncomplicated so that it can be performed by skilled but not highly specialized laboratory technicians. Without a practical method of detecting *Giardia* cysts in water, further research on removal or control of cysts is handicapped.

Water treatment

Outbreaks of waterborne giardiasis result from deficiencies in water treatment, contamination of water sources that receive no (or inadequate) treatment, or contamination of the water after treatment. These outbreaks generally involve small municipal water systems or semipublic water systems in recreational areas. There is no evidence that such outbreaks occur in well operated systems with conventional or adequate treatment processes.

Disinfection. Epidemiologic studies indicate that most outbreaks of waterborne giardiasis have been due to consumption of untreated surface water or surface water treated with chlorine disinfection alone, as was the case at Aspen, Colo.,[33] and Rome, N.Y.[35] Data indicate that

chlorine disinfection with normal dosage and contact time is insufficient to destroy *G. lamblia* cysts.[46] High dosages of chlorine and longer contact times are required in water treatment systems in which sedimentation and filtration cannot be employed.

The early concepts of the high resistance of *G. lamblia* cysts to disinfection were based on the results for *Entamoeba histolytica*, for which viability was determined by eosin exclusion. If a cyst remains colorless when exposed to eosin dye, it is considered viable; when it is penetrated and stained by the dye, it is considered dead. Results of recent studies suggest that eosin exclusion consistently gave higher estimates of cyst viability than can be found by in vitro excystation. Hoff[62] recommended that the results of early disinfection studies using eosin exclusion be disregarded. The recently developed in vitro excystation method may be applied to disinfection studies on *Giardia* cysts.

Using excystation, Jarroll et al[63] studied the effects of chlorine dosage on *G. lamblia* cyst viability under a variety of conditions. Cyst survival was generally found to increase as buffer pH increased. Water temperature, chlorine concentration, and contact time were important in cyst survival. Increasing chlorine dosage at the same temperature and increasing water temperature resulted in corresponding increases in inactivation rates. At 25°C, exposure to 1.5 mg chlorine/L for 10 min killed all cysts at pH 6, 7, and 8. At 15°C, exposure to 2.5 mg chlorine/L for 10 min killed all cysts at pH 6, but at pH 7 and 8 small numbers of cysts remained viable after 30 min but not after 60 min. At 5°C, exposure to 1 mg chlorine/L for 60 min failed to kill all the cysts at pH 6, 7, and 8; exposure to 2 mg chlorine/L killed all cysts at pH 6 after 60 min, all cysts at pH 7 after 10 min, and all cysts at pH 8 after 30 min. These results demonstrate that *G. lamblia* cysts can be destroyed by chlorine.

Rice et al[64] reported similar results, which indicated that *G. muris* cysts were as resistant as or more resistant than *G. lamblia* cysts. Since *G. muris* cysts are easily obtained, this organism can be used as a model for studies of human parasitic protozoa.

Ultraviolet radiation has been considered as a potential alternative to chlorine for disinfection in small water systems. Rice and Hoff[65] found that *G. lamblia* cysts were highly resistant to germicidal ultraviolet radiation at a maximum dosage of 42 000 μW-s/cm^2. The percentage of survival, compared with the control excystation, varied from 73 to 95 percent. From this, one can conclude that ultraviolet irradiation cannot be used as a cysticide.

Jarroll et al[19] tested six disinfectants that are used for treating small quantities of water for their ability to destroy *G. lamblia* cysts. Excystation was the criterion of viability. The disinfectants used were water-sterilizing tablets (A), Globaline tablets (B), EDWGT (C), 2 percent tincture of iodine (D), saturated iodine (E), and bleach (F). The tests were performed on two different water samples (clear and cloudy) and at two temperatures (3° and 20°C).

At 20°C all six disinfectants completely destroyed the cysts in both waters. However at 3°C, in cloudy water, disinfectant E was less than completely effective; and in clear water, four disinfectants (B, D, E, and F) failed to inactivate all of the cysts. The failure of these disinfectants was due to insufficient halogen residuals or to short contact times, and not to an extreme resistance of the cysts to the halogens. The study showed that very high levels of chlorine and iodine will kill *G. lamblia* cysts. A total chlorine residual of 6.2 mg/L after a 30-min contact time at pH 7.9 and 3°C was sufficient to inactivate *G. lamblia* cysts in both clear and cloudy waters.

Filtration. Studies on the removal of asbestos, bacteria, *Giardia* cysts, and viruses by coagulation, sedimentation, and filtration have been reviewed by Logsdon et al.[66,67] Proper coagulation, the key to effective filtration, depends on water temperature, pH, the constituents of the raw water, dosages of coagulant and coagulant aid, mixing time, and mixing intensity. Some of these factors are beyond the control of plant operators.

Slow sand filters, granular media filters, or diatomaceous earth (DE) filters may be used for filtration. Slow sand filters, generally used by small water supply systems, are not frequently found in the United States. Slow sand filtration is a biological rather than a physical–chemical process and is effective for removing microorganisms and particulates. About 90 percent of 7- to 12-μm particles (the size of *G. lamblia* cysts) are removed by slow sand filters.[66] Research into the removal of *Giardia* cysts by slow sand filters is being conducted at Colorado State University under the sponsorship of the US Environmental Protection Agency.[67]

Rapid sand filtration is widely used in the United States by large cities as well as by small communities. For effective granular media filtration, water must be properly coagulated. Baylis et al[68] reported that *Entamoeba histolytica* cysts were effectively removed (99.99 percent) from water by coagulation and rapid sand filtration. During World War II, US Army data also showed that many more *Entamoeba histolytica* cysts passed through rapid sand filters at 6.4 to 9.6 gpm/sq ft (4.4 to 6.6 mm/s) when poorly coagulated or uncoagulated water was filtered.[69]

Recently, Logsdon et al[70,71] found that *G. muris* could break through dual-media filters when raw water was inadequately coagulated or not coagulated at all. In addition, cysts might pass through the filter during times when the coagulant dose was interrupted, when the filtration rate was abruptly increased, or when a breakthrough of turbidity occurred. Nevertheless, a dual-media filter operating at 4 gpm/sq ft (2.7 mm/s) with proper coagulation of the raw water removed 99 percent or more of cysts. Similar results were obtained in a filtration study of *G. lamblia* cysts at the University of Washington, where a rapid sand filter with optimum coagulation removed >99 percent of cysts. Only 15 percent of cysts were removed from uncoagulated water.[67]

Logsdon et al[67] emphasized that AWWA's water quality goal of 0.1 ntu was a good one. To assure protection of public health, filtration plants should produce water with a turbidity of 0.1 ntu rather than 1.0 ntu.

Diatomaceous earth filtration for water treatment was first investigated during World War II.[69] Since then, it has been shown to be effective in removing algae, bacteria, and viruses. The US Army data[69] indicated that DE filtration could remove essentially all *E. histolytica* cysts and cyst-sized particles. Although removal of particles in the 7–20-μm size range is not necessarily related to turbidity removal, pressure diatomite filtration in Duluth, Minn., produced a filtered water turbidity of <0.1 ntu.

Recent research showed that DE filtration was just as effective for removing *G. lamblia* cysts as for removing those of *E. histolytica*.[70,71] Diatomaceous earth filtration was effective for removing 9-μm-diameter radioactive beads (used as models for *G. lamblia* cysts), *G. muris* cysts, and *G. lamblia* cysts. To assure the most effective filter performance, the filter should be precoated with a thickness of 0.2 lb/sq ft (1.0 kg/m²) of DE, and body feed should always be used. Current research at the University of Washington confirmed these results. Logsdon and Lippy[67] reported that cyst removals in four DE filter runs ranged from 99.03 percent to >99.87 percent in 12 separate determinations.

Filtration of surface water supplies is required to provide a second barrier against the transmission of waterborne diseases. It also helps to reduce the load on the disinfection process, thereby increasing the efficiency of disinfection, which is the primary means of disease control.

Conclusions

Research is needed on the viability, isolation, and enumeration of *G. lamblia* cysts; the removal of cysts by coagulation, flocculation, sedimentation, and filtration; and the inactivation of cysts by disinfection. Water treatment processes should result in the inactivation of 100 percent of *G. lamblia* cysts.

References

1. SCHMIDT, G.D. & ROBERTS, L.S. *Foundations of Parasitology.* The C.V. Mosby Co., St. Louis, Mo. (1977).
2. MOFFET, H.L. *Clinical Microbiology.* J.B. Lippincott Co., Philadelphia, Pa. (2nd ed., 1980).
3. LEVINE, N.D. *Giardia lamblia*: Classification, Structure, Identification. In: Waterborne Transmission of Giardiasis. EPA-600/9-79-001 (1979).
4. MEYER, M.C. & OLSON, O.W. *Essentials of Parisitology.* Wm. C. Brown Co., Dubuque, Iowa (2nd ed., 1975).
5. GARCIA, L.S. & ASH, L.R. Diagnostic Parasitology. *Clinical Laboratory Manual.* The C.V. Mosby Co., St. Louis, Mo. (1975).
6. JAWETZ, E.; MELNICK, J.L.; & ADELBERG, E.A. *Review of Medical Microbiology* (12th ed., 1976).
7. MARKELL, E.K. & VOGE, M. *Medical Parasitology.* W.B. Saunders Co., Philadelphia, Pa. (1981).
8. CHATTERJEE, K.D. *Parsitology: Protozoology and Helminthology in Relation to Clinical Medicine* (10th ed., 1975).
9. TOMBES, A.S.; LANDFRIED, S.S.; & WILLIAMS, L.D. Surface Morphology of *Giardia* Cysts Recovered from a Variety of Hosts. In: Waterborne Transmission of Giardiasis. EPA-600/9-79-001 (1979).
10. RENDTORFF, R.D. & HOLT, D.J. The Experimental Transmission of Human Intestinal Protozoan Parasites. IV. Attempts to Transmit *Endamoeba coli* and *Giardia lamblia* Cysts in Water. *Amer. Jour. Hygiene,* 60:3:327 (Nov. 1954).
11. DANZIGER, M. & LOPEZ, M. Numbers of *Giardia* in Feces of Infected Children. *Amer. Jour. Tropical Medicine & Hygiene,* 24:2:237 (Mar. 1975).
12. SCHMERIA, M.J.; JONES, T.C.; & KLEIN, H. Giardiasis: Association with Homosexuality. *Annals Internal Medicine,* 88:6:801 (June 1978).
13. DAVIES, R.B.; KUKUTAKI, K.; & HIBLER, C.P. Cross Transmission of *Giardia.* EPA-600/51-82-013 (Jan. 1983).
14. YARLEY, J.H. & BAYLESS, T.M. Giardiasis. *Gastroenterol.,* 52:2:301 (Feb. 1967).
15. RENDTORFF, R.C. The Experimental Transmission of Human Intestinal Protozoan Parasites. II. *Giardia lamblia* Cysts Given in Capsules. *Amer. Jour. Hygiene,* 59:2:209 (Mar. 1954).
16. WOLFE, M.S. Managing the Patient with Giardiasis: Clinical, Diagnostic and Therapeutic Aspects. In: Waterborne Transmission of Giardiasis. EPA-600/9-79-001 (1979).
17. KAMATH, K.R. & MURUGASU, R. A Comparative Study of Four Methods for Detecting *Giardia lamblia* in Children with Diarrheal Disease and Malabsorption. *Gastroenterol.,* 66:1:16 (Jan. 1974).
18. BEZJAK, B. Evaluation of a New Technique for Sampling Duodenal Contents in Parasitologic Diagnosis. *Amer. Jour. Digestive Disorders,* 17:0:848 (Sept. 1972).
19. JARROLL, E.L.; BINGHAM, A.K.; & MEYER, E.A. *Giardia* Cyst Destruction: Effectiveness of Six Small-Quantity Water Disin-

fection Methods. *Amer. Jour. Tropical Medicine & Hygiene,* 29:1:8 (Jan. 1980).

20. HEALY, G.R. The Presence and Absence of *Giardia lamblia* in Studies on Parasite Prevalence in the USA. In: Waterborne Transmission of Giardiasis. EPA-600/9-79-001 (June 1979).

21. HARTER, L.; FROST, F.; & JAKUBOWSKI, W. *Giardia* Prevalence Among 1- to 3-Year-Old Children in Two Washington State Counties. *Amer. Jour. Public Health,* 72:4:386 (Apr. 1982).

22. HOFFMAN, S.L. ET AL. Intestinal Parasites in Indochinese Immigrants. *Amer. Jour. Tropical Medicine & Hygiene,* 30:2:340 (Mar. 1981).

23. KEITTIVUT, B. ET AL. Prevalence of Schistosomiasis and Other Parasitic Diseases Among Cambodian Refugees Residing in Ban Kaeng Holding Center, Prachinburi Province, Thailand. *Amer. Jour. Tropical Medicine & Hygiene,* 31:5:988 (Oct. 1982).

24. LAWRENCE, D.N. ET AL. Epidemiologic Studies Among Amerindian Populations of Amazonia. III. Intestinal Parasitoses in Newly Contacted and Acculturating Villages. *Amer. Jour. Tropical Medicine & Hygiene,* 29:4:530 (July 1980).

25. SOLE, T.D. & CROLL, N.A. Intestinal Parasites in Man in Labrador, Canada. *Amer. Jour. Tropical Medicine & Hygiene,* 29:3:364 (May 1980).

26. CRAUN, G.F. Waterborne Outbreaks of Giardiasis. In: Waterborne Transmission of Giardiasis. EPA-600-9-79-001. (1979).

27. CRAUN, G.F. & McCABE, L.J. Review of the Causes of Waterborne-Disease Outbreaks. *Jour. AWWA,* 65:1:74 (Jan. 1973).

28. McCABE, L.F. & CRAUN, G.F. Status of Waterborne Diseases in the US and Canada. *Jour. AWWA,* 67:2:95 (1975).

29. CRAUN, G.F.; McCABE, L.J. & HUGHES, J.M. Waterborne Disease Outbreaks in the United States: 1971–1974. *Jour. AWWA,* 68:8:420 (Aug. 1976).

30. CRAUN, G.F. Outbreaks of Waterborne Disease in the United States: 1971–1978. *Jour. AWWA,* 73:7:360 (July 1981).

31. LIPPY, E.C. Waterborne Disease: Occurrence is on the Upswing. *Jour. AWWA,* 73:1:57 (Jan. 1981).

32. VEAZIE, L.; BROWNLEE, I.; & SEARS, H.J. An Outbreak of Gastroenteritis Associated with *Giardia lamblia.* In: Waterborne Transmission of Giardiasis. EPA-600/9-79-001 (June 1979).

33. MOORE, G.T. ET AL. Epidemic Giardiasis at a Ski Resort. *New England Jour. Medicine,* 281:8:402 (Aug. 1969).

34. GLEASON, N.N. ET AL. A Stool Survey for Enteric Organisms in Aspen, Colorado. *Amer. Jour. Tropical Medicine & Hygiene,* 19:3:480 (May. 1970).

35. SHAW, P.K. ET AL. A Community-Wide Outbreak of Giardiasis with Documented Transmission by Municipal Water. *Annals Internal Medicine,* 87:4:426 (Oct. 1977).

36. JURANEK, D. Waterborne Giardiasis. In: Waterborne Transmission of Giardiasis. EPA-600/9-79-001 (June 1979).

37. KIRNER, J.C.; LITTLE, J.D.; & ANGELO, L.A. A Waterborne Outbreak of Giardiasis in Camas, Washington. *Jour. AWWA,* 70:1:35 (Jan. 1978).

38. LIPPY, E.C. Tracing a Giardiasis Outbreak at Berlin, New Hampshire. *Jour. AWWA,* 70:9:512 (Sept. 1978).

39. LIPPY, E.C. Water Supply Problems Associated With a Waterborne Outbreak of Giardiasis. In: Waterborne Transmission of Giardiasis. EPA-600/9-79-001 (June 1979).

40. LOPEZ, C.E. ET AL. Waterborne Giardiasis: A Community-Wide Outbreak of Disease and High Rate of Asymptomatic Infection. *Amer. Jour. Epidemiol.,* 112:4:495 (Oct. 1980).

41. JONES, W.E. & EDELL, T.A. Giardiasis in Vail, Colorado. In: Waterborne Transmission of Giardiasis. EPA-600/9-79-001 (June 1979).

42. Center for Disease Control. Giardiasis in Travelers. *Morbidity & Mortality Weekly Rept.,* 19:455 (1970).

43. WALZER, P.D.; WOLFE, M.S.; & SCHULTZ, M.G. Giardiasis in Travelers. *Jour. Infectious Diseases,* 124:2:235 (Aug. 1971).

44. LOPEZ, C.E. ET AL. Giardiasis in American Travelers to Madeira Island, Portugal. *Amer. Jour. Tropical Medicine & Hygiene,* 27:6:1128 (Nov. 1978).

45. BARBOUR, A.G.; NICHOLS, C.R.; & FUKUSHIMA, T. An Outbreak of Giardiasis in a Group of Campers. *Amer. Jour. Tropical Medicine & Hygiene,* 25:3:384 (May 1976).

46. AWWA Committee Report. Waterborne Disease in the United States and Canada. *Jour. AWWA,* 73:10:528 (Oct. 1981).

47. KARAPETYAN, A.E. Methods of *Giardia lamblia* Cultivation. *Tsitologia,* 2:379 (1960).

48. KARAPETYAN, A.E. In Vitro Cultivation of *Giardia duodenalis. Jour. Parasitol.,* 48:2:337 (Apr. 1962).

49. FORTESS, E. & MEYER, E.A. Isolation and Axenic Cultivation of *Giardia* Trophozoites from the Guinea Pig. *Jour. Parasitol.,* 62:5:689 (Oct. 1976).

50. MEYER, E.A. *Giardia lamblia:* Isolation and axenic cultivation. *Exp. Parasitol.,* 27:1:179 (Feb. 1976).

51. MEYER, E.A. The Propagation of *Giardia* Trophozoites in Vitro. In: Waterborne Transmission of Giardiasis. EPA-600/9-79-001 (June 1979).

52. GILLIN, F.D. & DIAMOND, L.S. Axenically Cultivated *Giardia lamblia:* Growth, Attachment and the Role of L-cysteine. In: Waterborne Transmission of Giardiasis. EPA-600/9-79-001 (June 1979).

53. SAHA, T.K. & GHOSH, T.K. Invasion of Small Intestinal Mucosa by *Giardia lamblia* in Man. *Gastroenterol.,* 72:3:402 (Mar. 1977).

54. ANDREWS, J.S. JR.; ELLNER, J.J.; & STEVENS, D.P. Purification of *Giardia muris* Trophozoites by Using Nylon Fiber Columns. *Amer. Jour. Tropical Medicine & Hygiene,* 29:1:12 (Jan. 1980).

55. FEELY, D.E. & ERIANDSEN, S.L. Isolation and Purification of *Giardia* Trophozoites From Rat Intestines. *Jour. Parasitol.,* 67:1:59 (Jan. 1981).

56. BINGHAM, A.K. ET AL. Induction of *Giardia* Excystation and the Effect of Temperature on Cyst Viability as Compared by Eosin-Exclusion and in Vitro Excystation. In: Waterborne Transmission of Giardiasis. EPA-600/9-79-001 (June 1979).

57. LUCHTEL, D.L.; LAWRENCE, W.P.; & DE WALLE, F.B. Electron Microscopy of *Giardia lamblia* Cysts. *Appl. & Envir. Microbiol.,* 40:4:821 (Oct. 1980).

58. RIGGS, J.L. ET AL. Detection of *Giardia lamblia* by Immunofluorescence. *Appl. & Envir. Microbiol.* 45:2:698 (Feb. 1983).

59. JAKUBOWSKI, W. Methods for Detection of *Giardia* Cysts in Water Supplies. In: Waterborne Transmission of Giardiasis. EPA-600/9-79-001 (June 1979).

60. *Standard Methods for the Examination of Water and Wastewater.* APHA, AWWA, and WPCF. Washington, D.C. (1980).

61. JAKUBOWSKI, W. ET AL. Large-Volume Sampling of Water Supplies for Microorganisms. *Jour. AWWA,* 79:12:702 (1978).

62. HOFF, J.C. Disinfection Resistance of *Giardia* Cysts: Origins of Current Concepts and Research in Progress. In: Waterborne Transmission of Giardiasis. EPA-600/9-79-001 (June 1979).

63. JARROLL, E.L.; BINGHAM, A.K.; & MEYER, E.A. Effect of Chlorine on *Giardia lamblia* Cyst Viability. *Appl. & Envir. Microbiol.,* 41:2:483 (Feb. 1981).

64. RICE, E.W.; HOFF, J.C.; & SCHAEFFER, F.W. III. Inactivation of *Giardia* Cysts by Chlorine. *Appl. Microbiol.,* 43:1:250 (Jan. 1982).

65. RICE, E.W. & HOFF, J.C. Inactivation of *Giardia lamblia* Cysts by Ultraviolet Irradiation. *Appl. & Envir. Microbiol.,* 42:3:546 (Sept. 1981).

66. LOGSDON, G.S. & FOX, K. Getting Your Money's Worth From Filtration. *Jour. AWWA,* 74:5:249 (May 1982).

67. LOGSDON, G.S. & LIPPY, E.C. The Role of Filtration in Preventing Waterborne Diseases. *Jour. AWWA,* 74:12:649 (1982).

68. BAYLIS, J.R.; GULLANS, O.; & SPECTOR, B.K. The Efficiency of Rapid Sand Filters in Removing the Cysts of Amoebic Dysentery Organisms from Water. *Public Health Repts.,* 50:46:1567 (Nov. 1936).

69. US Army. Efficiency of Standard Army Purification Equipment and Diatomite Filters in Removing Cysts of *Endamoeba histolytica* from Water. War Dept. Rept. 834. Washington, D.C. (July 3, 1944).

70. LOGSDON, G.S.; SYMONS, J.M.; & HOYE, R.L. Water Filtration Techniques for Removal of Cysts and Cyst Models. In: Waterborne Transmission of Giardiasis. EPA-600/9-79-001 (June 1979).

71. LOGSDON, G.S. ET AL. Alternative Filtration Methods for Removal of *Giardia* Cysts and Cyst Models. *Jour. AWWA,* 73:2:111 (Feb. 1981).

About the author: *Shun Dar Lin, a scientist in the Water Quality Section, Illinois State Water Survey, P.O. Box 697, Peoria, IL 61652, has 23 years of experience in research as well as three years in teaching. A graduate of National Taiwan University, Taipei (BS, civil engineering), the University of Cincinnati, Ohio (MS, sanitary engineering), and Syracuse University, N.Y. (PhD, sanitary engineering), Lin is a member of AWWA, ASCE, WPCF, and the* Standard Methods *committee.*

Reprinted from *Jour. AWWA,* 77:2:40 (Feb. 1985).

Methods for
Detecting Giardia

GIARDIA METHODOLOGY FOR WATER SUPPLY ANALYSIS

Frank W. Schaefer, III
Microbiology Branch
Health Effects Research Laboratory
U. S. Environmental Protection Agency
Cincinnati, Ohio 45268

Eugene W. Rice
Microbiological Treatment Branch
Municipal & Environmental Research Laboratory
U. S. Environmental Protection Agency
Cincinnati, Ohio 45268

In the United States giardiasis is the most reported parasitic infection. Giardia lamblia, the etiologic agent, is a protozoan flagellate parasite. It causes a range of pathology in the population from a-symptomatic at one extreme to severe diarrhea associated with mal-absorption, flatulence, and weight loss at the other extreme. Giardia lamblia is known to occur in two forms: the trophozoite which is a vegetative stage and the cyst which is a resistant, dormant stage. The cyst stage is the primary means of transmission (1,2). As few as 10 cysts are required to produce an infection in man (3).

During the past 15 years, numerous waterborne outbreaks of this pathogen have occurred in the United States. The areas in which these outbreaks have occurred have been characterized by pristine water with low coliform counts. Since the water was considered pure, in many cases it was not filtered but was subjected only to chlorination (4). Complicating this problem is the fact that reservoir hosts for this organism are being identified. These include the dog, beaver, and muskrat (1,2,5,6,7).

We are interested in detecting Giardia lamblia in drinking water supplies. Once detected, the Agency would like to know if the cysts detected are viable and are capable of infection. Also of interest is how much disinfectant is required to kill the organism. The focus of this paper is on detection of cysts in large volumes of water.

In 1978 the United States Environmental Protection Agency hosted a symposium entitled Waterborne Transmission of Giardiasis. In those proceedings, a method was published for the detection of Giardia cysts in water supplies (8). This technique was updated and recently was published in the 15th edition of Standard Methods for Examination of Water and Wastewater (9). The method basically involved filtering water through a standard orlon fiber sampling device. The fibers were removed and blended to shake the cysts free of the fibers. To try to separate the cysts from sediment and debris, the fiber extract was flocculated and passed through nylon screens successively down to a mesh of 7 µm. This procedure was nearly 100% effective in removing cysts from seeded water samples, but it was only 30 to 50% effective in its recovery of cysts from the fibers (8).

In 1980 a Giardia workshop composed of academic and governmental authorities from around the United States was held at the United States Environmental Protection Agency. Publication of certain details of this body's deliberations are expected in 1982 (10). One of the concerns of this conference was that no two laboratories were processing their samples the same way even with the knowledge of the published method. In addition, certain aspects of the published method were criticized. These included fiber blending, flocculation, and screening. A consensus was reached on changing the published method by this group. The hope was that in making these changes recovery efficiencies would improve, and that processing uniformity would result. The revisions of the method include washing the fibers by hand rather than blending them and elimination of the flocculation and screening steps. The declaration of a positive identification of a Giardia cyst was also more clearly defined. Obtaining the sample is the only aspect of the original procedure that remains unchanged.

A. SAMPLE COLLECTION

Sample collection requires an apparatus consisting of an inlet hose connected to a plastic filter holder containing a yarn-wound orlon filter (1 or 7 μm porosity) (9), outlet hose, water meter, and a limiting orifice flow control device with a flow rate of 3.15×10^{-5} m^3/sec (9). Aseptic technique should be used during sample collection to prevent cross-contamination. Line pressures between 100 and 500 kPa are satisfactory. When line pressures are inadequate or do not exist, the apparatus must be connected to a pump. While the volume of water to be sampled depends upon the water turbidity and the intent of the investigator, 1,900 liters collected over 18 to 24 hours is suggested. Records of the time, meter reading, and location should be noted at the beginning and end of collection. The inlet hose should be kept above the level of the outlet hose to prevent filter backwashing. After the sample collection is complete, the apparatus should be carefully drained. The plastic filter holder is opened and the filter is aseptically trans-ferred to a sealing plastic bag which is appropriately labeled. To guard against leaks, the filter is placed inside a second sealable plastic bag. Temporary storage and transportation of the filters back to the laboratory for analysis is always on wet ice. Freezing of filters should never be permitted, because the cysts are disrupted sufficiently so that they do not respond to the concentration steps below.

B. SAMPLE PROCESSING

The sediment volume obtained from different filters varies according to the turbidity of the water. For sediment volumes less than 5 ml, the zinc sulfate method of concentrating Giardia cysts is adequate and may be used. When sediment volumes are greater than 5 ml, the zinc sulfate method becomes impractical due to cyst distortion. Before all the slides can be scanned, any cysts that might be present become distorted due to the prolonged exposure to the zinc sulfate. In the case of sediment volumes above 5 ml, sucrose floatation must be done instead.

Processing should begin within 48 hours of receiving the sample. Each filter is individually processed using aseptic technique and wearing rubber gloves to protect the processor. The filter is transferred from the plastic bag to an enamel, stainless steel, or other appropriate pan. Any residual water and/or debris from the transport bag is rinsed into a collection container before being discarded. The end of the orlon fiber is found, and the entire fiber is unwound. The unwound fiber is divided roughly into fourths which are individually washed in successive 1 liter aliquots of distilled water, until the fibers appear clean. The clean orlon fibers are wrung out into a collecting beaker by placing them in an 8 x 8 inch interlocking seal polyethylene bag which has had one corner snipped off to allow drainage. Depending on the turbidity of the water that was sampled, as much as 16 liters of fiber washings may result.

At this point the fiber washings can either be sedimented 3 to 4 hours to overnight at 4°C; or alternatively may be concentrated by centrifugation at 600 x g for 5 minutes. In either event, the supernatant is discarded and the sediment is processed further. The sediment is diluted to 800 ml with distilled water. To this is added 54 ml of 37% formalin. The final volume is brought to 1,000 ml with distilled water. This yields an approximate 2% formalin concentration which is designed to stabilize the Giardia cysts.

The sample again is either sedimented overnight at 4°C or is concentrated by centrifugation for 5 minutes in 250 ml bottles at 600 x g. If possible, the supernatant is aspirated until around 50 ml of sediment and supernatant remain.

Zinc Sulfate Floatation

The remaining sediment and supernatant are transferred to a 50 ml graduated conical centrifuge tube which is spun 5 minutes at 600 x g. As much of the supernatant as possible is aspirated from the pellet.

If no more than 1 ml of sediment exists, then the sediment is vortexed directly with 7 to 10 ml zinc sulfate solution (Sp. Gr. 1.18) (11). If more than 1 ml sediment exists, then it is subdivided into 1 ml amounts or less and is suspended in zinc sulfate. Each 1 ml-sediment-zinc sulfate suspension is transferred to a 15 ml conical centrifuge tube. To each 15 ml conical centrifuge tube one drop of Lugol's iodine (1) is added to stain the Giardia cysts. Each conical centrifuge tube is filled with additional zinc sulfate solution and is spun at 800 x g for 5 minutes without braking at the end in a swinging bucket rotor. After removal from the centrifuge, each tube is filled to the brim with zinc sulfate solution without allowing runover. A number one 22 x 22 mm coverslip is touched to the meniscus after allowing the preparation in each tube to remain static for 2 to 3 minutes. The coverslip is immediately removed after touching the meniscus, and the adherent drop side is placed down onto a clean glass slide. The coverslip is sealed to the slide with vaspar (1 part paraffin and 1 part petroleum jelly) to prevent drying of the preparation during microscopic examination.

Sucrose Floatation

The supernatant is aspirated until 10 to 25 ml supernatant remain above the pellet.

Five to 10 ml of sediment are suspended in 70 to 90 ml of distilled water. This suspension is layered on top of 70 ml of 1.5 M sucrose solution (Sp. Gr. 1.18) contained in a 250 ml glass centrifuge bottle.

The bottles prepared in this fashion are balanced and centrifuged at 800 x g for 5 minutes without braking at the end. A swinging bucket rotor is required for the centrifugation step. Three-fourths of the aqueous supernatant is aspirated. The remaining fluid bilayer is poured into another 250 ml glass centrifuge bottle which is half full of distilled water. The sediment, which remains after pouring off the fluid bilayer, is discarded.

The centrifuge bottles prepared in the previous step are centrifuged at

600 x g for 2 minutes. The supernatants are all discarded, and the pellets are pooled in a 50 ml conical centrifuge tube. Some distilled water dilution of the pellet is required to affect the pooling. Two drops of Lugol's iodine (1) is added to the centrifuge tube to stain the protozoan cysts. The centrifuge tube is vortexed to evenly mix the Lugol's iodine solution with the pellet. The contents of the conical centrifuge tube are spun at 600 x g for 2 minutes. The resultant pellet is washed 2 more times with distilled water by centrifugation at 600 x g for 2 minutes to insure removal of all residual sucrose.

After the final wash, the supernatant is aspirated down leaving between 0.5 and 1.0 ml of fluid above the pellet. The pellet is resuspended in a small volume of distilled water and is observed drop by drop for Giardia cysts. The slides are prepared with 22 x 22 mm coverslips which are sealed to the slide with vaspar (1 part paraffin and 1 part petroleum jelly) to retard drying.

MICROSCOPIC OBSERVATION

The whole coverslip is scanned on a brightfield microscope using 10X, 25X, and 40X objectives as needed. Every coverslip from the zinc sulfate procedure and every drop of suspension from the sucrose procedure must be checked for cysts, for even with these concentration procedures the Giardia cysts, if present at all, are usually there at low suspension densities. Sometimes only one cyst is recovered.

A sample is reported as positive only if the Giardia cyst(s) are the right shape and size and have two demonstrable internal anatomical characters. The sizes range from 10-20 µm long by 5-15 µm wide (1). The internal anatomical cyst characters which can be viewed include nuclei, 2-4; axonemes; and claw-hammer shaped median bodies.

At present this procedure is not recommended for routine use, because it is still inefficient and time consuming. At best it is helpful for epidemiological and outbreak studies.

REFERENCES

1. LEVINE, NORMAN D. Protozoan Parasites of Domestic Animals and of Man. Burgess Publishing Co., Minneapolis, MN (1966). p. 119, 391.

2. HEWLETT, ERIK L.; ANDREWS, JR., JOHN S.; RUFFIER, JUANITA; & SCHAEFER, III, FRANK W. Experimental Infection of Mongrel Dogs with Giardia lamblia Cysts and Cultured Trophozoites. J. Infect. Dis., 145:1:(In Press) (Jan. 1982).

3. RENDTORFF, ROBERT C. The Experimental Transmission of Human Intestinal Protozoan Parasites. II. Giardia lamblia Cysts Given in Capsules. Am. J. Hyg., 59:3:209-220 (1954).

4. CRAUN, GUNTHER F. Waterborne Giardiasis in the United States: A Review. Am. J. Public Health, 69:8:817-819 (1979).

5. SHAW, PETER K.; BRODSKY, RICHARD E.; LYMAN, DONALD O.; WOOD, BRUCE T.; HIBLER, CHARLES P.; HEALY, GEORGE R.; MACLEOD, KENNETH I.; STAHL, WALTER; & SCHULTZ, MYRION G. A Communitywide Outbreak of Giardiasis with Evidence of Transmission by a Municipal Water Supply. Ann. Intern. Med., 87:4:426-432 (1977).

6. DAVIES, ROBERT B. & HIBLER, CHARLES P. Animal Reservoirs and Cross-Species Transmission, Waterborne Transmission of Giardiasis (Walter Jakubowski & John C. Hoff, editors) EPA 600/9-79-001. National Technical Information Service, Springfield, VA (1979). pp. 104-126.

7. FROST, FLOYD; PLAN, BYRON; & LIECHTY, BILL. Giardia Prevalence in Commercially Trapped Mammals. J. Environ. Health, 42:5:245-249 (1980).

8. JAKUBOWSKI, WALTER & ERICKSEN, THEADORE H. Methods for Detection of Giardia Cysts in Water Supplies, Waterborne Transmission of Giardiasis (Walter Jakubowski & John C. Hoff, editors) EPA 600/9-79-001. National Technical Information Service, Springfield, VA (1979). pp. 193-210.

9. GREENBERG, ARNOLD E.; CONNERS, JOSEPH J.; & JENKINS, DAVID (editors). Standard Methods for the Examination of Water and Wastewater. American Public Health Association, Washington, D. C. (15th edition, 1981). pp. 842-847.

10. ERLANDSEN, STANLEY L. & MEYER, ERNEST A. (editors). Giardia and Giardiasis. Plenum Press, New York (1982). (In Press).

11. MELVIN, DOROTHY M. & BROOKE, MARION M. Laboratory Procedures for the Diagnosis of Intestinal Parasites. U. S. Government Printing Office, Washington, D. C. (1975). pp. 95-108.

Reprinted from Proc. AWWA WQTC, Seattle, Wash. (Dec. 1981).

A REFERENCE METHOD FOR DETECTING GIARDIA CYSTS IN WATER

Walter Jakubowski, Microbiologist
Chief, Parasitology & Immunology Group

Microbiology Branch
Toxicology & Microbiology Division
Health Effects Research Laboratory
U. S. Environmental Protection Agency
Cincinnati, Ohio 45268

INTRODUCTION

The waterborne transmission of giardiasis has been firmly established with 80 outbreaks having been reported in the United States from 1965 through 1983 (1,2). The outbreaks are continuing and more than 350,000 people in Pennsylvania were on boil-water orders in 1984 because of concern about giardiasis (3). Until 1975, the evidence for waterborne transmission was entirely epidemiological--no one had successfully detected Giardia cysts in water samples. In that year, the Centers for Disease Control (CDC) used a swimming pool filter, containing sand as the filtering medium, to collect 10 samples representing more than one million liters of raw Rome, New York water (4). Portions of the sediment from two of the 10 samples caused infection in beagle puppies and one cyst was detected microscopically. The sampling equipment was bulky and heavy, requiring transport by truck or air freight, and the field collection of the filter backwash into 55-gallon drums was difficult and cumbersome.

The CDC discontinued use of their swimming pool filter method in 1976 when the USEPA Health Effects Research Laboratory (HERL) in Cincinnati developed a portable sampling device and associated methodology (5). With the HERL methodology, cysts were demonstrated for the first time in the raw water and distribution system of a water supply during the Camas, Washington outbreak in May, 1976. The method involved concentrating particulates from water samples on microporous yarn-wound orlon filters, extracting the filters, purification steps to separate the cysts from interfering debris, and microscopic examination of the concentrate.

It should be recognized that a cultural method for the detection, isolation, and enumeration of Giardia organisms from water is not currently available. Cultures of Giardia trophozoites can be established from clinical specimens and maintained in the laboratory (6,7). However, this procedure is far from routine and it may not be practical, even with relatively pure source material from duodenal aspirates, until additional information is available on the nutritional requirements of the organisms (8). Developing a cultural method for Giardia cysts in water samples may be an even more formidable task than it is for trophozoites in clinical specimens because of the wide variety of competing organisms, including bacteria, viruses, and aquatic protozoa, that must be controlled. Currently, detection methodology for Giardia in water is dependent upon microscopic examination or upon animal feeding of concentrates. Because of the difficulty, expense or unavailability of suitable animal models, the methodology has historically focused on microscopic detection methods.

"STANDARD" METHODS

The method developed by HERL, variously known as the EPA method, the HERL method, or the orlon filter method, subsequently appeared in the 15th edition of Standard Methods (9). This method is NOT a standard method. It appeared in Section 912, Detection of Pathogenic Micro-organisms. Methods included in this section are intended as a starting point for the investigator with a need to sample for one or more of the included pathogens. For many of these pathogens, the methodology may be rapidly changing and there may be a lag period of two or more years from the writing of the method to publication in Standard Methods. As indicated above, Giardia methodology is new, having been in existence for less than 10 years, and by the time the 15th edition became available, there had been significant changes in the methodology.

The original impetus for developing a method was to assist in outbreak investigations. The qualitative methodology that was developed did prove useful in some such investigations. However, the need now has shifted to a quantitative monitoring method that may be used to satisfy a possible maximum contaminant level (MCL) or treatment regulation. Additionally, a method of this type might be useful in determining whether or not a variance or exemption would be appropriate for a given water supply.

A REFERENCE METHOD

Recognizing the shifting priorities and the need for some form of standardization, HERL convened a small workshop on Giardia methods in the fall of 1980 (10). Participants in the workshop consisted of investigators who had been working to some extent with methods for detecting Giardia cysts in water samples. The product of the workshop was a consensus method and it was suggested that laboratories with a need or desire and the ability to perform the test use this method as a starting point or reference. This reference method will appear in the 16th edition of Standard Methods. The details of the method are presented below and a flow chart is given in Fig. 1.

a. Sampling:

Collect samples of the raw water source, at the treatment plant before disinfection, and from the distribution system. The total number of samples will depend on the study objectives and available resources. Collect a sample using an apparatus consisting of an inlet hose, plastic filter holder with 23-cm long yarn-wound filter (1-μm porosity), outlet hose, water meter, and a limiting orifice flow control device with a flow rate of 6.3×10^{-5} m^3/sec (1 GPM). The yarn may be orlon, polypropylene, or other suitable material that does not release fibers during subsequent processing. Components of the apparatus need not be sterile, but thoroughly drain and rinse the equipment between samples. Use aseptic technique during sample collection to protect the sample collector and to prevent sample cross-contamination. A line pressure of 100 to 130 kPa (15-18 psi) is satisfactory; if water under pressure is unavailable, use a pump. The volume of water to be sampled depends on the intent of the investigation; a minimum sample size of 380 L is suggested.

To collect a sample, connect inlet hose to an appropriate sampling tap. The direction of water flow is from the outside to the inside of the yarn-wound filter cartridge. Record time and meter reading and open sampling tap. Collect sample, turn off sampling tap, record time and meter reading, and disconnect sampling apparatus. Take care to maintain opening on the inlet hose above the level of opening on outlet hose to prevent filter backwashing loss of particulate matter. Drain residual water as completely as possible from the sampling apparatus. When unit is completely drained, open filter holder, aseptically remove filter cartridge, place it in a labelled plastic bag, and seal bag. If additional samples are to be collected with the same filter device, thoroughly rinse the influent hose and filter holder with the water to be sampled before installing another filter cartridge. Refrigerate samples or place on wet ice as soon as possible after collection. Transport to the laboratory on wet ice and process as soon as practical but within 48 hours. Do not freeze. Minimize shipping and storage times.

b. Sample Processing:

1. Extraction. Handle filter samples aseptically. Using a razor knife, separate the filter fibers from the support core by making a longitudinal cut down the entire length of the filter cartridge. Cut the fibers into two or more approximately equal portions consisting of an outer and an inner layer. There often will be a fairly distinct line of demarcation based on the depth to which sample particulate matter has penetrated the filter cartridge. Alternatively, locate the end of yarn on the outside of the cartridge, unwind the fiber, and divide into equal portions. Extract each portion separately with 1 L distilled water by kneading or by shaking for approximately 10 minutes. Express fluid from the fiber portions and combine all fluid. If the fiber mat retains significant amounts of particulate matter, repeat the extraction process on each portion with fresh liter volumes of water until the fibers appear clean. If further processing is not possible at this point, preserve the extract by adding a sufficient volume of 37% (V/V) formaldehyde to produce a final concentration of 2% (V/V) formaldehyde. Refrigerate the preserved extract until concentration.

2. Extract Concentration. Concentrate the filter extract by letting the extract settle in the refrigerator overnight or by centrifuging at 600 x g for 5 minutes. If available, use an algal centrifuge. Decant or aspirate the supernatant fluids and resuspend the pellets in a volume of 2% (V/V) formaldehyde equal to the total volume of the combined sediments. If little or no sediment is present, suspend the sediments in 10 ml of 2% (V/V) formaldehyde. [The algal centrifuge produces a total volume of 7 ml of supernatant fluid and sediment. Add 3 ml 2% (V/V) formaldehyde and wash by centrifugation.] Recentrifuge the combined sediments for 5 minutes at 600 x g. If the sediment volume is equal to or less than 1 ml, continue processing by decanting or aspirating the entire supernatant. Discard supernatant. If the sediment volume is >1 ml, resuspend it in the supernatant fluid, transfer a volume of suspension equivalent to 1 ml of sediment to a 15 ml tube and again centrifuge for 5 minutes at 600 x g. Decant and discard the supernatant.

Add 2 to 3 drops of Lugol's iodine [dilute stock solution 1:5 with sp. g. 1.20 zinc sulfate ($ZnSO_4$) solution] to the pelleted material and mix with an applicator stick. Add 5 ml $ZnSO_4$ solution (sp. g. 1.20) and mix on a vortex mixer. Bring volume up to the top of the tube with additional $ZnSO_4$ swinging bucket rotor, centrifuge for 3 minutes at 650 x g. Touch a clean, grease-free, #1 22 x 22 mm coverslip to the meniscus and let tube and contents remain static for 2 to 3 minutes after placement. Be careful to handle the coverslip by the edges. Carefully lift coverslip straight up, place on a slide, and seal coverslip with vaspar.

3. <u>Microscopic Examination</u>. Scan entire coverslip area with a 10X objective on a bright-field microscope using reduced light. Identify suspect organisms using a 45X objective and an ocular micrometer. Identification must be made by an individual with a demonstrated proficiency for recognizing and differentiating intestinal protozoa. Take care not to confuse <u>Giardia</u> with yeast, diatoms, <u>Coccidia</u>, and other organisms. Positive identification of <u>Giardia</u> requires observing structures of the right size and shape and having at least two internal morphological characteristics. Examine sediment from the entire sample in the above manner before declaring a sample to be negative.

c. Reporting Results:

Report dimensions of <u>Giardia</u> cyst-like structures larger than 8 μm and less than 19 μm. In addition, report any internal morphological features observed. Photomicrographs of suspect cysts may be useful in documenting or corroborating identification. Record and report non-<u>Giardia</u> microorganisms detected.

d. Cultivation:

No in vitro method for cultivating <u>Giardia</u> from the cyst stage is currently available. Trophozoites may be cultured; however, it is less likely that trophozoites would be present in water supplies in detectable numbers or that they would survive long.

DISCUSSION

The above method, that will appear in the 16th edition of <u>Standard Methods</u>, again is <u>NOT</u> a standard method. It is a tentative method and should be used as a reference or starting point. This method, or modifications of it, have been used by a small number of laboratories in the last four years. The method is workable in the hands of trained individuals but it is not without problems: it is time and experience intensive, it fails to provide information on the viability and source of detected cysts, and extrapolating to get quantitative results is difficult and questionable. These problems are inherent in a detection technique that is dependent upon microscopic examination of specimens.

Work is in progress at HERL and elsewhere on fluorescent antibody (FA) staining techniques (11,12). FA techniques offer the promise of simplifying the detection of cysts in concentrates and they may be useful in differentiating animal and human-source cysts (12). However, currently available FA techniques do not solve the problem of determining viability. Recent reports on the use of the Mongolian gerbil as an animal model (13,14) may allow determining the viability or infectivity of cysts, but the expense and other disadvantages of animal assay systems would remain. <u>Giardia</u> methodology is evolving and the challenge is to develop simpler and better approaches to identifying, detecting, isolating, and enumerating <u>Giardia</u> in water supplies.

REFERENCES

1. Lippy, E. C. and Logsdon, G. S., "Where Does Waterborne Giardiasis Occur, and Why?", In Proceedings of the 1984 Specialty Conference on Environmental Engineering, Pirbazari, M. and Devinny, J. S. (eds.), pp. 222-228, ASCE, New York, NY, 1984.

2. Centers for Disease Control "Water-related Disease Outbreaks", 1983 Annual Summary, U. S. DHHS, PHS, Atlanta, GA, September, 1984.

3. AWWA, "Giardiasis Outbreaks Show Steady Increase", Mainstream, 28:1, 13, July, 1984.

4. Shaw, P. K., et al., "A Communitywide Outbreak of Giardiasis with Evidence of Transmission by a Municipal Water Supply", Ann. Intern. Med., 87:426-432, 1977.

5. Jakubowski, W. and Ericksen, T. H., "Methods for Detection of Giardia Cysts in Water Supplies", In Waterborne Transmission of Giardiasis, Jakubowski, W. and Hoff, J. C. (eds.), U. S. Environmental Protection Agency, 600/9-79-001, Cincinnati, OH, June, 1979.

6. Meyer, E. A., "Giardia lamblia: Isolation and Axenic Cultivation", Exp. Parasitol., 39:101-105, 1976.

7. Gordts, B., et al., "Routine Culture of Giardia lamblia Trophozoites from Human Duodenal Aspirates", The Lancet, p. 137, July 21, 1984.

8. Meyer, E. A., "Culture of Giardia lamblia", The Lancet, p. 527, September 1, 1984.

9. APHA-AWWA-WPCF, "912 H. Pathogenic Protozoa", In Standard Methods for the Examination of Water and Wastewater, 15th edition, 1981.

10. Jakubowski, W., "Detection of Giardia Cysts in Drinking Water, State-of-the-Art", In Giardia and Giardiasis, Erlandsen, S. L. and Meyer, E. A. (eds.), Plenum, New York, NY, 1984.

11. Sauch, J. A., "Detection and Identification of Giardia Cysts Using Immunofluorescence and Phase Contrast Microscopy", In Advances in Water Analysis and Treatment, 12th Annual AWWA Water Quality Technology Conference, 2B-3, Denver, CO, December, 1984.

12. Riggs, J. L., Nakamura, K. and Crook, J., "Identifying Giardia lamblia by Immunofluorescence", In Proceedings of the 1984 Specialty Conference on Environmental Engineering, Pirbazari, M. and Devinny, J. S. (eds.), pp. 234-238, ASCE, New York, NY, 1984.

13. Belosevic, M., et al., "Giardia lamblia Infections in Mongolian Gerbils: An Animal Model", Jour. Inf. Dis., 147:222-226, 1983.

14. Faubert, G. M., et al., "Comparative Studies on the Pattern of Infection with Giardia spp. in Mongolian Gerbils", Jour. Parasitol., 69:802-805, 1983.

Reprinted from Proc. AWWA WQTC, Denver, Colo. (Dec. 1984).

Figure 1. Giardia Reference Method Flow Chart

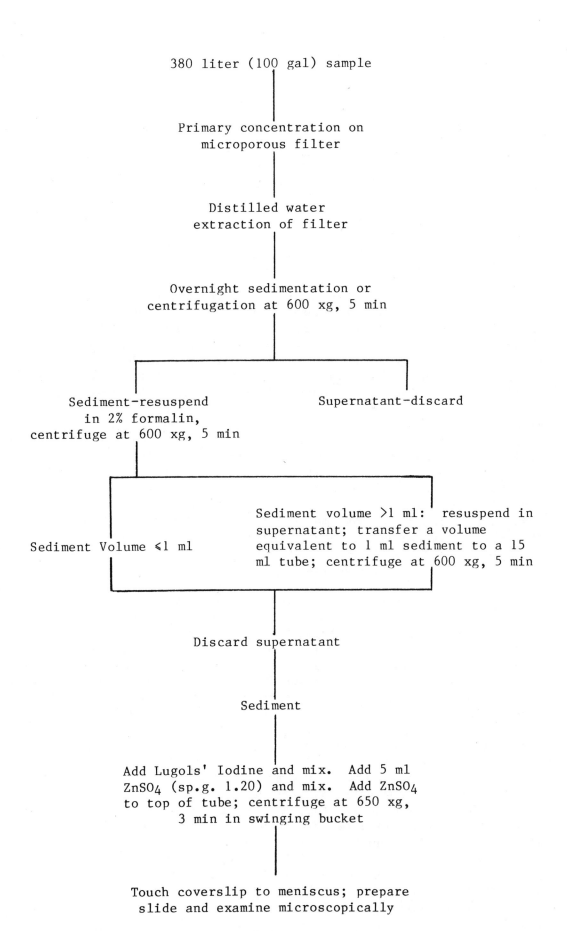

380 liter (100 gal) sample

Primary concentration on
microporous filter

Distilled water
extraction of filter

Overnight sedimentation or
centrifugation at 600 xg, 5 min

Sediment-resuspend
in 2% formalin,
centrifuge at 600 xg, 5 min

Supernatant-discard

Sediment Volume ≤1 ml

Sediment volume >1 ml: resuspend in
supernatant; transfer a volume
equivalent to 1 ml sediment to a 15
ml tube; centrifuge at 600 xg, 5 min

Discard supernatant

Sediment

Add Lugols' Iodine and mix. Add 5 ml
$ZnSO_4$ (sp.g. 1.20) and mix. Add $ZnSO_4$
to top of tube; centrifuge at 650 xg,
3 min in swinging bucket

Touch coverslip to meniscus; prepare
slide and examine microscopically

DETECTION AND IDENTIFICATION OF GIARDIA CYSTS USING IMMUNOFLUORESCENCE AND PHASE CONTRAST MICROSCOPY

Judith A. Sauch
Microbiologist

Microbiology Branch
Toxicology & Microbiology Division
Health Effects Research Laboratory
U. S. Environmental Protection Agency
Cincinnati, Ohio 45268

INTRODUCTION

The flagellated protozoan, Giardia, recognized now as the most common intestinal parasite in the U. S., can be transmitted via municipal water supplies and mountain streams (1). Coliform counts are unreliable as indicators for the presence of Giardia cysts in water supplies because the cysts are much more resistant to disinfection (2). In the absence of a cultural method, detection and identification of these protozoans has been accomplished by direct microscopic observation of sample concentrates (3). In this method, the observer must locate and identify Giardia cysts among numerous microorganisms of similar size and shape and against a background of debris.

Figure 1 is a phase contrast microphotograph of one field in a typical sample. An average sample would be composed of thousands of such fields. Within each field, the observer must quickly scan each object and decide if it meets the size [larger than 8 μm and smaller than 19 μm] (3) and shape requirements for a Giardia cyst. This is very time consuming and fatiguing. Fatigue and any distortion of a cyst from normal morphology may result in the observer not recognizing it and subsequently declaring the sample free of Giardia. In practice it is undesirable to use a test for regulatory purposes that can easily yield false negative results.

An immunofluorescence detection method was developed by the author and used to generate Figure 2. This is exactly the same field as seen in Figure 1. Using this method, the Giardia cyst in this microscopic field is now easily detectable since it glows bright green against a black background. Contaminating microorganisms and debris are either not visible or show up as dull red or light yellow objects. The observer can confirm the identity of any Giardia cyst-like objects found by switching objectives and using phase contrast at high magnification to locate characteristic internal morphological features (1-4 nuclei, median bodies, and axonemes).

When the amount of solids in a sample is small, it has been possible using the fluorescent antibody technique to scan the concentrate from an entire water sample of 100 or more gallons in as little as 20 minutes. The time required to scan the concentrate from a sample will vary with the quantity of particulate matter present in the concentrate. In the author's laboratory, it was determined that to examine all of the concentrate from a given water sample would take about 30 times as long using direct observation by phase as compared to the immunofluorescent method.

The focus of this paper will be to provide the reader with an introduction to the immunofluorescent method of detecting microorganisms. A paper describing the technical details for detection of Giardia cysts in water samples is in preparation.

IMMUNOFLUORESCENCE TECHNOLOGY

OVERVIEW

The Direct Fluorescent Antibody Assay. If a mammal is inoculated with proteins from a different organism, the mammal's immune system responds by producing a class of molecules known as antibodies. Each antibody molecule is capable of specifically binding to the foreign molecule (antigen) that induced its production. Thus, if a rabbit is inoculated with a highly purified preparation of Giardia cysts, in a few weeks blood serum from that rabbit will contain large amounts of antibodies which are capable of specifically binding to Giardia cysts.

Antibodies to Giardia cysts can be purified from serum and then be chemically cross-linked to a fluorescent dye. If a sample containing Giardia cysts (Figure 3A) were exposed under the proper conditions to this fluorescent-labelled specific antibody, washed free of the excess antibody and then mounted for microscopic examination, these cysts would fluoresce upon exposure to ultraviolet (UV) light. With an epifluorescent microscope (UV illumination projected through the objective lens onto the specimen) the microorganisms could then be observed to glow brightly against a black background. In the ideal situation, if the antibody used is specific enough and other test parameters have been optimized, no other microorganisms or debris in the sample would fluoresce. The assay just described is known as a direct immunofluorescent or direct fluorescent antibody assay.

The Indirect Fluorescent Antibody Assay. A modified and more versatile approach is used to label microorganisms in the indirect fluorescent antibody assay (Figure 3B). In this technique, the fluorescent dye is not cross-linked to the specific antibody (anti-Giardia cysts). Instead, a second antibody-fluorescent dye complex, the conjugate, attaches to the specific antibody. Thus, the specific, or primary, antibody acts as a bridge binding the Giardia cyst antigens and the fluorescent conjugate together. The microscopically visible end result is virtually identical for both approaches. However, the indirect method allows many specific antibody preparations to be used in a laboratory without the need to individually cross-link each one to a fluorescent dye. Only one fluorescent conjugate directed against rabbit antibody need be stocked.

The rest of this section consists of comments on the major technical considerations involved in an indirect fluorescent antibody assay. Details on fluorescent antibody technology can be found in references 4-8.

REAGENTS

Antigens and Antibodies. Only the purest possible antigen (Giardia cyst) preparations (9) should be used to stimulate antibody production. Well-purified antigen preparations can yield high titer (relative potency) specific antibody that has very minimal cross-reactivity with other microorganisms and debris in water samples.

In the author's laboratory, rabbits have been used because many animals can be easily housed and each can produce large volumes of high quality antisera. To produce the antiserum, several healthy adolescent Giardia-free rabbits housed in a well-controlled clean environment should be inoculated with antigen. Antiserum harvested from these animals should be titered, tested for major cross-reactivity and stored frozen. Detailed information on antibody production and purification can be found in references 6, 7, and 10.

The Conjugate and the Fluorochrome. Conjugates for the indirect assay consist of the fluorochrome (the label) chemically cross-linked to a species specific anti-globulin antibody (Figure 3B). The suitability of commercially-prepared conjugate from different sources for use in an indirect assay must be evaluated in each assay. In the author's experience, the ideal conjugate should be usable at a high dilution, cause little or no background fluorescence, and retain its brightness during storage and sample examination.

Fluorescein, as fluorescein isothiocyanate, has such a high efficiency of converting exciting UV light into emitted visible light that it has been almost universally adopted as the principle fluorochrome in conjugates used for immunofluorescent labeling.

The characteristic intense apple green fluorescence of fluorescein is not commonly encountered in water samples and is easily differentiated from the primarily red and light yellow autofluorescence of many microorganisms. Rhodamine is an alternate fluorochrome that fluoresces red and is widely used when another color is required for identification of a second antigen in a given sample. Sources for commercially-available fluorescein and rhodamine conjugates are listed in Linscott's Directory (11).

The Counterstain. Counterstains, such as Evans Blue, are often used in fluorescent antibody assays to improve the phase contrast or brightfield image and to minimize both non-specific binding of the conjugate and autofluorescence (6). Optimal concentrations of counterstains must be determined for use with a given specific antibody/conjugate combination, as too high a concentration results in decreased intensity of the conjugate's fluorescence.

Evaluation of Reagents. The relative concentrations of each of the reagents used in an indirect fluorescent antibody assay must be optimized. A "checkerboard" titration is performed in which systematically varied concentrations of the specific antibody, conjugate, and counterstain are used in actual assays in all reasonable combinations. The optimal combination is that which yields: (1) intense fluorescent staining of the cysts when specific antibody is used, (2) only background levels of staining when unimmunized rabbit serum is substituted for specific antibody, (3) easily visible internal morphology by phase contrast, and (4) only barely perceivable levels of nonspecific staining of other microorganisms and debris when specific antibody is used.

Mounting Media. Selection of the mounting medium for microscopic examination is very important and depends on: the substrate upon which the immunofluorescent method was performed (i.e., glass slides, membrane filters), storage time of the sample after labeling with antibody and conjugate, and the necessity to examine the sample by phase contrast microscopy. The best mounting fluid for a particular system will not autofluoresce, will not inhibit fluorescence emission, and will allow a high contrast phase image of the microorganisms.

Exposure of the fluorescein conjugate to exciting UV light also causes rapid fading of the apple green fluorescence. However, this fading can be minimized by the incorporation of either n-propyl gallate (12) or phenylenediamine (13) into the mounting media. Optimal concentrations of these chemicals must be determined for each system as too high a concentration may completely inhibit fluorescence emission.

INSTRUMENTATION AND METHODOLOGY

The Manifold. If fluorescent antibody assays are to be run on samples supported on membrane filters, it is the author's experience that a multiport vacuum manifold be used, since this allows several samples and controls to be assayed simultaneously. The manifold should be equipped with individual vacuum shut-off valves under each filter support. In addition, it is necessary to regulate the overall degree of vacuum in the manifold so as to maintain only a minimal pressure differential across the membrane filters.

It is convenient to perform immunofluorescent assays after applying controlled amounts of sample concentrates to the surfaces of membrane filters. These are held in place on the manifold by removable weights that form wells over the membranes. The filter surfaces are then sequentially flooded with the reagents which are drawn through the membrane filter by vacuum after the appropriate incubation times. Filters should be chosen that have pore sizes small enough to retain the microorganism on the filter surface and they should be composed of a material compatible with the mounting media and observation conditions.

Positive and negative controls are included each time the manifold is used to assay water samples for Giardia cysts. Positive fluorescence and good internal morphology of G. lamblia cysts (the positive control) assures the operator that the assay has been performed correctly. The absence of fluorescing cysts on membrane filters where no cysts or samples were purposely added (negative control) indicates the lack of cross membrane contamination.

The Microscope. Since its introduction, epifluorescent (incident light, vertical) illumination has replaced transmitted light as the illumination system of choice for immunofluorescent assays. In epifluorescent illumination (Figure 4) exciting light, usually from a high pressure mercury vapor lamp (50 or 100W), passes through an exciting filter, is deflected downward by a dichroic beam splitting mirror into and through the objective to the stained specimen. Emitted and some reflected light are deflected by the dichroic mirror to the observer through a barrier filter which absorbs unwanted and harmful UV light. This results in a high level of fluorescence intensity with little bleaching of the fluorochrome in areas not yet examined. Most major manufacturers now produce microscopes with epifluorescent illumination systems and frequently it is possible to convert an existing microscope to this mode of operation.

It is also desirable to use microscope objectives and condensers that allow any Giardia cysts located by their fluorescence to also be observed by phase contrast. This will permit detection of any characteristic internal morphological features that are classically used to confirm a _Giardia_ cyst's identity.

CONCLUDING STATEMENTS

It should be appreciated that to initially develop such an assay involves many complexities only implicit in the foregoing explanation. This was especially true when the author developed an indirect fluorescent antibody assay for _Giardia lamblia_ cysts, as the cysts cannot be generated in vitro. It was necessary, therefore, that all work be done using purified primary isolates from human feces.

Fortunately, once developed, fluorescent antibody assays are relatively easy to perform. This is reflected in their widespread use for clinical applications, usually with commercially available reagents. The author believes that should the necessary reagents become commercially available, immunofluorescent detection of _Giardia lamblia_ cysts in water samples could become routine.

REFERENCES

1. Craun, G. F., "Waterborne Giardiasis in the United States: A Review", Am. J. Public Health, 69:817-819, 1979.

2. Hoff, J. C., Rice, E. W. and Schaefer, F. W., "Disinfection and the Control of Waterborne Giardiasis", In Proceedings of the 1984 Specialty Conference on Environmental Engineering, Pirbazari, M. and Devinny, J. S. (Eds.), pp. 239-244, ASCE, New York, 1984.

3. Jakubowski, W., "Detection of _Giardia_ Cysts in Drinking Water. State-of-the-Art", In Giardia and Giardiasis, Erlandsen, S. L. and Meyer, E. (Eds.), Plenum, New York, 1984.

4. Cuello, A. C. (Ed.), "Immunohistochemistry", John Wiley and Sons, New York, pp. 47-82, 1983.

5. Fetaenu, A., "Labelled Antibodies in Biology and Medicine", McGraw-Hill International, New York, 2nd Edition, 1978.

6. Goldman, M., "Fluorescent Antibody Methods", Academic Press, New York, 1968.

7. Weir, D. M. (Ed.), "Immunochemistry", Volume 1 of Handbook of Experimental Immunology, Blackwell, London, 3rd Edition, Chap. 15, 1978.

8. Wick, G., Traill, K. and Schauenstein, K. (Eds.), "Immunofluorescence Technology", Elsevier, New York, 1982.

9. Sauch, J., "Purification of _Giardia muris_ Cysts by Velocity Sedimentation", Appl. and Environ. Microbiol., 48:454-455, 1984.

10. Marchahalonis, J. J. and Warr, G. W. (Eds.), "Antibody as a Tool", John Wiley and Sons, New York, 1982.

11. Linscott's Directory of Immunological and Biological Reagents, 3rd Edition, 1984-1985.

12. Giloh, H. and Sedit, J. W., "Fluorescence Microscopy: Reduced Photobleaching of Rhodamine and Fluorescein Protein Conjugate by N-Propylgallate", Science, 217:1252-1255, 1982.

13. Johnson, L. D. and deToqueira Araujo, G. M., "A Simple Method of Reducing the Fading of Immunofluorescence During Microscopy", J. Immunol., 43:349-350, 1981.

Reprinted from Proc. AWWA WQTC, Denver, Colo. (Dec. 1984).

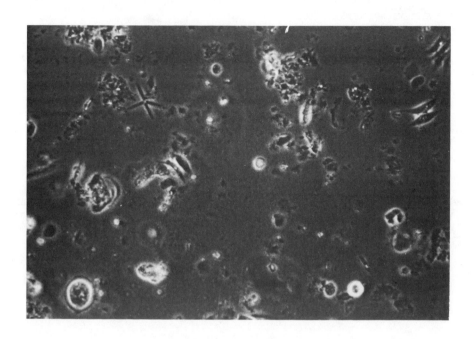

FIGURE 1.   PHASE CONTRAST IMAGE OF A WATER SAMPLE
(200X MAGNIFICATION)

FIGURE 2.   FIELD IN FIGURE 1 ILLUMINATED BY ULTRAVIOLET LIGHT
SHOWING A <u>GIARDIA</u> CYST-LIKE OBJECT NEAR CENTER

FIGURE 3. DIAGRAMATIC REPRESENTATION OF IMMUNOFLUORESCENT METHODS

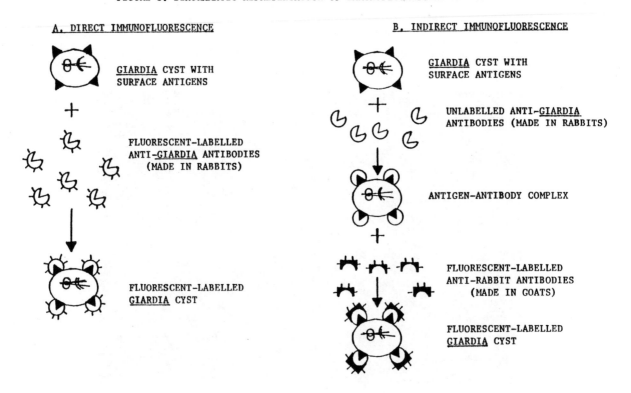

A. DIRECT IMMUNOFLUORESCENCE

GIARDIA CYST WITH SURFACE ANTIGENS

FLUORESCENT-LABELLED ANTI-GIARDIA ANTIBODIES (MADE IN RABBITS)

FLUORESCENT-LABELLED GIARDIA CYST

B. INDIRECT IMMUNOFLUORESCENCE

GIARDIA CYST WITH SURFACE ANTIGENS

UNLABELLED ANTI-GIARDIA ANTIBODIES (MADE IN RABBITS)

ANTIGEN-ANTIBODY COMPLEX

FLUORESCENT-LABELLED ANTI-RABBIT ANTIBODIES (MADE IN GOATS)

FLUORESCENT-LABELLED GIARDIA CYST

FIGURE 4. EPIFLUORESCENT ILLUMINATION

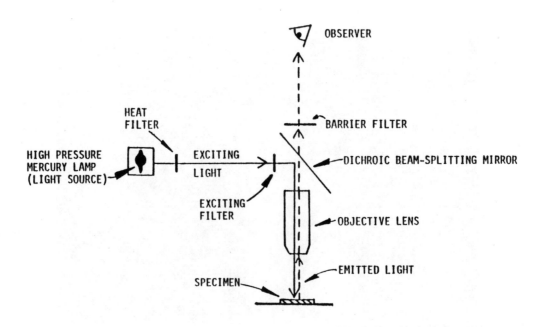

OBSERVER

BARRIER FILTER

DICHROIC BEAM-SPLITTING MIRROR

HEAT FILTER

HIGH PRESSURE MERCURY LAMP (LIGHT SOURCE)

EXCITING LIGHT

EXCITING FILTER

OBJECTIVE LENS

EMITTED LIGHT

SPECIMEN

GIARDIA METHODS WORKSHOP

Walter Jakubowski, Moderator
Health Effects Research Laboratory
U.S. Environmental Protection Agency
Cincinnati, Ohio 45268

John L. Cleasby, Coordinator
Department of Civil Engineering
Iowa State University
Ames, Iowa 50011

Panel Members

Frank W. Schaefer, III, Parasitologist, Health Effects Research Laboratory, U.S. Environmental Protection Agency, Cincinnati, Ohio 45268

Charles P. Hibler, Professor, Colorado State University, Fort Collins, Colorado 80523

E. Alan Meyer, Professor, Oregon Health Sciences University, Portland, Oregon 97201

George J. Vasconcelos, Microbiologist, Region X Laboratory, U.S. Environmental Protection Agency, Manchester, Washington 98353

Judith A. Sauch, Immunologist, Health Effects Research Laboratory, U.S. Environmental Protection Agency, Cincinnati, Ohio 45268

John L. Riggs, Research Scientist, Department of Health Services, Berkeley, California 94704

George R. Healy, Parasitologist, Centers for Disease Control, Atlanta, Georgia 30333

INTRODUCTION

The objectives of the workshop were to familiarize participants with the use and limitations of Giardia cyst detection methodology; to critically review the tentative Giardia method appearing in the 16th edition of Standard Methods for the Examination of Water and Wastewater; to share the experiences of the panelists and audience in working with methods for detecting Giardia cysts in water, and to make recommendations for improving the methodology. Most of the panel members had participated in a workshop held in September, 1980 from which a consensus method had been developed. That consensus method formed the basis for the 16th edition tentative method.

The format for the current workshop involved a brief introduction by a panel member of a specific portion of the methodology, discussion among panel members and the audience to expand various aspects of that topic, and finally, a summary and recommendations concerning that portion of the methodology. The reader may refer to a paper by W. Jakubowski, which appears in this handbook for a description of the 16th edition tentative method. Additional information on Giardia immunofluorescence methods may also be found in a paper by J. Sauch elsewhere in this handbook.

The summaries below were developed by the moderator from a verbatim transcript of the workshop. A serious attempt was made to make the summaries as complete and as thorough as possible, and the moderator accepts all responsibility for the accuracy of their content and editorial quality.

F. Schaefer - Sample collection, volume, transport and storage: Every water sampling situation is different, varying with whether raw or finished water is being sampled, and requiring decisions to be made about how to sample. For example, if you are sampling a raw water from a very turbid river, it is possible to collect too large a sample. Under these circumstances, the filter will contain so much silt and debris that large extract volumes will be produced (possibly in excess of 25 liters) resulting in pellet volumes exceeding 10-15 ml after concentration. Large pellet volumes create difficulties in purifying the cysts from the debris by flotation procedures. On the other hand, when sampling a filtered finished water supply, or other low turbidity water, very large volumes can sometimes be sampled with production of small pellets. As a guideline, it is suggested that the minimum sample size be 100 gallons and the maximum be 500 gallons. Another point to be stressed is the requirement for scrupulously cleaning sampling equipment in order to prevent cross-contamination between samples. Giardia cysts can be sticky and adhere to laboratory apparatus and sampling equipment. Equipment should be thoroughly scrubbed before sampling, and, if using the same filter holder to sample more than one water supply on a given trip, it should be thoroughly washed and rinsed with the next water to be sampled. The use of gloves and aseptic technique is also recommended to reduce the possibility of cross-contamination. Samples should be transported to the laboratory on ice as soon as possible. Under no circumstances should the sample be frozen. Freezing the sample can cause formation of ice crystals within the cysts, which could then result in their disruption upon thawing. Upon arrival in the laboratory, samples should be processed as soon as possible, preferably within 48 hours.

Discussion

Different laboratories are using various filter types for sample collection. Some laboratories are using yarn-wound filters of the Commercial Filter Corporation type, made from orlon or polypropylene. Other laboratories are using AMF/CUNO type filters with a cotton or polypropylene base. At least one laboratory is using the Balston epoxy fiberglass filter tube, and there has been one published report on the use of Millipore membrane filters. Most laboratories are using the depth-type filters, as opposed to membrane filters, because of the advantages afforded in filtering large volumes. No rigorous, controlled comparative evaluations of filter types have been done to determine if one filter type is superior to another. The selection seems to be based upon laboratory preference, and perception of the ease with which trapped particulates can be recovered from the filter material. At Dr. Hibler's laboratory, filters made of polypropylene, orlon or cotton are processed. They prefer the polypropylene filters because they look cleaner after washing. In addition, they can readily process 1000 gallons of "clean" western waters. Dr. Hibler also emphasized the need for cleanliness in preparation of sampling equipment, recommending the use of chlorination, soap and hot water. Participants agreed that depth-type filters should have a porosity not greater than 1 μm since some investigators have found cysts to pass through filters with nominal porosities as low as 3 μm. For processing yarn-wound filters, cutting or unwinding the yarn is a matter of personal preference.

There was some discussion on the need to regulate flow or pressure during sample collection. Some laboratories collect samples at a pressure of 15 PSI or less, and others control flow at 1 to 1-1/2 GPM. Still others use no pressure or flow regulation. With depth-type filters, high pressure can increase the nominal porosity and there is the possibility of cyst disruption through shear force. Until data become available, it would appear prudent to collect samples at the lowest feasible pressure differentials.

It is known that chlorine can distort and inactivate cysts, possibly rendering them unidentifiable. Data are now being acquired on the concentration of chlorine and contact times required to produce different degrees of distortion. This is important because it may take most of a day, if not more, to collect a single sample. Cysts collected on the filter early in the sampling could be subjected to chlorine contact for several hours, resulting in distortion that produces false negative results. There was reluctance on the part of panelists to endorse dechlorination for sampling of disinfected supplies because of the added expense, equipment and effort required to do so. However, it was agreed that chlorine can produce cyst distortion.

The question of maintaining the sampling unit filter and holder cold during sampling was raised. One panelist had occasion to sample in a warm, humid environment without access to ice or refrigeration during sampling and transport. Cysts were found, but it is not known if losses might have occurred. It would seem reasonable to keep the filter at least in the shade during sampling, and cold if possible.

Recommendations

Controlled laboratory studies are needed in order to make defensible recommendations on filter types and porosities, flow rates and pressure differentials, and transport and holding times. In the meantime, depth-type or membrane filters with a porosity of 1 μm or less, and capable of sampling the required volume, should be used. The filter composition should be of a material that will allow ready recovery of the cysts and that will not interfere with purification of cysts from the debris. Sample volumes should be a minimum of 100 gallons, and the maximum volume would depend upon the quantity of particulate matter in the sampled water. When sampling disinfected waters, dechlorination is advisable. Pressure differentials, flow rates and transport and holding times should be minimized as much as possible. Samples should be protected from sunlight and heat during collection and transport.

C. Hibler - Sample processing to concentrate state: As already mentioned we work with polypropylene, cotton and orlon filters, but we prefer the polypropylene. We split the filter to the core and hand wash it in distilled or deionized water doing as thorough a job as possible. The volume of wash water may be 0.5 to 2 gallons, depending upon how much dirt is on the filter. Formalin is added to the wash water to a concentration of 2% V/V to preserve the sample and prevent growth of other protozoa, algae, etc. The wash water is then refrigerated for 24 hours and the supernatant is siphoned down to the organic sediment. With some samples you can get down to almost nothing; with some eastern waters the organic sediment-containing portion may be greater than 200 ml. If there is a need for rapid results the wash water can be centrifuged instead of using refrigeration settling. The sediment portion is centrifuged and the pellet is split if more than 1 ml of pellet is produced. The pellet is then overlayed or underlayed using zinc sulfate and centrifuged for 3 to 8 minutes. The interface material is collected and washed by centrifugation. It can then be concentrated by flotation using zinc sulfate in 15 ml tubes with a coverslip placed on the meniscus. Or it can be filtered through a 5 µm Nucleopore membrane. The material on the membrane is washed off and examined under the microscope using an iodine stain. If the debris in the sample is primarily inorganic, a zinc sulfate or potassium citrate solution with a specific gravity of 1.28 may be used to clean it up. If it is primarily organic, a specific gravity of 1.09 is used. If the debris is a mixture of inorganic and organic, a specific gravity of 1.2 is used. The important thing is to separate the cysts from the inorganic and organic debris prior to microscopic examination.

Discussion

This part of the methodology seems to be where the "art" plays a big role. There are differences of opinion, and very little data, on the medium to use for washing filters, on chemicals used in flotation procedures and on details for effecting separation of cysts from debris (e.g., underlaying vs. overlaying) and collecting the concentrate for examination (e.g., coverslip vs. recovering all of the floated phase). Some laboratories are using distilled or deionized water buffered at pH 7.0 as the washing medium and some are using potassium citrate or citrate/tween 20 combinations. The proponents of this latter method find it effective in cleaning the debris from orlon filters and say that it may help to release "sticky" cysts from filters and prevent cyst loss to glassware. When using detergent solutions or wetting agents with orlon filters, the experience has been that fewer washings and rinsings are needed in order to get the fibers clean and white to the eye. Consequently, a smaller volume of wash extract is produced and processing time is decreased if centrifugation rather than refrigeration sedimentation is used. The proponents of distilled or deionized water have not encountered sticky cysts and say that the use of detergents may interfere with recovery of cysts when centrifugation flotation is performed with a coverslip touching the meniscus and top of the tube. There did not appear to be much concern with isotonicity of the wash medium since the cyst is a resistant form and does occur in a hypotonic environment, i.e., water. However, osmotic effects do occur during flotation when cysts are exposed to the chemical solutions used in this procedure. Some prefer zinc sulfate, others prefer sucrose or potassium citrate or sucrose/percoll. All of the solutions can produce cyst distortion or changes in the specific gravity of cysts upon prolonged contact. Zinc sulfate seems to be the chemical most widely preferred by parasitologists. Percoll may exert less of an osmotic effect but it is expensive.

Recommendations

Comparative data are needed on the effectiveness of various wash media and on the suitability of different flotation chemicals. Distilled or deionized water, as recommended in the 16th edition tentative method, or phosphate buffered saline at pH 7.0 may be used as the wash medium. The addition of detergents, wetting agents or other substances should be compared and evaluated against these media. Any of the flotation chemicals listed above may be used for purification but caution should be exercised to minimize contact time of cysts with the concentrated solutions.

   E.A. Meyer - Examination of concentrate, criteria for identification, reporting results: This portion of the methdology overlaps somewhat with the previous discussion by Dr. Hibler in that examination of the concentrate involves getting the purified cysts from the centrifuge tube or membrane filter to the microscope. The tentative method calls for taking a 15 ml centrifuge tube with less than 1 ml of pellet (I would prefer less than 0.25 ml) from the sample, adding zinc sulfate at a specific gravity of 1.20 to the top of the tube and placing a clean coverslip on the meniscus. The coverslip is allowed to remain in place for several minutes to hopefully allow the cysts to stick to it. The coverslip is gently placed on a microscope slide and examined. This coverslip procedure of removing cysts from the meniscus is subject to wide variation and potential large losses of Giardia cysts. In our laboratory, we have found that the volume of meniscus material recovered on the coverslip varies from 0.25 ml to 0.75 ml. An alternative method is to recover all of the floated material with a pipet. The tentative method then suggests scanning the entire coverslip area using a 10X objective. The intent here is to recommend low power and I would suggest a total scanning magnification, including the ocular, on the order of 100X. Either bright-field or phase microscopy may be used for examination, depending upon what is available and the experience of the observer.

   Concerning interfering organisms and debris, Dr. Hibler suggested some ways of dealing with the specimen when there is much inorganic or organic debris present. There are many types of organisms including yeasts, diatoms, algae and protozoa that can be confused with Giardia on the basis of size and shape. The major problem organisms are flagellated protozoa and pollens. Criteria for identification include seeing an oval organism in the size range of 8 to 19 μm and observing two or more internal morphological characteristics. I have seen Giardia cysts as small as 5 μm in length, but, generally, they are larger. With iodine stain the cyst is brownish and refractile. Lying diagonally along the length of the organism are axonemes or fibrils that are the internalized portions of the flagellae. The cysts could have up to four nuclei, usually at one end of the cyst. Median bodies are the final internal characteristic that may be observed. To identify these internal characteristics, a total magnification of 450X or greater should be used. This method cannot tell you if the cysts are viable. Finally, for reporting, it may be a good idea to photograph what you have seen and to record dimensions of organisms in the right size range with notations of internal morphological characteristics that were observed.

Discussion

   One panelist suggested that it might be helpful to acclimate yourself to examination of concentrates by retuning your eye each day through observation of at least ten known cysts from a human source and recording measurements. Photography may not be too helpful in documenting the finding of a Giardia cyst because the internal morphological characteristics usually occur in several planes of focus. The quantity of concentrate pellet used in the flotation affects the recovery of cysts and recovery may drop significantly at pellet volumes of 0.5 ml and greater. The cleanliness of the coverslip is very important and even new coverslips are greasy and should be cleaned with alcohol prior to use. For identification of <u>Giardia</u>, no other known organism has median bodies but it may not be possible to observe these in as much as 50% of the cysts encountered. Concerning viability, it may be possible, after gaining considerable experience, to make an educated guess that a cyst is nonviable based on its appearance, i.e., shrunken, shriveled, distorted, muddy internal characteristics, etc. However, the viability status of a cyst that appears to be morphologically excellent cannot now be determined by microscopic examination alone. There are no surrogate indicators for <u>Giardia</u> cysts in water, but it may be possible to determine if the type of plant debris present is of beaver or muskrat origin. Also, in reporting results, information on the effectiveness of water treatment plant filtration can be gained by examining raw and finished water. Even if <u>Giardia</u> cysts are not found, if there is debris in the finished water in the same size range as cysts, then there is a potential problem.

Recommendations

Quantitative data are needed on the efficiency of cyst recovery when floating cysts from concentrate pellets up to 1 ml in volume. The reliability of the coverslip method versus examination of all floated material needs to be determined. Scanning of the coverslip at a total magnification of about 100X is recommended with confirmation at 450X or greater. Either bright-field or phase microscopy may be used but the investigator should become experienced by examining known <u>Giardia</u> cyst-containing specimens. Most cysts will be in the size range of 8 to 19 µm but it should be be realized that cysts smaller the 8 µm can occur. At least two internal morphological characteristics must be observed in order to make positive identification.

G.J. Vasconcelos - Equipment requirements, analysis time, costs: Costs for sampling and processing equipment and supplies, and estimates for personnel time are shown in Tables 1-4. The sampling time will vary with the volume and type of water being sampled. Personnel time for sampling includes assembly and disassembly of the equipment. Ranges are given for the cost of sample processing equipment and examination in Table 2. These costs are estimates and were taken from recent catalogs. Personnel time listed in Table 3 is based on experience at the USEPA Region X laboratory in Manchester, WA. A range of summary costs and times is given in Table 4. Per sample costs at individual laboratories will vary with number of samples processed, quality and quantity of equipment and with labor costs.

Table 1. Sampling Costs and Time

A.	Sampling Device with Filter	$131
B.	Sampling Device with Regulator, Pressure Gauge and Backflow Preventor	$161
C.	Sampling Time (range)	
	1. Filtration time	4-24 hrs
	2. Personnel time to attach and disconnect device	1-2 hrs

Table 2. Processing and Examination Equipment and Supply Costs

A. Major Equipment
    1. High speed/capacity centrifuge $3,200-5,000
        - Swinging bucket rotor and cups $850-3250
    2. Bright-field microscope $2,500-6,500
    3. Large capacity refrigerator $1,000-2,500
    4. Hydrometers, tube mixers, stage and
        ocular micrometers, baths, etc. $350-850

B. Minimal Supplies and Chemicals
    1. Expendables - plasticware, bags,
        slides, reagents, etc. $500
    2. Non-expendables - glassware, etc. $570

C. Optional Equipment
    1. Microscope combinations
        - Bright-field/phase contrast $3,750-8,500
        - Bright-field/phase/FA $6,000-12,000
        - Bright-field/phase/FA/DIC   > $10,000
    2. Refrigerated, high speed/capacity
        centrifuge w/wind-shielded swinging
        bucket $5,700-10,000
    3. Algal (Foerst) centrifuge $1,500-2,000
    4. Microscope camera system $1,250-6,000

Table 3. Processing and Examination Time

A. Extraction
    1. Reference method   1-2 hrs
    2. Reference method w/tween 20 and
        sodium citrate wash   1-3 hrs

B. Extract Concentration
    1. Refrigerator settling   24 hrs
    2. Centrifuging   1-3 hrs
    3. Algal (Foerst) centrifuge   (not known)
    4. Processing combined sediments
        - volume = to or < 1 ml   1 hr
        - volume = to or > 1 ml   2 hrs
    5. Sucrose/percoll flotation w/vol >1 ml   2 hrs

C. Microscopic Examination
    1. Range per slide (of 75 examined)   10-58 min
    2. Average time per slide   23 min
    3. Average number of slides examined/sample   3
    4. Average total time per sample   69 min

Table 4. Cost and Time Summary for Sampling and Analysis

	Low	High	With Optional Equipment
A. Equipment and supplies	$9,100	$19,331	$28,581
B. Personnel time/sample	5.5 hrs	11.5 hrs	
C. Labor cost/sample at $22/hr	$121	$253	
D. Equipment cost amortized over 5 yrs	$34	$73	$114
E. Supplies cost over 5 yrs at 50 samples/yr	$21	$24	$33
F. Cost per sample	$176	$350	$400

Discussion

Costs for individual orlon filters are about $4 each when purchased in case lots. Dr. Hibler's laboratory is currently charging $40 per sample for analysis if refrigeration sedimentation of the extract is used. He estimates that the total cost per sample is actually closer to about $100. A private water company representative utilizing Dr. Hibler's services estimates his costs at about $125 per sample in addition to the analysis costs.

J. Sauch - Fluorescence techniques: FA techniques offer an advantage over light microscopic examination in that, instead of looking for a bright brown object against the dull brown objects in the background, now we're looking for bright apple green fluorescing objects against a black background with perhaps, very dull yellow or red autofluorescing objects. Advantages over light microscopy are time and the ease of finding the cysts. Unfortunately, fluorescent antibody techniques require specialized equipment and reagents and an initial commitment in personnel time to develop the method and to manufacture reagents. In setting up the assay, you have to consider the antigen to make the antibody. What are you going to use for antigen? Where are you going to get cysts? The purer they are, the lower the possibility of cross-reactivity. One of the disadvantages of this method is that there is a slight possibility of cross-reactivity due to organisms that share antigens. Secondly, how are you going to prepare the antibody? What animals are you going to use? Right now, no commercial manufacturer makes Giardia antibodies. We've used rabbits to prepare antibodies because they're very easy to house in a small area, and each rabbit is capable of producing a large amount of serum with very high titer antibody. There are some manufacturers who may make the antibody for you and they may even conjugate it to do a direct type of assay. Additional considerations in using the FA/phase method are the counterstain to use, the membrane filter type and how to mount the filter. The counting media must not autofluoresce or quench the fluorescence of the conjugate. Thirdly, specialized equipment is needed. An epifluorescent microscope with vertical illumination will probably be your biggest expense. Vertical illumination is important because the illumination is only upon the section of the slide that you are looking at. This prevents bleaching of the rest of the sample.

To detect cysts in a sample, I look for fluorescing objects of the correct size and shape. I run positive controls with every assay and I've attuned my eyes to look for a certain size, shape and fluorescence with the positive controls (G. lamblia cysts). My identification procedures are based upon the size, shape and fluorescence as well as internal morphological characteristics. I use the same criteria that the reference method uses: nuclei, axonemes or median bodies upon examination by phase. It is important to use light microscopy for identification because of the potential, even though slight, for cross-reactivity. Negative controls should also be run to guard against sample cross-contamination or laboratory contamination of glassware or equipment. The time required to prepare the concentrate for examination by FA is about 1.5 hours. Depending upon the quantity of particulate matter in the sample, it could take anywhere from 20 minutes to examine a very clean sample to a couple of days for a very dirty water sample.

Discussion

In the FA/phase technique, FA is used to detect cyst-like objects and confirmation of identity is made by switching to phase microscopy and observing the internal morphologic characteristics. The condition of the cyst cannot be determined by FA alone and identification is not based solely on the FA reaction. Organisms which fluoresce under FA but in which internal characteristics cannot be discerned are reported as "Giardia cyst-like objects". Cysts exposed to chlorine at high dosage levels or for long contact times will react with FA reagents and glow under UV, but if the internal morphology cannot be seen under light microscopy, the sample will not be reported as Giardia-positive. When using FA techniques, additional time is required to prepare the specimen for examination under the microscope. However, the level of expertise required to detect cysts, as well as the amount of time required to scan the specimen, may be reduced. The actual amount of time involved in using either the tentative method or FA techniques will vary from laboratory to laboratory with the experience and training of the personnel. Available microscopic detection and identification techniques, including FA, do not distinguish living from dead cysts. The USEPA is currently funding work to develop metabolic stains for viability that might be used in conjunction with FA techniques.

Recommendations

FA techniques for Giardia are in an early stage of development. Additional work is needed to characterize Giardia antigens; to develop guidelines for antibody production, purity and use, and to develop criteria for cyst identification.

J. Riggs - Antibody identification, species differentiation: We first worked with polyclonal antibody which is produced when you inoculate an animal with an antigen, in this case, whole cysts from a human source. The antibody is recovered from the serum of the inoculated animal, labelled or conjugated with a fluorescent dye and used in the fluorescence technique. Polyclonal antibodies leave a lot to be desired because, if the inoculated animal has an antibody against an organism that is in, for example, a filter from a water system, the antibody will react to that and produce fluorescence. The purity of the antigen inoculated into the animal is another source of background fluorescence. It is actually a specific fluorescence because the animal will make antibodies against bacteria or other organisms or proteins that are in the "purified antigen". What we have done is to produce monoclonal antibodies against an antigen or antigens in the Giardia cyst. Monoclonal antibodies are produced through a tissue culture procedure using mouse hybridoma cells. In this procedure, single cells that produce antibodies to specific antigens in the Giardia cyst can be isolated and cloned. These cells can be grown in culture and they will continue to produce the antibody that the cell was first primed against. We have used monoclonal and polyclonal antibodies to try to separate human Giardia from non-human Giardia. We prepared monoclonal antibody against human-source cysts and conjugated the antibody with rhodamine, a red fluorescing dye. We reacted the monoclonal antibody with a mixture of human and mouse-source cysts. Only the human-source cysts reacted with the monoclonal antibody and they stained red. After the reaction with monoclonal antibody had taken place, we washed the excess antibody away and then reacted the mixture of cysts with a polyclonal antibody conjugated with fluorescein, a green fluorescing dye. The polyclonal antibody couldn't react with the human-source cysts because they were already covered with monoclonal antibody. But the polyclonal antibody could react with the murine or non-human type cysts and they stained green. This allows us to differentiate and say that in this specimen both human and non-human Giardia cysts were present.

Discussion

The monoclonal antibodies produced by Dr. Riggs' laboratory were developed from cysts obtained from a single human stool specimen. His laboratory has looked at over 1,000 human stool specimens and the monoclonal antibody reacted with all of them. Specimens from two beavers have been examined and one was positive with both the poly-clonal and monoclonal antibodies indicating that the beaver was infected with human and non-human type Giardia. It is not known if this monoclonal antibody would react with all human-source Giardia or if it would react with all cysts in a given specimen. There appears to be a biological difference in cysts from different human sources in that some will produce infection in Mongolian gerbils with resul-tant cyst excretion while others will cause infection in gerbils but without cyst production. Even with cysts from the same donor, there is variability in the percent of cysts that will excyst. Panelists cautioned that parasites may be subject to antigenic modification dependent upon the host. Panelists also suggested that monoclonal antibodies be developed using cysts from different human and animal sources and that they be cross-checked against each other. Dr. Riggs indicated that his laboratory is in the process now of trying to pro-duce monoclonal antibodies against cysts from other sources.

Recommendations

Additional groundwork is needed with respect to both polyclonal and monoclonal antibodies including characterization of cyst antigens, production of cysts in culture and standardization of cyst prepara-tions and reagents. Antibodies should be produced to cysts from a variety of human and animal sources and thoroughly evaluated before their use in identifying species of origin can be validated.

G. Healy - Quality control, performance evaluation, laboratory certi-fication: Quality controls in parasitology are less defined and less rigid than they are in other areas of microbiology. Reliable diagnosis is based on accurate identification of organisms and competent performance of techniques for recovering and demonstrating the organisms. Proper quality control involves:

1. Trained Personnel - minimum of 150 hours of lectures and laboratory, periodic refresher courses, participation in proficiency testing program(s).

2. Equipment - microscopes, centrifuges, refrigerators, fume hood, slides, coverslips, stains, solutions.

3. Use of Appropriate Techniques - wet mounts, concentrates, stained slides.

4. Reference Materials - manuals, books, library of well-defined specimens from commercial or professional courtesy origin.

5. Reference Collection - maintaining a number of organisms in formalin and/or polyvinyl alcohol fixative.

6. Periodic Use of Reference Material for Accuracy of Routine Performance.

In 1975, because of some severe problems in diagnosis in New York City, the Congress passed the Clinical Laboratory Improvement Act, which mandated that laboratories had to have a certain level of proficiency. The College of American Pathologists, the American Association of Bioanalysts and the Centers for Disease Control are the three principle proficiency testing groups. In clinical parasitology, the situation is different than it is for water bacteriology or water microbiology in that everything is done by consensus. Specimens are sent out to a number of laboratories that are identified as reference or referee laboratories. The performance of other laboratories is then compared against what the reference laboratories have reported to be present in the specimens. Proficiency testing and performance evaluation are going to be very important in the area of water parasitology not only because of Giardia. Cryptosporidium is another protozoan intestinal parasite that may be transmitted through drinking water. The question of laboratory certification is another complex problem even in clinical parasitology, but it is something that also needs to be addressed in environmental parasitology.

Discussion

QA/QC, performance evaluation and laboratory certification are important topics in water parasitology and the need for them will probably increase. The initial focus should be on assessing qualitative performance since quantitation of cysts with available methods is very difficult and open to question. Detecting and identifying cysts is one problem and determining the viability of detected cysts is yet another problem. QA/QC will be needed as well for viability determinations. Recently developed animal models include the Mongolian gerbil and suckling mice but much additional work needs to be done to determine infective dose, conditions for establishing infection and the ability of cysts from different sources to routinely produce infection in animal models. Training is another area that needs to be addressed. The USEPA does not conduct formal training programs in Giardia methods. Since so many of the procedures are more art than science, hands-on workshops might be advisable. These may be conducted under the auspices of the American Society for Microbiology or through some other suitable mechanism. Alternatively, videotape or slide presentations may be of value.

Recommendations

Steps should be taken to form work groups or committees to develop QA/QC, performance evaluation and laboratory certification guidelines in the area of water parasitology, and to develop a training program in Giardia methods.

Concluding Discussion

At the conclusion of the workshop, it was re-emphasized by a participant from Colorado that a negative finding of Giardia using currently available methods does not mean that a given water supply is safe or not at risk. This is true because of variability in method efficiency and because a stream or water supply might be negative on the day of sampling and positive on a day when a sample was not collected. More useful information might be provided by looking at the differential in particle numbers and sizes between raw and filtered waters to determine the efficiency of treatment. It was agreed that there are better ways to monitor the efficiency of filtration than by performing a Giardia analysis. However, there are many unfiltered water supplies in the United States and until and unless a regulation is promulgated requiring the filtration of all surface water supplies, there apparently is a need for a method to specifically detect Giardia cysts and not just particulates in the same size range. The USEPA does not require monitoring for Giardia and the methods work that has been done is related to assisting in outbreak investigations and gathering data on occurrence and health effects of the organism.

Reprinted from Proc. AWWA WQTC, Denver, Colo. (Dec. 1984).

EPA Project Summary

Determination of the Use of Solid Particle Samplers for *Giardia* Cysts in Natural Waters

William S. Brewer

William S. Brewer is with Wright State University, Dayton, OH 45434.

Frank W. Schaefer III is the EPA Project Officer (see below).

The complete report, entitled "Determination of the Use of Solid Particle Samplers for Giardia Cysts in Natural Waters," (Order No. PB 83 246 090) will be available only from:

National Technical Information Service
5285 Port Royal Road
Springfield, VA 22161
Telephone: 703-487-4650

The EPA Project Officer can be contacted at:

Health Effects Research Laboratory
U.S. Environmental Protection Agency
Research Triangle Park, NC 27711

Parasitic flagéllates in the genus *Giardia* are distributed worldwide and are now the most commonly reported intestinal parasites in the United States and Britain. Twenty-three waterborne outbreaks of giardiasis affecting over 7000 people occurred in various states in the United States between 1965 and 1977. Because of this significant increase in the incidence of waterborne outbreaks of giardiasis, efforts have been made to develop reliable and/or sensitive methods to determine the presence or absence of *G. lamblia* cysts in water supplies. The primary objective of this study was to improve the current methodology for concentrating, recovering and detecting cysts of *G. lamblia* in water supplies.

Two sampling processes for the concentration of cysts were examined. One process was diatomaceous earth filtration while the second was that of cyst concentration onto charged particles. Cysts of *G. muris* were used to determine the retention efficiency of ion-exchange resins and each type of diatomaceous earth filter examined. Cyst desorption efficiencies were evaluated for ion-exchange resins that best retained cysts, while backwashing parameters were optimized for diatomaceous earth filters. Results of cyst retention experiments indicated that two processes, anion-exchange concentration of cysts and diatomaceous earth filtration, had the potential to be developed into field methods. Comparison of these two processes at low cyst inoculum concentrations (1×10^3 cysts/liter) indicated that a greater number of cysts could be recovered from the diatomaceous earth filters.

When 40 liter samples of tap water containing between 1.0×10^4 and 7.0×10^5 cysts were passed through diatomaceous earth filters, 5.2 to 31.1% of the cysts were recovered in the backwash. As a result, the diatomaceous earth filter was comparable to microporous filtration and may have application in sampling finished water supplies. However, its utility in raw sampling was limited since turbidity severely reduced the recovery efficiency of cysts.

This Project Summary was developed by EPA's Health Effects Research Laboratory, Research Triangle Park, NC, to announce key findings of the research project that is fully documented in a separate report of the same title (see Project Report ordering information at back).

Introduction

The primary objective of this study was to improve the current methodology for concentrating, recovering, and detecting cysts of *G. lamblia* in water supplies. Two sampling processes for concentration of *G. lamblia* cysts were examined. One process was diatomaceous earth filtration while the second process examined was that of cyst concentration onto charged particles. The first process was based on the hypothesis that cysts could be efficiently trapped on the surface of diatomaceous earth filters and subsequently recovered through backwashing the filter with a small volume of water. The second process was based on the hypothesis that cysts could be attracted to charged surfaces, since they have been shown to have a charge of approximately -25 mV at pH 5.5 and to increase their electronegativity as the pH rises to 8.0. In addition, charge-attraction techniques have been applied to concentration of viruses and bacteria from water, to the concentration of trypanosomes, and to the concentration of *Plasmodium* in clinical samples.

Parasitic flagellates in the genus *Giardia* are distributed worldwide and are now the most commonly reported human intestinal parasites in the United States and Britain. The cycle of this parasite is composed of two stages: the cyst stage and the trophozoite stage. Transmission of *Giardia* most often occurs when viable cysts are ingested directly or through water contaminated with feces. The average incubation period for human giardiasis is 8 days, with a range of 3-42 days. While most infections are asymptomatic, some people have a short-lasting acute diarrheal disease, nausea, and anorexia. A small percentage develop an intermittent or protracted course characterized by diarrhea, cramping, abdominal pain, bloating, and flatulence. Diarrhea with or without overt malabsorption may last months or even years. As a measure of the significance of giardiasis in the United States, 23 waterborne outbreaks affecting over 7000 people have been caused by this infection between 1965 and 1977. Epidemic outbreaks have been reported in New York, New Hampshire, Pennsylvania, Colorado and Washington. The majority of these outbreaks were attributable to public consumption of minimally treated water which was fecally contaminated. In addition, sporadic epidemics have occurred among infants and children in hospital nurseries, custodial and residential institutions, and in day-care centers where personal hygiene standards were not stringent. Water supplies were most likely contaminated by untreated human waste or by aquatic mammal waste since asymptomatic carriers and aquatic mammals have been identified as the major reservoirs.

Because of the significant increase in the incidence of waterborne outbreaks of giardiasis reported, efforts have been made to develop reliable and/or sensitive methods to determine the presence or absence of *G. lamblia* cysts in water supplies. The major problem associated with developing sampling technology is that cysts are assumed to be present in low numbers, therefore necessitating the need for large-volume water sampling. During the past few years, several approaches have been taken to concentrate and detect cysts in water supplies. Developed methods can be divided into three major categories: (a) membrane filtration, (b) particulate filtration, and (c) microporous filtration.

Moore et al. (1969) utilized membrane filtration to examine water and sewage samples for *Giardia* cysts. One- and two-liter samples were passed through cheesecloth and then filtered through 0.45 μm porosity membrane filters. Sediment on the filter surface was brushed into water, centrifuged and preserved in 10% formalin for microscopic examination. No cysts were observed in 10 water samples. Barbour and his coworkers (1976) used the method of Chang and Kabler to filter 22 liters of stream water following an outbreak. However, no *Giardia* cysts were found. Luchtel and colleagues (1980) utilized 293 mm diameter 5.0 μm pore size Nuclepore* filters to concentrate formalin-fixed *G. lamblia* cysts from 20 liter tap water samples. Recovery rates of approximately 75% were found under such conditions.

The Center for Disease Control developed a large-volume sampling technique. The method used a swimming pool filter in which sand was the sampling medium. During an outbreak in Rome, New York, a total volume of 1.1 x 10^6 liters (28,000 gal.) of water was collected at an average flow rate of 76 liters/minute (20 gal./min) through the filter daily for 10 days. The filter backwash was collected each day in two 210-liter drums and coagulated with alum. The resulting sediment was collected and aliquots were fed to beagle puppies and examined microscopically. Two of ten samples fed to dogs produced infection and a single cyst was observed in one sample. In 1976, the U.S. Environmental Protection Agency (EPA) developed a cyst concentration technique involving the filtration of a large volume of water (100 gal. or more) through a microporous orlon fiber filter. This method has been tentatively adopted as the "method of choice" for concentrating cysts from water supplies. However, the reliability and validity of the technique has yet to be fully evaluated. Limited laboratory evaluations have indicated that cyst recovery was only in the range of 3-15%, with a mean of 6.3%. The interpretation of positive or negative field data is uncertain. In the sampling methodology described above, efficient cyst concentration from water was possible under certain conditions. However, processing of the concentrate led to significant losses of cysts. Therefore, a major problem in detecting *Giardia* cysts in contaminated water supplies using any of the above techniques has been the quantitative recovery of cysts from the filter medium.

Summary

The feasibility of using diatomaceous earth filters or ion-exchange resins to concentrate cysts of *G. lamblia* from water samples and subsequently recover those cysts in a quantitative manner was evaluated with water samples experimentally contaminated with *G. muris* cysts. A series of ion-exchange resins were initially selected for evaluation. Anionic resins selected included DEAE-cellulose and two polystyrene, divinyl-benzene-crosslinked resins, Dowex 1-X4* and Dowex 1-X8. Cationic resins evaluated included Dowex 50W-X4 and Dowex 50W-X8. The relative ability of each resin to capture cysts from water samples was determined by passing cysts in buffered, distilled water through ionically-charged columns and determining the fraction of cysts retained at varying inoculum concentrations. DEAE cellulose retained all the cysts passed through the column regardless of the inoculum concentration; whereas, Dowex 1-X4, Dowex 1-X8, Dowex 50W-X4, and Dowex 50W-X8 retained 101, 71, 75, and 97% of the inoculated cysts, respectively. Statistical analysis of the data indicated that there was no significant difference between the ability of DEAE cellulose, Dowex 1-X4 and Dowex 50W-X8 to capture cysts from water samples. Therefore, these resins were selected for desorption studies. Cysts concentrated on the surfaces of ion-exchange resins were eluted with 40 ml of buffer optimized to a particular pH and ionic strength (pI) and the percent recovered from each resin was compared. Approximately 49% of the cysts concentrated on Dowex 1-X4 resins could be recovered throughout the inoculum range tested; however, an average of only 38% of the inoculated cysts were recovered from the Dowex 50W-X8 columns. No cysts were recovered from DEAE cellulose resins regardless of the elution pH or pI.

The capacity of three types of diatomaceous earth filter, Celite 505, Hyflo-SuperCel, and Celite 560, to concentrate cysts was evaluated by passing distilled water samples containing between 6.0 cysts/liter and 1.62 x 10^4 cysts/liter through each column. No significant differences between the diatomaceous earth filters were observed, with retention ranging between 66 and 100%. However,

---

*Mention of trade names or commercial products does not constitute endorsement or recommendation for use.

significant differences between columns were noted when each was backwashed to recover the cysts captured on the surface. Best cyst recoveries were observed when the filters were backwashed with 2000 ml of distilled water at a flow rate of 2 liters/minute. An average of 13% of the cysts concentrated on Celite 560 columns could be recovered. However, because of the concentration of small diatomaceous earth particles or "fines" in the backwash of both Celite 505 and Hyflo-SuperCel columns, no cysts could be detected under the microscope. Therefore, Celite 560 was selected for comparative studies with Dowex 1-X4.

The utility of either Dowex 1-X4 or diatomaceous earth (Celite 560) columns as *Giardia* cyst sampling devices was compared by sampling 40 liter distilled water samples contaminated with cysts and comparing the number of cysts recovered from each sampler. The mean number of cysts recovered was not significantly different over the range of inoculum concentrations used; however, analysis of the data indicated that there was a higher probability of recovering cysts from water samples containing low concentrations of cysts (1.0-1.5×10^3 cysts/liter) when the diatomaceous earth filter was used. Cyst recoveries from water samples passed through diatomaceous earth averaged 13%, similar to that observed when microporous filtration was evaluated. However, the efficiency of the diatomaceous earth filter was markedly decreased when turbid water was sampled. When the recovery of cysts inoculated into turbid water samples (24 FTU) was compared to that of turbid-free samples, an 85-86% reduction was observed.

Conclusions

Based on the objectives and results of this study the following conclusions can be drawn: 1) Results of resin capacity and cyst retention experiments indicated that two processes, anion-exchange concentration of cysts and diatomaceous earth filtration, had the potential to be developed into field methods. 2) Analysis of the data with respect to inoculum concentration indicated that a greater number of cysts were recovered from the diatomaceous earth filters when dilute samples (1×10^3 cysts/liter) were filtered. Based on these data diatomaceous earth filters were considered the best choice. 3) Comparison of the efficiency of the diatomaceous earth filter to the reported efficiency of the EPA method of microporous filtration (13.0 and 6.7%, respectively) indicated that the two were similar. However, it appeared that the diatomacous earth filter was more severely affected by the composition of the water sample than the microporous filter would be. 4) The diatomaceous earth filter may have application to finished water supplies; however, its use on raw water samples is limited at the present time.

Recommendations

1. The results of this study indicate that *Giardia* cysts could be efficienctly concentrated on either the surface of diatomaceous earth or on the surfaces of charged particles. However, subsequent research should be directed toward the efficient recovery of concentrated cysts. Further research should be carried out with weak ion-exchange resins that operate efficiently in a narrower pH range or with charge-modified filters similar to those used to concentrate viruses.

2. One of the major problems encountered in this study was the lack of a sensitive detection technique. Quantification of *Giardia* cysts is presently based upon microscopic identification and counting. Experimental studies should be carried out to develop accurate detection of small numbers of cysts mixed with other microorganisms and debris from aquatic habitats. Immunofluorescent techniques similar to those used for the detection of bacteria should be investigated.

3. The behavior of *Giardia* cysts on certain resins observed in this study raised some question concerning the biochemical and physiological nature of cysts. Little information is available on these subjects. Basic research on the nature of the cysts themselves would lead to rational decisions on applicable sampling methods.

References

Barbour, A.G., C.R. Nichols, and T. Fukushina. 1976. An outbreak of giardiasis in a group of campers. Am. J. Trop. Med. Hyg. 25: 384-389.

Luchtel, D.L., W.P. Lawrence, and F. B. DeWalle. 1980. Electron microscopy of *Giardia lamblia* cysts. Appl. Environ. Microbiol. 40: 821-832.

Moore, G.T., W.M. Cross, D. McGuire, C.S. Mallohan, N.N. Gleason, G.R. Healy, and L.H. Newton. 1969. Epidemic giardiasis of a ski resort. N. Eng. J. Med. 281: 402-407.

Giardiasis Outbreaks and
Giardia Occurrence

Waterborne Disease Outbreaks—1946–1980: A Thirty-Five-Year Perspective

Edwin C. Lippy and Steven C. Waltrip

The 672 outbreaks of waterborne disease that were reported in the United States between 1946 and 1980 affected more than 150 000 persons. An analysis of the data from these outbreaks provides information on how often and where outbreaks occurred, as well as on what caused them.

Information on waterborne disease outbreaks has been collected since 1920 and published in summary fashion for the periods 1920-36,[1] 1938-45,[2] 1946-60,[3] 1961-70,[4] and 1971-78.[5] Outbreak information on file by year dating back to 1946 was recently entered into an automatic data processing system that provided rapid recall of data for analyses of occurrence, distribution, and trends and for other statistical interpretation. The data for 1946–80 are presented in this article to give a 35-year perspective on waterborne illness associated with drinking water.

Information on waterborne illness has been submitted voluntarily to the US Public Health Service (USPHS) since 1920 by state agencies concerned with intestinal illness. What originally began as a concern for morbidity and mortality resulting from typhoid fever and infant diarrhea and their relationship with food, milk, and water led to the development of important public health measures that had a profound influence on decreasing the incidence of enteric diseases. The voluntary reporting procedure was changed somewhat after the responsibility for it was transferred in 1966 from the National Office of Vital Statistics to the Centers for Disease Control (CDC), which began contacting state health agencies on a yearly basis for reports of disease outbreaks. In 1971 the US Environmental Protection Agency (USEPA) joined with CDC in a collaborative effort to improve reporting of waterborne illness by contacting state regulatory agencies responsible for water supply activities. A form was developed to provide detailed information on waterborne outbreaks, and its use was initiated in 1976 after two years of field testing. Reporting was further refined in 1978 when annual summaries of waterborne outbreaks were separated from those caused by food and published under the title "Water-Related Disease Outbreaks."[6]

Authors involved in writing summaries and articles on waterborne disease outbreaks have been critical in their appraisal of the effectiveness of voluntary reporting. They indicate that it is a passive surveillance system which detects only a fraction of the outbreaks and cases of illness that occur each year in the United States. In support of this criticism are the statistics developed from an improved surveillance system in one state that accounted for 48 of the 189 outbreaks or 25 percent of the total reported in the period from 1976 to 1980. Craun and McCabe,[7] in a review spanning 1946-60, estimated that one half of the outbreaks in community systems and only one third in noncommunity systems are reported. Information from a study done in a Rocky Mountain state under auspices of a USEPA contract showed that under intensified surveillance efforts, the annual number of outbreaks increased nearly fourfold in a three-year period compared with the occurrence in the previous three years. Surveillance and reporting efforts improved from 1971 to 1980 (Figure 1), but much more needs to be done.

The analyses that follow must be interpreted with caution because the intensity of surveillance varies among states. Agencies that conscientiously report outbreaks may cast doubt on the safety of drinking water in their states although this may not be warranted. If all states placed equal emphasis on surveillance, a more accurate representation of occurrence, distribution, and trends could be presented.

The data presented in this article are from outbreaks associated with water obtained from public or individual water systems and intended for drinking or domestic purposes. At least two cases of acute infectious disease or one case of an acute intoxicating illness must occur and originate from a common source to

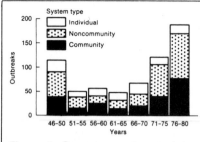

Figure 1. Occurrence of waterborne disease outbreaks by five-year increments

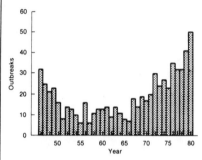

Figure 2. Annual occurrence of waterborne disease outbreaks

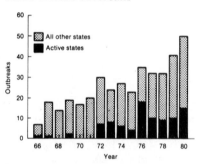

Figure 3. Waterborne disease outbreaks in two active states versus those in all other states

TABLE 1
Occurrence of waterborne disease outbreaks by type of water system

System Type	Outbreaks		Occurrence per 1000 Systems*
	1946–80	**1971–80**	
Community	237	121	1.9 (121/65)
Noncommunity	296	157	1.0 (157/150)
Individual	137	36	
Unknown	2	0	

*1971–80 data

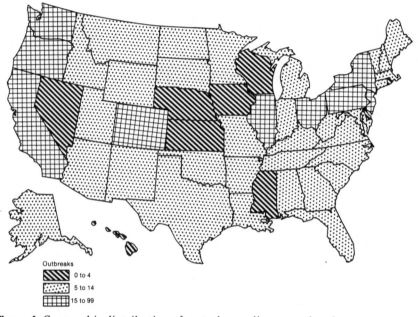

Figure 4. Geographic distribution of waterborne disease outbreaks

qualify as an outbreak. An infectious disease that produces symptoms affecting the upper and lower gastrointestinal tract is caused by a microbiological agent and is accompanied by an incubation period ranging from about 24 hours to several weeks. An intoxicating illness that produces symptoms associated with the upper gastrointestinal tract is caused by a chemical agent, and the incubation period from exposure to illness is brief—usually a matter of a few hours. For one case of illness to qualify as an outbreak associated with a chemical agent, the causative agent must be identified in a sample of water collected from the system.

A public water system normally consists of (1) a water supply from a surface source (e.g., river, lake, reservoir), a ground source (e.g., well, spring, or a mixture of the two); (2) treatment facilities (e.g., filtration, disinfection); and (3) a distribution network of mains and pipes that deliver water to the consuming public. Unfortunately, many systems

have no treatment facilities. By definition, public water systems also have at least 15 service connections and provide water to 25 persons. There are two types of public water systems. A community water system serves year-round residents; a noncommunity water system, which does not meet the residence requirement, serves the traveling public at institutions, camps, parks, or motels. Where public water systems are not available, people develop their own household water supplies—individual systems.

Public water systems currently number about 215 000, with 65 000 community systems serving a resident population of 195 million persons. Noncommunity systems number 150 000, and about 10 million individual systems serve 35 million persons.

Occurrence, distribution, and trends

Outbreaks. Figure 2 illustrates the annual frequency of the 672 waterborne outbreaks reported for the 35-year period commencing in 1946. The trend in num-

ber of outbreaks has been increasing since 1966. The trend is even better exemplified in Figure 1 and is especially notable since 1971, when the current system of reporting was established. Figure 2 shows a peak in 1980; the 50 reported outbreaks were the third highest number of outbreaks since record collection began in 1920. The record of 60 outbreaks was reported in 1941. Cursory examination of Figures 1 and 2 indicates that outbreaks were under control for the 1951–70 period, whereas the opposite may be said about the outbreaks for 1971–80. A four-year cycle in annual occurrence began in 1972, with peaks following in 1976 and 1980. There is no obvious scientific explanation for the cycle; however, if it and the current trend continue, 95 outbreaks can be expected in 1984. Again, Figure 2, as well as subsequent analyses, must be tendered with cautious interpretation. Although Figures 1 and 2 show large increases in the number of reported outbreaks for the 1971–80 period compared with those in previous years, relatively few outbreaks were reported by the 50 states. The 50 outbreaks reported in 1980 averaged one per state. If a few states become aggressive in investigating and reporting waterborne outbreaks, the trend can easily be affected. The influence of two such states is shown in Figure 3.

The geographic distribution of outbreaks is shown in Figure 4. Delaware was the only state that did not report an outbreak. States in which the reported outbreaks were low are clustered in the upper Midwest or Plains states, and those in which the number was high are clustered in the eastern United States (New York, Pennsylvania, New Jersey, and Ohio) and in the western United States (Washington, Oregon, and California). The combination of a low population density and a long distance between sources of water supply and wastewater discharge locations may explain the low-outbreak cluster, whereas opposite conditions may explain the high-outbreak clusters. The clusters may also be due to the effort devoted to surveillance of outbreaks. New York has been active in investigating and reporting since the 1940s, and Pennsylvania began a more active effort in the 1970s (Figure 3).

The gross distribution shown in Figure 4 was corrected for population density by expressing the information as a rate—outbreaks per number of community water systems—for each state (Figure 5). Outbreaks in noncommunity and individual systems were not included in the distribution. In Figure 5 the rate of occurrence is based on the number of outbreaks that occurred over the 35-year period for each state and the number of water systems that were counted during a nationwide inventory in 1963.[8] The 1963 date represents a near midpoint in the set of outbreak data.

Of those states in the high-outbreak category of the gross distribution shown in Figure 4, only New York, Colorado, and Oregon remain in the high-outbreak category of the distribution based on the number of water systems (Figure 5). These three remaining states share the high category in the rate distribution with the states of Vermont, New Hampshire, Connecticut, Rhode Island, South Carolina, Wyoming, Idaho, and Alaska. The preponderance of the states in the low-rate category are located in the mid-continental United States. The topic of geographic outbreak clustering is also considered in the section on "Water System Deficiencies."

A temporal distribution of outbreaks, which is given in Figure 6, shows that a peak is evident in the summer months. Summer is also the period of the year for vacations, including visits by many people to recreational areas that are primarily served by noncommunity water

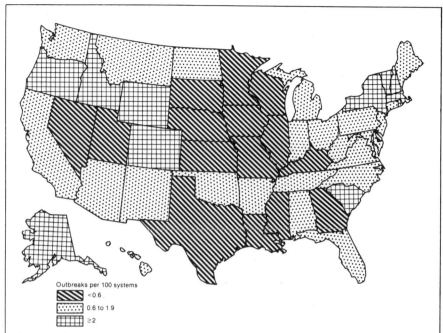

Outbreaks per 100 systems
<0.6
0.6 to 1.9
≥2

Figure 5. Geographic distribution of waterborne disease outbreaks per 100 community water systems

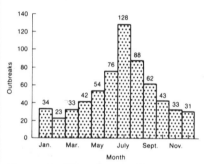

Figure 6. Waterborne disease outbreaks by month *(month not known for 25 outbreaks)*

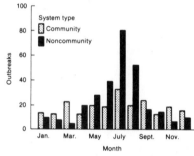

Figure 7. Waterborne disease outbreaks for community and noncommunity systems

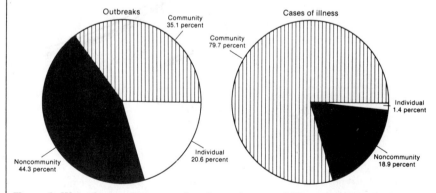

Figure 8. Waterborne disease outbreaks and cases of illness by type of system

systems. Figure 7 was developed to compare the monthly distribution of outbreaks by type of water system. The pronounced peaks shown in this figure indicate that noncommunity systems experience problems in the summer. A similar, clearly defined, temporal distribution is not apparent for community systems.

Outbreaks in noncommunity systems outnumbered those in community systems, as shown in Table 1 and Figure 8. However, when the rate of occurrence was based on the number of outbreaks per 1000 community or noncommunity systems, the rate for community systems was nearly twice that for noncommunity systems (Table 1). Rates were computed

69

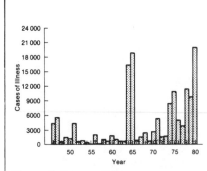

Figure 9. Cases of illness

TABLE 2
Causative agents of waterborne disease

Agent	Outbreaks	Cases of Illness
Bacterial		
Campylobacter	2	3 800
Pasteurella	2	6
Leptospira	1	9
Escherichia coli	5	1 188
Shigella	61	13 089
Salmonella	75	18 590
Total	146	36 682
Viral		
Parvoviruslike	10	3 147
Hepatitis	68	2 262
Polio	1	16
Total	79	5 425
Parasitic		
Entamoeba	6	79
Giardia	42	19 734
Total	48	19 813
Chemical		
Inorganic	29	891
Organic	21	2 725
Total	49	3 616
Unknown	350	84 939
Grand total	672	150 475

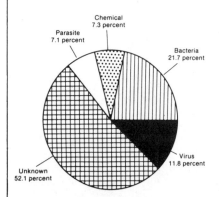

Figure 12. Waterborne disease outbreaks by causative agent

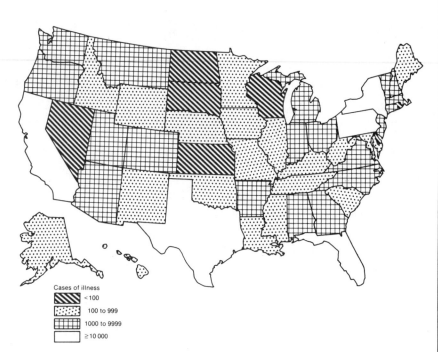

Figure 10. Geographic distribution of cases of illness

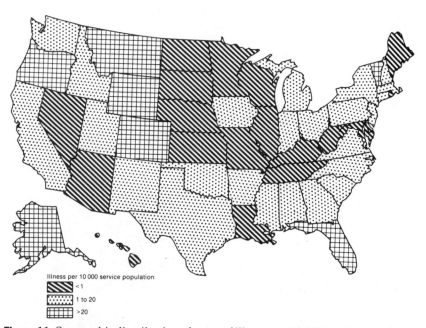

Figure 11. Geographic distribution of cases of illness per 10 000 service population in community water systems

only for the 1971–80 period, to take advantage of information recently collected on the number of water systems in the United States through the Federal Reporting Data System (FRDS). Because national inventories prior to the FRDS did not discriminate between community and noncommunity systems, rates for 1946–80 could not be calculated.

Compared with the outbreak rate for noncommunity systems, the greater rate for community systems is probably due to better reporting of outbreaks; clusters of illness would be more readily recognized and reported in a resident population than in a nonresident population. Travelers exposed to contaminated water from a noncommunity system may be-

come ill some time after exposure (the incubation period) and at a distance from the point of exposure. Associating their illness with contaminated water served at a motel, campground, or restaurant that they visited two to three days previously would be difficult. If an association would be perceived, the separation by time and distance might tend

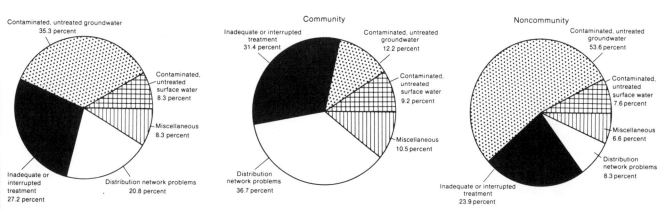

Figure 13. Waterborne disease outbreaks by deficiency in public water systems

Figure 14. Waterborne disease outbreaks by deficiency in community and noncommunity water systems

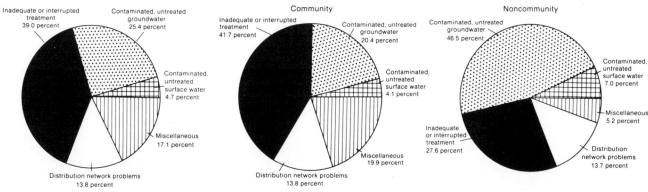

Figure 15. Cases of illness by deficiency in public water systems

Figure 16. Cases of illness by deficiency in community and noncommunity water systems

to make people want to forget the event and not become involved with reporting an illness potentially related to contaminated drinking water. Therefore, although the outbreak rates are greater for community systems than for noncommunity systems, the recognition and reporting of outbreaks in noncommunity systems must be acknowledged as affecting the rates.

Illness. The annual occurrence of cases of illness associated with waterborne disease outbreaks are shown in Figure 9. The total number of cases during the period was 150 475. More than 20 000 cases occurred in the peak year of 1980, with four major outbreaks accounting for 14 000 of these cases. The prominent peaks in two other years, 1964 and 1965, each resulted from one major outbreak that caused 16 000 cases of illness. Excluding these two years, illness could be labeled as insignificant for a 25-year period. Since 1974, however, the number of cases has increased dramatically and been accompanied by a similar increase in outbreaks (Figures 1 and 2).

The geographic distribution of illness is shown in Figures 10 and 11. Figure 10 represents a gross distribution of all cases. Figure 11 shows a distribution that was corrected for population density

by expressing illness as a rate based on the number of people served by community water systems. Those cases attributable to outbreaks in noncommunity and individual systems were not considered in the rate. In Figure 11 the rate is based on the number of cases per 10 000 service population counted during the USPHS national inventory in 1963.

The gross distribution emphasizes the large outbreaks that occurred in California (Riverside with 16 000 cases in 1965), Texas (Georgetown with 8000 cases in 1980), Florida (Gainesville with 16 000 cases in 1964), Pennsylvania (Sewickley with 5000 cases in 1975 and Bradford with 3500 cases in 1979), and New York (99 outbreaks, the largest of which was in Rome with 4800 cases in 1974). In Figure 11, the use of a rate to portray illness changed the distribution pictured in Figure 10, with only Florida remaining in the high-rate category and being joined by Oregon, Montana, Wyoming, Colorado, Vermont, and Alaska.

Figure 12 indicates that bacterial agents were a major factor among outbreaks. In more than half of the outbreaks, however, an agent was not determined. Microbiological agents are rarely identified in water (<1 percent of outbreaks) but are determined through col-

lection and assay of stool or blood specimens. The lack of success in identifying agents in water stems mainly from the lack of local capability to isolate, culture, and identify microorganisms from an environment in which they are normally reduced in number by dilution or die away as opposed to medical specimens collected from a human subject in which agents are propagated. Another reason for poor success with environmental samples is that outbreaks are normally recognized one to two weeks after the exposure, which allows ample time for the water system to be purged of contamination through normal usage of water. Conversely, opportunities for identifying the agent in medical specimens are better because of secondary transmission of illness through person-to-person pathways. A person originally exposed to the agent in his drinking water may transmit the disease some days later to a contact who then becomes a positive candidate for investigators. Because some agents survive freezing, contaminated water drawn from the tap and frozen as ice cubes for later use can also be analyzed and used to supplement the analysis of medical specimens.

Chemical agents can affect aesthetic qualities of water. Unusual taste, odor,

TABLE 3
Comparison of sources of water supply [7] for community water systems with low and high outbreak rates

Outbreak Rate	State	Systems Using Surface Sources	Systems Using Ground Sources	Ratio of Ground to Surface Sources
Low	Delaware	5	36	7.2
	Georgia	80	327	4.1
	Illinois	118	786	6.7
	Iowa	33	666	20.2
	Kansas	73	400	5.5
	Kentucky	130	123	0.9
	Louisiana	43	235	5.5
	Minnesota	23	578	25.1
	Mississippi	3	230	76.7
	Missouri	87	341	3.9
	Nebraska	3	427	142.3
	Nevada	9	43	4.8
	South Dakota	11	237	21.5
	Texas	172	960	5.6
	Utah	32	249	7.8
	Wisconsin	21	404	19.2
	Total	843	2661	*
High	Alaska	25	11	0.4
	Colorado	92	196	2.1
	Idaho	32	147	4.6
	Massachusetts	54	168	3.1
	New Hampshire	50	52	1.0
	New York	296	653	2.2
	Oregon	123	197	1.6
	Rhode Island	8	35	4.4
	South Carolina	63	148	2.3
	Vermont	41	116	2.8
	Wyoming	27	69	2.6
	Total	811	1792	†

*Median ratio = (6.7 + 7.2)/2 = 6.95
†Median ratio = 2.3

or discoloration leads to complaints that prompt collection of samples and investigation by local authorities. Outbreaks and illness caused by chemicals are probably better recognized and reported than are those caused by microbial agents. Table 2 lists microbiological and chemical agents responsible for outbreaks from 1946 to 1980.

Cases of illness are classified according to type of water system in Figure 8. Although noncommunity systems have the largest percentage of the total number of outbreaks, cases of illness in community systems greatly outnumber those occurring in other water systems. Simply stated, outbreaks in community systems are more serious in terms of illness because of the availability of a resident population to become infected. The severity of illness in community or noncommunity systems is further exemplified by comparing cases per outbreak. Community systems averaged 506 cases of illness per outbreak, and noncommunity systems averaged 96 cases, which gives a severity factor of five.

The number of deaths resulting from waterborne outbreaks has greatly diminished over the years. From 1920 through 1945, 960 deaths were reported (mostly from typhoid fever), for an annual rate of 37. During the 35-year period since 1946, the rate has been reduced to 1 death per year. An underreporting problem probably exists in terms of deaths associated with gastrointestinal illness, which can stress a weakened cardiovascular system or other critical body function. That is, even though mortality occurs in the weakened and distressed population affected by gastrointestinal illness, the death is not attributed to gastrointestinal illness.

Water system deficiencies

Deficiencies in water system design, operation, and maintenance that contribute to waterborne outbreaks are of importance to regulatory agencies and to the water utility industry, including purveyors, equipment manufacturers, consulting firms, and laboratories. Overwhelming percentages of a certain deficiency are indicative of a breakdown in an approach by a regulatory agency or the industry. Unfortunately, the analyses that follow show that the causes of outbreaks for 1946 to 1980 do not markedly differ from those presented in summaries for 1920 to 1945, which may indicate that drastic changes are needed in attitudes of regulatory agencies and the industry.

Water system deficiencies that caused or contributed to outbreaks were categorized under five major headings as follows:

1. Use of contaminated, untreated surface water;
2. Use of contaminated, untreated groundwater;
3. Inadequate or interrupted treatment;
4. Distribution network problems; and
5. Miscellaneous.

Use of contaminated, untreated surface water and groundwater is self-explanatory. Inadequate or interrupted treatment includes breakdown or failure of equipment, insufficient chlorine contact time, and an overloaded process. Distribution network problems include cross-connections, improper or inadequate main disinfection, and contamination of open distribution reservoirs. The miscellaneous category encompasses deliberate sabotage, events that systems are not expected to cope with, and undetermined causes of outbreaks.

Water system deficiencies that caused or contributed to outbreaks are graphically depicted by the five categories in Figure 13. Outbreaks in individual systems were omitted so that deficiencies in public water systems, which are regulated by a federal law, could be emphasized. Use of untreated, contaminated groundwater and poor practices in treatment and distribution of water are deficiencies that deserve attention. More than 80 percent of the outbreaks were associated with these deficiencies. The deficiencies in community and noncommunity systems are compared in Figure 14. Two thirds of the outbreaks in community systems were caused by deficiencies in treatment and distribution of water, whereas three fourths of the outbreaks in noncommunity systems were related to use of untreated groundwater or to poor treatment practices.

The overwhelming statistic gleaned from Figure 14 and Table 1 is that the 229 outbreaks in noncommunity systems were caused by use of untreated, contaminated groundwater or by inadequate or interrupted treatment. Disinfection is generally the only treatment provided in these systems. The 229 outbreaks represent nearly one half of all outbreaks reported for public water systems that are regulated by law; i.e., 237 outbreaks in community systems plus 296 in noncommunity systems equals 533 outbreaks, and 229/533 is approximately one half. If disinfection was in place where needed and applied properly, many outbreaks could be prevented. Some of the common problems that deserve attention from regulatory agencies and the water utility industry include:

1. Noncommunity systems suffer from lack of proper design, construction, and

TABLE 4
Comparison of community water systems providing treated and untreated surface water [7] in states with low and high outbreak rates

Outbreak Rate	State	Systems Providing Treated Surface Water	Systems Providing Untreated Surface Water
Low	Delaware	5	0
	Georgia	79	1
	Illinois	118	0
	Iowa	33	0
	Kansas	73	0
	Kentucky	127	3
	Louisiana	43	3
	Minnesota	23	0
	Mississippi	3	0
	Missouri	87	0
	Nebraska	3	0
	Nevada	9	0
	South Dakota	11	0
	Texas	166	6
	Utah	21	11
	Wisconsin	21	0
	Total*	822	24
High	Alaska	13	12
	Colorado	77	15
	Idaho	22	10
	Massachusetts	48	6
	New Hampshire	47	3
	New York	275	21
	Oregon	118	5
	Rhode Island	8	0
	South Carolina	63	0
	Vermont	21	11
	Wyoming	25	2
	Total†	717	85

*Low rate = 822/24 = 34.2
†High rate = 717/85 = 8.4

TABLE 5
Comparison of community water systems providing treated and untreated groundwater [7] in states with low and high outbreak rates

Outbreak Rate	State	Systems Providing Treated Groundwater	Systems Providing Untreated Groundwater	Ratio of Treated to Untreated Groundwater
Low	Delaware	22	13	1.6
	Georgia	155	171	0.9
	Illinois	575	211	2.7
	Iowa	387	272	1.4
	Kansas	399	1	399.0
	Kentucky	92	31	3.0
	Louisiana	112	123	0.9
	Minnesota	287	291	1.0
	Mississippi	123	107	1.1
	Missouri	150	191	0.8
	Nebraska	39	387	0.1
	Nevada	13	30	0.4
	South Dakota	75	148	0.53
	Texas	524	436	1.2
	Utah	68	181	0.4
	Wisconsin	217	173	1.3
	Total	3238	2767	*
High	Alaska	9	2	4.5
	Colorado	98	98	1.0
	Idaho	40	106	0.4
	Massachusetts	63	101	0.6
	New Hampshire	14	38	0.4
	New York	279	366	0.8
	Oregon	63	128	0.5
	Rhode Island	19	16	1.2
	South Carolina	49	99	0.5
	Vermont	10	105	0.1
	Wyoming	30	39	0.8
	Total	674	1098	†

*Median ratio = (1.0 + 1.1)/2 = 1.05
†Median ratio = 0.6

operation and receive minimal attention from regulatory agencies to correct these problems.

2. One microbiological sample is required during each quarter a noncommunity system is operational. A sample collected from the system during start-up after the off-season may not show the same results as one collected during heavy visitation and system overload.

3. Coliform samples that are collected periodically from a challenged system with problems and show negative results provide a false sense of security. Though it is easy to inactivate coliforms with chlorination, more resistant pathogens not detected by the testing procedure may survive.

4. The key to disinfection is reliability. Reliability is enhanced by yoked-up cylinders; switchover devices; auxiliary power; dosage applied in response to output; automatic residual recording; loop-controlled feed; adequate contact time; pH, temperature, and turbidity considerations; standby equipment or spare parts; and reports to substantiate operation.

A simple correction can often enhance reliability. For example, during investigation of a recent outbreak, examination of the chlorination facility showed that hypochlorite was pumped from a day tank in response to float operation in a reservoir. The operator duly noted in the record every day that the chlorine solution level in the day tank was "O.K." After an unsuccessful wait for the pump to cycle during the inspection, it was discovered that the electrical contacts in the relay box were so corroded that the pump was not receiving the signal. The solution to this problem was a simple correction in record-keeping: the recorded level of the chlorine solution would be used to indicate whether the pump was operating. It took only a few minutes for the operator to realize that if he regularly measured a 250–350-mm (10–12-in.) decrease in the solution level on a daily basis and suddenly found only a 75-mm (3-in.) decrease, something was wrong.

Cases of illness attributed to deficiencies in public water systems are shown in Figure 15 and are further classed according to community and noncommunity systems in Figure 16. Inadequate or interrupted treatment was responsible for the greatest number of cases of illness in community systems, whereas use of contaminated, untreated groundwater caused the most illness in noncommunity systems.

The states with low and high outbreak rates (Figure 5) were tabulated, and comparisons were made based on the source of water supply and treatment provided. The comparisons were limited to community water systems identified in the 1963 USPHS inventory. In Table 3, ratios of systems using groundwater to those using surface water were computed for states with low and high outbreak rates, and the medians of the ratios were compared. The results indicate that water systems in states with a low outbreak rate depended more on groundwater than on surface water as a source of supply. This is a reasonable expectation in that groundwater generally has better microbiological quality than does surface water. This does present a conflict, however, in that more outbreaks are attributed to use of groundwater (for noncommunity systems). The conflict may be explained by the premise that community water systems are better operated, with disinfection in place where it is required.

Table 4 shows a comparison between states with low and high outbreak rates and the number of systems providing treated and untreated surface water. Treatment was defined in the 1963 USPHS inventory as "any action taken upon the water"; therefore, for example, communities that treated surface water with aeration were included. Comparison of the ratios indicates that states with a low rate of outbreaks had a much higher ratio of systems treating water to those not treating than did states that had a high rate of outbreaks. The fourfold difference in the ratios for low and high rates reinforces the need for adequate treatment of surface sources.

A similar comparison for the treatment of groundwater is shown in Table 5. Comparison of the ratios indicates more systems provided treatment of groundwater in the states with low outbreak rates.

Summary

In the 35-year period from 1946 to 1980, 672 waterborne disease outbreaks affecting more than 150 000 persons were reported. The increase in the annual occurrence of outbreaks that began in 1967 was probably due to better reporting from a few states. If this is true, the statistics for the United States during this 35-year period are greatly understated. The reporting of outbreaks is improving, if an increase in the number of outbreaks is used as an indicator. If the data for 1976–80 are used as an attainable level of sensitivity under the current reporting procedure, the average annual frequency of outbreaks is 38. Applying this value for annual occurrence to the 35-year period increases the period total to 1330 or nearly double the number reported. A similar analogy applied to illness shows a frequency of 10 000 cases per year, which applied to the 35-year period results in 350 000 cases of illness—considerably more cases than the number reported.

Clustering was evident when outbreaks in community systems were geographically distributed and rate-based
according to the number of water systems in a respective state. Water systems in states with low outbreak rates were compared with those in states with high outbreak rates by using source of supply and treatment as indicators. Those states with low outbreak rates had a greater ratio of systems that depended on groundwater as a source of supply and that provided treatment of surface water and groundwater.

Outbreaks exhibit temporal distribution, with peak occurrences in June, July, and August. This distribution was more pronounced when outbreaks in community systems were compared with those in noncommunity systems. The number of outbreaks in noncommunity systems greatly exceeded those in community systems for the summer months. Noncommunity systems serve recreational areas that experience heavy visitation during the summer, causing maximum demand on water systems and overloads to sewerage–sewage facilities. Noncommunity systems suffer from design, construction, operation, and maintenance problems and normally receive less attention from regulatory agencies in terms of monitoring and inspection to correct the problems.

Although outbreaks in noncommunity systems outnumber those in community systems, the frequency of outbreaks is almost twice as great in community systems. However, outbreaks in noncommunity systems are more difficult to detect because they serve the traveler who may become ill 1600 km (1000 mi) from the point of exposure to contaminated drinking water and would not associate the illness with a place he visited two days earlier. Although underreporting of outbreaks occurs for community systems, it is thought to be a much more serious problem for noncommunity systems.

Cases of illness occurring in community water systems far outnumber those in noncommunity systems. Regulatory agencies should concentrate their efforts on protecting the source of water supply, ensuring adequate and reliable treatment, and improving distribution network practices in community systems to achieve the greatest overall reduction of waterborne illness.

Microbiological agents that cause waterborne outbreaks are rarely isolated from the water system. Investigators are more successful with identification of agents in specimens from cases where the agent is propagated and excreted in large numbers. For more than half of the outbreaks, no agent is identified, probably because of the lapse in time between occurrence of illness and the beginning of an investigation that includes collection of specimens.

The deficiencies in water systems that caused and contributed to water-
borne outbreaks during this 35-year period differed little from those reported for the previous 26 years (1920–45). The glaring deficiencies were that disinfection was not in place where it was needed and not properly operated where it was in place. Changes in regulatory and industry approaches are indicated and may include scrutiny in approval, frequent inspection, operator training, and expanded monitoring from regulatory agencies with improvements in design, operation, and equipment from industry.

Acknowledgments

The authors thank V. Tilford, secretary, Health Effects Research Laboratory, Cincinnati, Ohio, and B. Lippy. journalism student, Ohio University, Athens, Ohio, for typing, proofreading, and editing the manuscript.

References

1. GORMAN, A.E. & WOLMAN, A. Waterborne Outbreaks in the United States and Canada and Their Significance. *Jour. AWWA*, 31:2:225 (Feb. 1939).
2. ELIASSEN, R. & CUMMINGS, R.H. Analysis of Waterborne Outbreaks, 1938–45. *Jour. AWWA*, 40:5:509 (May 1948).
3. WEIBEL, S.R. ET AL. Waterborne-Disease Outbreaks, 1946–60. *Jour. AWWA*, 56:8:947 (Aug. 1964).
4. TAYLOR, A. ET AL. Outbreaks of Waterborne Disease in the United States, 1961–1970. *Jour. Infectious Diseases*, 125:3:329 (Mar. 1972).
5. CRAUN, G.F. Outbreaks of Waterborne Disease in the United States: 1971–1978. *Jour. AWWA*, 73:7:360 (July 1981).
6. Water-Related Disease Outbreaks—Annual Summary. Centers for Disease Control, Dept. of Health & Human Services, Atlanta, Ga. (1978).
7. CRAUN, G.F. & MCCABE, L.J. Review of the Causes of Waterborne-Disease Outbreaks. *Jour. AWWA*, 65:1:74 (Jan. 1973).
8. Statistical Summary of Municipal Water Facilities in the United States, Jan. 1, 1963. Publ. 1039. USPHS, Washington, D.C. (1965).

About the authors: *Edwin C. Lippy is a graduate of Clemson University (Clemson, S.C.), with degrees in civil engineering and water resources engineering. His work has been published pre-*viously in JOURNAL AWWA, *the* American Journal of Epidemiology, *and the* American Journal of Public Health. *He is a sanitary engineer and Steven C. Waltrip is a biological technician at the A.W. Breidenbach Environmental Research Center, USEPA, 26 West St. Clair Street, Cincinnati, OH 45268.*

Reprinted from *Jour. AWWA*, 76:2:60 (Feb. 1984).

A Waterborne Outbreak of Giardiasis in Camas, Wash.

J. C. Kirner, J. D. Littler, and L. A. Angelo

An account of the investigation launched in the wake of an outbreak of giardiasis reveals how and why the outbreak occurred and describes the steps taken to prevent a reoccurrence.

In late April and early May of 1976, local physicians in Camas, Wash., reported the occurrence of approximately 25 cases of giardiasis during a 2–3 week period. Subsequently, an investigation was launched by local, state, and federal health officials.

The results of the investigation revealed new areas of concern for the water-utility industry. The disease-causing organism was identified in the infected individuals and traced back through the water system to infected beavers active in the watershed. This was the first substantiated case of wild animals contaminating a human population with *Giardia. Giardia lamblia* cysts were concluded to have passed through the system's mixed-media pressure filters because of media loss, media disruption, and inadequate pretreatment. Approximately 10 per cent of the population (600 people) showed clinical signs of the infection. Initially, the distribution of the infected individuals implicated the water system.

Emergency measures were taken to prevent further spread of the infection. The source of supply was switched from a surface to a deep-well source; the system was flushed; and a "boil water order" was imposed in areas that could not be flushed immediately.

Investigations were undertaken to (1) determine the manner of disease transmission, (2) identify the source of the

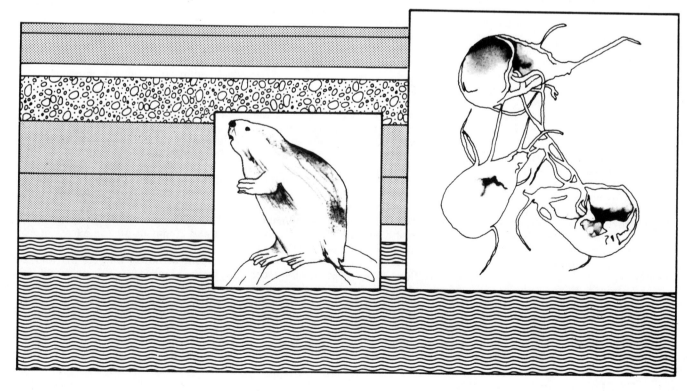

TABLE 1
Results of Filter Sampling Technique, May 1976—EPA Filter

Location	Day	Volume Filtered		Giardia Found
		gal	m³	
Butler Reservoir	15	100	0.38	−
Boulder Creek	15	25	0.09	−
Jones Creek	15	72	0.27	−
Lower Prune Hill Reservoir	16	23	0.08	−
Upper Prune Hill	16	225	0.85	+ (50/225)
Butler Reservoir	15	*	*	+
Water Treatment Plant (raw water)	20	2250	8.52	−
Water Treatment Plant (raw water)	20	162	0.61	+
Well No. 1	20	1000	3.79	−
Well No. 4	20	1255	4.75	−
Boulder Creek Headworks	21	2978	11.27	+
Boulder Creek Headworks	21	268	1.01	−
Jones Creek Headworks	21	6750	25.55	−

*Sediment (unfiltered)

disease, and (3) determine what corrective action should be taken. The investigations included an epidemiological study, a system sanitary survey, literature review, and the application of sampling techniques known to have been effective in detecting *Giardia lamblia* in previous outbreaks.

The System

The city of Camas is located 21 km (13 mi) east of Portland, Oregon on the north shore of the Columbia River and has a population of 6000. Its primary industry is paper production. The city utilizes both surface-water and deep-well sources of supply. Most of its water supply, for a large portion of the year, is from the surface-water source. Since this is a gravity-flow system, the slightly higher treatment costs are offset by not having to pump the water up to distribution system pressures. (Fig. 1, Schematic of Camas Water System.)

The surface-water source consists of Boulder and Jones Creeks. The two creeks come from adjoining watersheds comprising a total area of 17 sq km (6.5 sq mi). Both watersheds have extremely limited human activity. There are no dwellings within their bounds, and mechanical access is limited by exceptionally poor quality dirt roads and rugged terrain. Wildlife within the watershed is varied and consists of many of the warm-blooded animals indigenous to the Pacific Northwest. Both watersheds are well isolated, have no human habitation, and are rarely used by humans above the water-system intakes. Raw-water turbidities during the period of concern rarely exceeded the existing state and federal standards for finished-water turbidity.

Raw water from both creeks is chlorinated in a transmission main located approximately 1.5 hr upstream from the water-filtration plant. The water-treatment plant (Fig. 2) is a direct filtration system consisting of injection of pretreatment chemicals immediately prior to the filter vessel and filtration using a multimedia pressure filter. The raw-water quality is generally excellent and the design filter rate of 7.9×10^{-2} m³/s (1.8 mgd) and 3.3×10^{-3} m/s (4.8 gpm/sq ft) is within acceptable limits. The treated water produced by the treatment plant continuously met turbidity and bacteriological standards as specified by the Safe Drinking Water Act (SDWA) during the period when the disease-causing organisms would have passed the treatment processes (Fig. 3).

The city also receives water from seven wells located in the lower elevations of the distribution system along the Washougal River. The wells augment the surface-water supply during periods of high demand or during periods of low flow in the surface-water source. The quality of water from the wells is excellent and meets all requirements of the Interim Primary Drinking Water Standards of SDWA, and chlorination is provided as a routine safety measure.

Distribution-system storage is provided in three locations serving three pressure zones. All distribution-system reservoirs are covered and are well protected against contamination.

The Infection

Giardia lamblia is a protozoa that infects the small bowel. Although the infection may be asymptomatic, its common symptoms are chronic diarrhea, abdominal cramps, bloating, fatigue, weight loss, and frequent loose pale stools that are greasy and malodorous. Large numbers of the causative organism (*Giardia lamblia*) can be harbored by both asymptomatic and symptomatic victims. Treatment is performed through the use of chemotherapy (flagyl or atabrine).[1] The mode of transmission for the infection is accepted to be fecal contamination of water or food and hand-to-mouth transfer of cysts from the feces of infected individuals.[1] The incubation period for the organism is 6–22 days, and the infectious dose is in the order of 10 cysts.[2] The only known method for detection of the organism is

microscopic examination of the carrying medium.

The organism exists in two forms: (1) as a free living trophozoite and (2) as an incapsulated cyst, which is the most commonly found form outside of the host organism. Only the cyst form is infectious for man. The trophozoite will be destroyed during passage through the early stages of the digestive process, while the cyst will survive the initial stages of digestion until it reaches the small intestine where environmental conditions support the emergence of the trophozoite form. The cyst ranges in size from 7 to 12 microns and can survive for many months in a water-distribution system.[3] Present-day knowledge regarding treatment for the cysts of *Giardia lamblia* is quite limited. The organism is suspected to be resistant to the normal levels of chlorination practiced on public water supplies. It is large enough that a high percentage of removal could be anticipated in a treatment facility utilizing pretreatment with mixed-media filtration.

Presently, research has not been performed to determine the effectiveness of chlorine or other disinfectants in the destruction of *Giardia lamblia* organisms. This is primarily due to the high costs of such studies; the organism cannot be cultured, and a bioassay using laboratory animals is the only way of determining its viability. The results of disinfection tests performed with *Endamoeba histolytica* cysts, which are considered to be a similar organism, have been extrapolated[4] indicating that for water temperatures of 5 C and a pH of 7.0, a chlorine residual in excess of 5 mg/l and a contact time in excess of 60 min is necessary in order to obtain a significant organism kill. At pH above 7.6, a much higher free chlorine residual is necessary since hypochlorous ion has very little cysticidal effect. The water temperature at the Camas water-treatment plant averaged about 6–9 C during April and the pH remained below 7.0. Therefore, it was initially assumed that the normal chlorination of 1.2 mg/l that

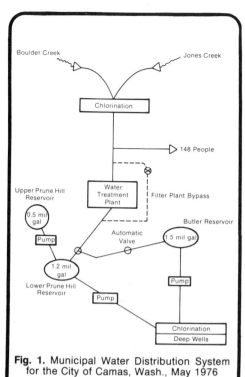

Fig. 1. Municipal Water Distribution System for the City of Camas, Wash., May 1976

Labels within figure:
Boulder Creek
Jones Creek
Chlorination
148 People
Upper Prune Hill Reservoir
0.5 mil gal
Pump
Water Treatment Plant
Filter Plant Bypass
Butler Reservoir
Automatic Valve
1.5 mil gal
1.2 mil gal
Lower Prune Hill Reservoir
Pump
Pump
Pump
Chlorination
Deep Wells

was practiced at Camas would not be sufficient to deactivate the *Giardia* cysts. Efforts during the sanitary survey were initially aimed at determining how the organisms could pass through the filtration equipment. Investigation of the disinfection system indicated several areas where failures in disinfection could have contributed to the survival of *Giardia* in the distribution system.

The Investigation

Epidemiological.[5] Confirmed cases initially reported to the county health officer indicated a predominance of cases in the northwest and north central portions of the community. Since these areas of the water-distribution system are most likely to receive surface water, this implicated the surface-water supply as a carrier of the disease-causing organism. Investigation of the initially confirmed cases by epidemiologists failed to indicate any common source of infection such as restaurants, community gatherings, or sources of food supply.

Since the cysts of *Giardia lamblia* are suspected to be highly resistant to chlorination, a preliminary survey of customers served by the city's raw-water line, which carries unfiltered disinfected water, was conducted. An ideal opportunity for a swift survey existed since customers on one side of the road were receiving water from the city's raw-water line while on the other side of the road residents received water from private wells. The survey indicated that 6

out of 36 individuals contacted who used city water showed symptoms compatible with Giardiasis. None of the 38 customers contacted who were on private wells showed symptoms of similar illness. Based on this preliminary investigation, epidemiologists from the state and the Communicable Disease Center decided to proceed with a full-scale epidemiological survey of the Camas area. The water system was considered to be the most likely source of the disease organism, and steps were taken to limit further exposure of the populace.

The complete epidemiological survey performed by the Communicable Disease Center and county health department personnel indicated an attack rate of 10–15 per cent throughout the population of 6000. The survey included determination of the attack rate and indicated a more widespread area of distribution than was previously suspected. The results of the survey supported the conclusion that the disease was waterborne.

Water system. The intake houses at Jones Creek and at Boulder Creek were inspected to determine if gross sources of contamination were evident at these locations. Each system consists of a coarse bar screen intake at the stream followed by a series of ¼-in. mesh screens inside the intake house. Neither intake house appeared to be an obvious source of contamination.

Aerial reconnaissance of the watershed indicated no human habitation within the drainage area and no human activity taking place at the time of the overflight. Further reconnaissance of the watershed by automobile revealed most roads to be in very poor condition and passable only by trailbike, horse, or four-wheel-drive vehicles. The roads appeared to have been used very infrequently. The reconnaissance provided only a general assessment of the area. Specific details of activity and terrain could not be adequately evaluated from the air or by automobile.

On-foot inspections of the Boulder and Jones Creek watersheds revealed no obvious sources of contamination but did lead to the observation of beaver signs within the stream. Since beavers had previously been implicated but not proven to be the source of a giardiasis outbreak in Utah, epidemiologists involved in the survey determined that beavers should be trapped and examined for *Giardia*.[6] The first beaver trapped in the watershed was infected with *Giardia*. This led the epidemiologists to conduct detailed bioassay studies to determine if the parasite carried by the beaver was infective for man. This was not determined to be the case until several months later. However, during the sanitary survey it was assumed that the beaver-borne parasite was infective

for man, and an attempt was made to determine how the organism had passed the treatment processes in sufficient number to cause the epidemic that occurred.

The initial plant survey revealed cross connections in the alum feedline, the filter-to-waste piping, and the surface washer. The valving on the waterline supplying the alum feed pump had the capability of providing water from either the raw or finished waterline. At the time of inspection the valve from the raw waterline appeared to be partially open and the main supply valve from finished water was open. This cross connection would have allowed flow from the raw waterline to the treated waterline. The filter-to-waste piping was cross-connected betweeen the filtered waterline and the backwash drainline. The filter-to-waste line was not used as a normal part of plant operation. When the check valve was tested and disassembled it was found to be in excellent condition. All cross connections were corrected.

Visual inspection of the media in both filter No. 1 and No. 2 indicated some loss in the quantity of media. The media in filter No. 1 was at a level approximately 19 cm (7½ in.) below the level of the wash-water arm and in filter No. 2 the media was approximately 34 cm (13½ in.) below the level of the surface-wash arm. At initial installation, the media was 5 cm (2 in.) below the level of the surface-wash arms. To further investigate the condition of the filters, an attempt was made to map the level of the fine garnet in both filters. It was found that the fine garnet in both filters was fairly level. Samples taken for analysis above the garnet indicated some loss of sand. A subsequent investigation by the city's consulting engineers indicated that, although the level of the fine garnet was fairly uniform in each filter, the coarse garnet had regions of mounding which could cause short-circuiting of the filters.

The surface water supply is normally disinfected with chlorine in the transmission line at approximately 1½-hr detention time from the filter plant. The normal chlorine feed rate provides approximately 1.2 mg/l of free chlorine residual at the treatment plant. This dosage was considered to be adequate. However, during the month of Apr. 1976 the chlorination plant was out of service because of mechanical difficulties during the periods of Apr. 5–8, 12–16, and 20–26 (Fig. 3). During the time that the chlorine plant was out of service chlorination was performed manually, but after review of the emergency chlorination procedures it was concluded that large amounts—possibly in excess of 76×10^3 m³ (20 mil gal) in 15 days—of water arrived at the treatment plant without adequate chlorination. Additional chlorine was added at the reservoirs following treatment, and an attempt was made to maintain a residual of 0.2 to 0.3 mg/l in the distribution system. Adjustments in chlorine dosage during this period were based on the field-monitoring of free chlorine residual. There was at least one occasion during this period when zero free chlorine residual was observed. It should be noted that not all water coming through the treatment plant goes directly to the reservoirs. Raw-water quality during the month of April remained in the immediate range of 1.0 tu with the exception of two nonconsecutive days during which the turbidity reached 5.0 and 7.0 tu, respectively.

The time differential between the chlorination plant failures and the majority of detected cases of giardiasis correlated closely to the incubation period for the diseases. However, the earliest cases of disease occurred in small numbers prior to the first chlorination failure, indicating that some viable cysts were passing the treatment processes during normal operation. The possible significance of disinfection failure is apparent when the period of chlorinator malfunction is compared to the peak of the epidemic as indicated by new cases per day (Fig. 4). A majority of disease onsets occurred following the failure of prechlorination and within the incubation period of giardiasis as measured from the time of that failure. There is reason to suspect that a partial deactivation of cysts was obtained through the chlorination practiced, and that possibly a very significant deactivation occurred predominantly as a result of the chlorination.

An alternative explanation of the time-distribution of disease outbreaks relates to the life cycle of beavers. During the spring of each year young beavers migrate within the watersheds they inhabit. If they enter areas that are suitable for them, then new colonies are formed. If they do not find habitable territory, they may return to their point of origin. Inspection of the Jones and Boulder Creek watersheds above the intakes revealed evidence of beaver activity; however, no beavers were trapped above the intakes despite the efforts of professional trappers. It appears possible that transient beavers moved through the upper reaches of the watershed during the spring, spread the infection, and then withdrew to the lower reaches of the watershed.[5] During inspection of the watersheds by utility staff the following year small beaver colonies were found to be located on very small feeder streams that were overlooked in the initial inspection. This points out one of the difficulties in locating such a source of contamination.

The treatment plant was initially designed and equipped for the removal of turbidity during the period when turbidity was considered to be an esthetic parameter. Chemical feed equipment for adding alum, polyelectrolytes, and soda ash was provided. The use of polyelectrolytes was terminated early because a high quality of finished water was easily maintained without their addition, and the only polyelectrolytes that were used on an experimental basis resulted in rapid clogging of the filter plus general operating difficulties.

Alum was routinely applied to the raw water at a rate of approximately 6–15 mg/l in liquid form entering the filters. However, there are several concerns as to the effectiveness of the coagulation process.

1. The raw water had an extremely low level of total alkalinity. A sample taken during this investigation had an alkalinity of 4 mg/l; therefore, there may not have been sufficient alkalinity in the raw water to provide a satisfactory coagulation reaction.

2. There was no routine calibration of the alum feed equipment.

3. Alum in the storage tank had, in the past, been diluted to an unknown degree because of the failure of a check valve that controlled water flow into the tank. Subsequent testing of the alum showed it to be about 50 per cent of normal concentration.

4. Detention time prior to filtration was only 6 min, and that is in the filter vessel above the filter.

5. Soda ash was applied to the finished water for the purpose of pH adjustment at the time of this survey. The plant was originally designed for the soda ash to be applied to the plant influent. Because of the low levels of total alkalinity in the raw water, the soda ash application is a necessity on the raw-water line.

During the review and evaluation of the unit processes at the water-treatment plant an attempt was made to determine the effectiveness of the filters in removing particles of a specific size. Since the outbreak occurred during a period when raw-water turbidities were not excessive and finished-water turbidities continuously met the requirements of SDWA, it was evident that Giardia cysts were present even when turbidity measurements indicated high quality water.

After water-treatment plant modifications were completed, samples of raw and finished water were analyzed using an automatic particle counter. Calculations were then made of the percentage of removal of particles in various size ranges. It was found that in the size range from 7 to 16 microns, the size of Giardia cysts, an average of 75 per cent removal of particles was obtained. The percentage of removal of all particles below 25 microns averaged 90 per cent.

Because the particle counter approach is affected by bits of dirt passing the filter or by alum floc forming after the filter, as well as by the biological organisms of concern, an attempt was made to locate an alternate analysis technique which would allow the distinction between biological organisms and bits of detritus. The membrane filter method for phytoplankton analysis was briefly utilized on a qualitative basis. The qualitative analysis indicated that even at a raw-water turbidity of 1.0 tu, when 100 ml of sample were filtered on a 25-mm membrane filter, a great portion of detritus particles and algae could be observed. Inspection of filtered water samples prepared in the same fashion, indicated a greatly reduced number of detritus particles and nearly complete removal of algae particles. An ocular micrometer was used to give added significance to the examination by allowing measurement of the particles passing the filters. The initial qualitative analysis using the membrane-filter procedure could not be followed with a quantitative analysis program because of limitations of time and equipment. This procedure could have considerable benefits in a water system employing or having access to the skills of a microbiologist.

Water quality. During the entire period in question a single unsatisfactory bacteriological sample was taken. Samples were being taken at an average rate of 8 per month. Throughout this period the water quality requirements of the Safe Drinking Water Act, including turbidity, were fully met.

Investigation of raw and finished water quality was made at several locations specifically for the organism Giardia lamblia. Two filtration techniques were utilized during this stage of the investigation, and both had previously been successful in capturing the organism. A filtration technique must be utilized to capture individual organisms, which must then be observed by microscope, as there is no technique currently available to culture the organism.

The technique that proved successful in detection of the organism was designed and operated by staff members of the EPA Health Effects Res. Lab. in Cincinnati.[7] The technique utilizes a cartridge-type filter unit with microscopic examination being performed on filter backwash material and filter homogenates. It is not possible at this time to present any meaningful extrapolations with reference to total number of organisms, as the sensitivity of the technique has not yet been determined. Research to identify its sensitivity is presently being conducted.

At Camas the technique was applied at the water intakes, filter plant, system reservoirs, and well supplies. The results

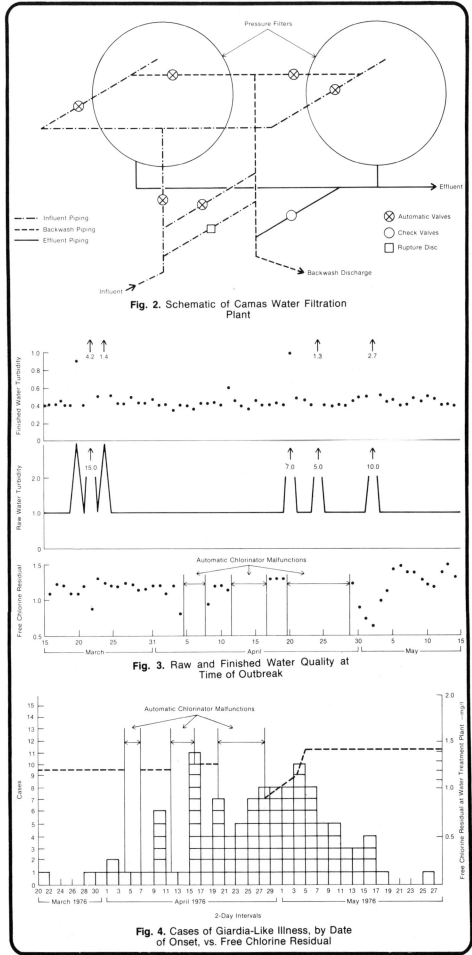

Fig. 2. Schematic of Camas Water Filtration Plant

Pressure Filters

Effluent

Backwash Discharge

Influent

-·-·- Influent Piping
- - - - Backwash Piping
——— Effluent Piping

⊗ Automatic Valves
○ Check Valves
□ Rupture Disc

Fig. 3. Raw and Finished Water Quality at Time of Outbreak

Finished Water Turbidity

4.2 1.4 1.3 2.7

Raw Water Turbidity

15.0 7.0 5.0 10.0

Free Chlorine Residual

Automatic Chlorinator Malfunctions

15 20 25 31 5 10 15 20 25 30 5 10 15

March   April   May

Fig. 4. Cases of Giardia-Like Illness, by Date of Onset, vs. Free Chlorine Residual

Cases

Automatic Chlorinator Malfunctions

Free Chlorine Residual at Water Treatment Plant —mg/l

20 22 24 26 28 30 1 3 5 7 9 11 13 15 17 19 21 23 25 27 29 1 3 5 7 9 11 13 15 17 19 21 23 25 27

March 1976   April 1976   May 1976

2-Day Intervals

for May 1976 are summarized in Table 1.

It is interesting to note that the highest number of organisms were located in system reservoir sediments. *Giardia* cysts are thought to have a density slightly greater than water.[8] This could have resulted in significant settling in reservoirs with long detention times. Although it is suspected that these cysts may have been present in the reservoirs for several weeks, it is not possible to estimate closely how long they were there, or what their viability was.

Conclusions

Based on the results of the investigation, the following conclusions were reached:

1. *Giardia lamblia* was the infective agent in the affected individuals, and it was waterborne.

2. Beavers active in the city's watershed were the only carriers of the organism detected; therefore, it is probable that they were the source of the infection.

3. The water-treatment-plant processes had been ineffective in removing the organism because of questionable filter media, ineffective pretreatment, and possibly disinfection failure.

4. The pretreatment processes were ineffective.

   a. The alum feed was poorly regulated and there was insufficient alkalinity in the raw water to allow adequate coagulation.

   b. Soda ash was added after the filters for pH adjustment and contributed no increase in alkalinity to the raw water. Soda ash was needed in the pretreatment process to provide effective pretreatment.

   c. Polyelectrolytes were not being utilized. They should be utilized as fully as possible.

   d. Chlorination in the manner being practiced by the city was either ineffective in disinfecting the raw-water supply where the infective organism was concerned, or failure of disinfection equipment allowed enough organisms to survive to cause a problem.

5. Turbidity and coliform count alone are inadequate parameters on which to judge the biological quality of filter effluent.

Remedial Action

In response to the need for system revisions, the city personnel have taken the following steps:

1. Regrading of the filter support media, and replacement of the filter material has been completed. It was first proposed to replace only that filter material that was lost, but economic analysis by the consulting engineer revealed that complete replacement, which could be

79

done more quickly, was only marginally more costly. This was due to the amount of work involved in regrading the existing material, and the difficulty in obtaining equipment to do it.

2. Pretreatment revisions have been made. Soda ash is now added upstream of the filters to boost the alkalinity. The alum, soda ash, and polymer feed rates have been optimized by a treatment specialist from the treatment-plant manufacturer.

3. The installation of standby chlorination equipment at the headworks is completed.

4. Monitoring of aluminum ion at the filter-plant outlet will be undertaken to determine the effectiveness of the filters in removing all alum floc and the effectiveness of pretreatment in bringing about the complete reaction of all alum added to the filters.

5. Installation of a bypass at the treatment plant will be made to allow water to be wasted without being routed to the distribution system.

6. Installation of a filter-to-waste line will be completed, with provision made for sampling of the filtered water.

7. A survey of mammals within the watershed has been conducted to determine the extent of giardial infection.

Implications

A system with a filtered surface-water source and well-water sources produced a waterborne epidemic of giardiasis despite excellent finished-water quality as normally evaluated (met SDWA requirements continuously). The following implications of this event are considered to have far-reaching effects:

1. Low turbidity and coliform in raw and finished waters from a treatment process are inadequate to ensure that all biological contaminants of a filtered water are subject to disinfection. In this case the influent turbidity was often as low as 1.0 tu and the effluent turbidity was generally less than 0.5 tu. The detention time provided after disinfection of the raw water was approximately 90 min before reaching the water treatment plant. The level of chlorination was approximately 1.2 mg/l. Even during periods of failure of the automatic chlorination equipment an attempt was made to maintain a chlorine residual of 0.2–0.3 mg/l in the distribution system.

2. It is essential that pretreatment be practiced at all times on a filtered water supply. It has been argued that if water can be filtered to an acceptable turbidity without pretreatment, then pretreatment to achieve more than this is unnecessary. The following analogy is offered in response to this statement. If 2.0 tu exist in the influent and the water leaving the plant has 0.5 tu, then 75 per cent of the material that entered the treatment plant has been removed. However, if sufficient alum is added to raise the influent turbidity to 10 tu and filter effluent shows 0.5 tu, then 95 per cent of the material that went to the filters has been removed. This analogy in itself justifies continual operation and monitoring of pretreatment facilities.

3. Pathogens can pass through a filter even when effluent turbidities are satisfactory. This indicates a need to make periodic determinations of the condition of the filter media. Mapping of the level of filter support gravels should be done on a regular basis (at least annually) to ensure that mounding is not occurring. The filter media down to the support gravel should also be monitored to ensure that deterioration or loss of media has not taken place.

4. Since it has been shown that a minimal level of turbidity passing the filters may result in a disease outbreak, consideration should be given to providing automatic plant shutdown when turbidity approaches a certain level. This level could be below the SDWA standard and could be the point at which turbidity levels begin to increase in a normal filter run. Such a safety device might be required for plants not having continuous operation and maintenance or for plants where organisms not susceptible to chlorination have been identified.

5. Monitoring aluminum ion in the filter-plant effluent may be an effective method for routine filter-performance evaluation. There are two cases in which significant levels of aluminum ion would pass the filters. Both cases are undesirable in filter-plant operation and indicate problems with pretreatment or filter media or both. In the first case there is inadequate reaction between alum and alkalinity in the water, and the aluminum ion would pass the filter without having participated in the coagulation process. In the second case a floc particle might pass the filter as a result of many factors, which would indicate a failure in the filter media to provide satisfactory removal. A colorimetric method is outlined in Standard Methods and would be applicable in treated waters not containing significant levels of iron, manganese, and fluoride.

Several questions relating to water-system management and watershed control are raised by the giardiasis outbreak in Camas. These questions are

1. Should filtration of all surface waters be required regardless of clarity or remoteness of the watershed?

2. Should backup sources be developed routinely? The availability of standby wells was of great value to the city of Camas.

3. Should periodic surveys of wildlife in remote watersheds be performed for bacterial and parasitic infestations?

In order to evaluate these implications fully and to provide protection against disease outbreaks of this nature, research is needed to answer the following questions:

1. What is the effectiveness of disinfectant dosage and detention time in deactivating *Giardia* cysts?

2. What is the distribution of *Giardia* and other potentially dangerous organisms in human and animal ecological systems?

3. What methods other than turbidity measurement can be used to evaluate more effectively the performance of filtration in water treatment?

Acknowledgements

The efforts of the following organizations and individuals contributed to the success of the study, analysis, and solutions for the giardiasis outbreak in Camas:

The Southwest Washington Health Dist. under the supervision of Donald A. Champaign; the Communicable Disease Center, Atlanta, Ga., represented by Aubert Dykes, Rodney Lorenz, and Bruce Woods; and the EPA Health Effects Res. Lab., Cincinnati, Ohio, represented by W. Jakubowski, E.C. Lippy, and Shih Lu Chang.

References

1. BENENSON, ABRAM S. *Control of Communicable Diseases in Man.* 12th edition, 1975.
2. RENDTDORFF, R.C. The Experimental Transmission of Human Intestinal Protozoan Parasites, II; Giardia Lamblia Cysts in Capsules. *American Jour. of Hygiene,* Vol. 59, 209-220 (1954).
3. RENDTDORFF, R.C. & HOLT, C.J. The Experimental Transmission of Human Intestinal Protozoan Parasites, IV; Attempts to Transmit Endamoeba Coli and Giardia Lamblia Cysts by Water. *American Jour. of Hygiene.*
4. Chang, Shih-Lu EPA Health Effects Research Laboratory, Cincinnati, Ohio, "Letter Report," Mar. 18, 1976.
5. DYKES, A.C. ET AL. Municipal Waterborne Giardiasis: Beavers Implicated as Reservoir. Unpublished manuscript.
6. BARBOUR, A.G.; NICHOLS, C.R.; & FUKUSHIMA, T. An Outbreak of Giardiasis in a Group of Campers. *American Jour. of Tropical Medicine and Hygiene,* Vol. 25, 384-389 (1976).
7. JAKUBOWSKI, W., ET AL. Large Volume Sampling of Water Supplies for Micro Organisms. USEPA Health Effects Research Laboratory, Cincinnati, Ohio, In press.
8. FAUST, RUSSELL JUNG. *Clinical Parasitology.* 7th edition, Lee and Feberger, Technical Appendix, p. 978.

An annual conference paper selected by the JOURNAL, authored by J. C. Kirner (Active Member, AWWA), regional engr., and J. D. Littler (Active Member, AWWA), dist. engr., both of the Washington State Dept. of Social and Health Services, Olympia, Wash., and by L. A. Angelo (Utility Rep., AWWA), publ. wks. dir., Camas, Wash.

Reprinted from *Jour. AWWA,* 70:1:35 (Jan. 1978).

Tracing a Giardiasis Outbreak at Berlin, New Hampshire

Edwin C. Lippy

Members of the water supply industry sometimes have to play Sherlock Holmes—in this case, to find causes and solutions to an outbreak of giardiasis in Berlin, N.H. The tale has all the elements of a good mystery, including a furry villain.

Giardiasis is an intestinal disease that during the past five years has gained the attention of public health authorities and the water supply industry. A review of the reported waterborne outbreaks of giardiasis that were linked epidemiologically to drinking water indicates the first one occurred in this country during the winter of 1965-66 in Aspen, Colo. Since then, outbreaks have been reported as shown in Table 1.[1]

While reporting is not completed for 1977, preliminary information indicates that at least five outbreaks occurred,

with approximately 1000 cases of the disease.

Geographically, the outbreaks took place in Colorado (8), Utah (2), Vermont (2), New Hampshire (1), New York (1), Tennessee (1), Idaho (1), California (1), and Washington (1). The largest outbreak occurred in Rome, N.Y., in 1974, affecting 5300 people,[2] and the second largest occurred in Camas, Wash., in 1976, with 600 cases.[3,4]

In 1976 EPA developed sampling methodology for detection of *Giardia* cysts in water,[5] which was successfully applied

during disease outbreaks in Vermont and Washington. In Washington the methodology enabled identification of the vehicle of transmission, an important objective of outbreak investigations. The isolation of the causative agent from the water system and from ill persons, in conjunction with supportive epidemiological evidence, is normally sufficient proof that the disease is waterborne. The methodology was used again in April of 1977 during an outbreak at Berlin, N.H.

Berlin is located in the northern part of the state in the White Mountains region

81

TABLE 1
Waterborne Giardiasis Outbreaks, 1969–1976

Occurrence	1969	1970	1971	1972	1973	1974	1975	1976	Total
Outbreaks	1	1	–	3	2	7	1	3	18
Cases of disease	19	34	–	112	28	5357*	9	639	6198

*A total of 4987 cases in 1974 were originally reported, but that figure is revised here to reflect recent data included in reference 2.

TABLE 2
Giardia Sampling Locations—Chlorine Residual Measurements in the Berlin Water System, 1977

Location	Chlorine Residual (4/23) Free mg/l	Chlorine Residual (4/23) Total mg/l	Chlorine Residual (4/24) Free mg/l	Chlorine Residual (4/24) Total mg/l	Chlorine Residual (4/25) Free mg/l	Chlorine Residual (4/25) Total mg/l
Androscoggin Finished	0.7	1.1	1.1	1.5	2.0	2.1
Ammonoosuc Finished	0.3	ND*	1.8	2	3+	3+§
City manager residence†	0	ND*	0.4	0.7	**	1.3
City Hall‡	0	ND*	1.1	1.5	**	1.7

*ND–not done
†Served from the Ammonoosuc source
‡Served mainly from the Androscoggin source
§Comparator scale limits–3 mg/l
**Out of reagent

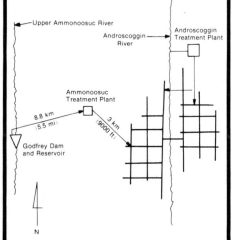

Fig. 1. Schematic Diagram of Berlin, N.H., Water System

and a few miles north of the Presidential Range. The city is situated at an altitude of 314 m (1030 ft) above mean sea level in the midst of the White Mountains National Forest and is noted for its scenic setting. Climatic conditions include an average summer temperature of 16 C (60 F), winter temperature of −10 to −7 C (15–20 F), annual rainfall of 0.9 m (36.5 in.), and snowfall of 2.54 m (100 in.). Berlin is a manufacturing and commercial center with industries employing 3500 people. The principal products are paper, lumber, canvas and rubber footwear, iron castings, knitted wear, machinery, chemicals, leather, conduit pipe, electrical components, business forms, and plastic products. The community has a population of 15 000.

The first case of giardiasis was identified by the laboratory at the Androscoggin Valley Hospital in Berlin. The patient, an 11-year-old girl, had been sick since October of 1976 and had been discharged from the hospital three times without cure. When a consulting physician ordered an ova and parasite test, a laboratory technician who had recently completed a course in parasitology identified the sample as positive for *Giardia*. Within three days, two more cases were identified, the hospital laboratory staff became concerned with the unusual frequency, and they notified the State Health Dept. By April 19, ten to fifteen cases had been identified, and an outbreak was suspected. EPA became involved when a State Health Dept. nurse discussed the situation with the Regional Water Supply Branch in Boston, which in turn requested technical assistance from the Health Effects Research Laboratory in Cincinnati. The notification occurred on Friday afternoon, April 22, and the first *Giardia* sampler was installed on the Berlin water system Saturday night, April 23.

The investigation began on Saturday evening with a meeting that included EPA, the city manager, representatives from the New Hampshire Water Supply and Pollution Control Commission, Berlin Water Works, and City Health Dept. The layout and operation of the water system were discussed, disease in the community described, and a course of action agreed upon.

The Berlin Water System

Berlin utilizes the Upper Ammonoosuc River and the Androscoggin River (which flows through the community) as independent sources of supply. Water from the two sources is treated separately at the older Ammonoosuc plant, which has a capacity of 8.1×10^3 m³/day (2.15 mgd) and at the new Androscoggin plant, with a capacity of 7.6×10^3 m³/day (2 mgd). Although the two plants provide water to identifiable areas within the distribution system, some areas receive a mixture.

Godfrey Dam impounds the Upper Ammonoosuc River with a reservoir of 99.4×10^3 m³ (26 mil gal) capacity and a reported safe yield of 7.6×10^3 m³ (2 mil gal). The watershed has a drainage area of 42 km² (20 sq mi) and is located within the boundaries of the White Mountain National Forest. The land is owned by the US government and under control of the National Forest Service. There are no known point-source waste discharges; however, hunting, fishing, and other forms of recreation are permitted. Water is discharged from Godfrey Dam through a bottom intake 4.9 m (16 ft) deep and transmitted approximately 8.8 km (5.5 mi) to a treatment plant located on the western fringe of Berlin (Fig. 1).

The Ammonoosuc plant, constructed in 1939–40, has eight pressure filters operated in parallel. The filters are 2.4 m (8 ft) in diameter and 5.5 m (18 ft) in length and are supported on concrete cradles. Filter effluent is chlorinated through addition of sodium hypochlorite (produced at the Androscoggin plant). Fluoride is also added to maintain a concentration of about 1 mg/l. The filters are backwashed every other day. The backwashing procedure requires drawing the water level down to within a few inches of the top of the filter bed; air scouring for approximately 5 min; backwashing with filtered water until the effluent is clean by visual determination; and closing the waste valve and introducing raw water onto the filters and eventually into the system following filtration.

The media in five of the filters (1, 3, 4, 6, and 7) consisted of 61 cm (24 in.) of graded anthrafilt [46-cm (18-in.) top layer, 0.5 mm diameter and 15-cm (6-in.) supporting layer, 1.0 mm diameter] supported by 46 cm (18 in.) of graded anthrafilt. Filters 2, 5, and 8 were rebuilt in 1974. The design of these filters differed in that the media were supported by a porous plate underdrain system, while the media of the other five filters rested on a concrete and grout bed. The porous plates had deteriorated and had been replaced with 0.9 m (36 in.) of graded gravel topped off with 0.45 to 0.61 m (18 to 24 in.) of anthrafilt.

Several obvious problems were noted at the plant.

1. Only one filter had a sampling tap; therefore the efficiency of individual filter operation could not be monitored on seven of the eight filters.

2. The plant was not designed for use of a conditioning chemical.

3. Turbidity monitoring equipment was not available.

4. Pressure filters are not easy to inspect, which creates an "out-of-sight, out-of-mind" attitude.

5. Backwash efficiency by bed expansion could not be determined.

6. Individual filter rates could not be controlled, so that in parallel flow operation the clean filters could operate beyond design rates. (Also, filters having blowholes, cracks, short-circuiting, and other flaws could operate beyond normal rates.)

7. Chlorinator operation depended upon a single electrical power feed that was not backed up with another source in case of power failure.

The filters were designed to operate at a rate of 0.05 m/s (1.3 gpm/sq ft), which yields a plant capacity of 8.1×10^3 m³/day (2.15 mgd).

The second source of supply, the Androscoggin River, has a drainage area of approximately 3367 km² (1300 sq mi), with an average flow of 67 m³/sec (2400 cfs). There were no known point-source waste discharges of any magnitude upstream of the plant intake. The recently constructed treatment plant became operational on Mar. 10, 1977, and was undergoing acceptance testing; as with any new facility, there were still several problems to be corrected. Some of the automated controls were not in operation, and an intake diffusion system was not working.

Water was pumped from an intake canal constructed in the bank of the Androscoggin River. Since the canal was designed to receive filter backwash directly, as well as the supernatant from a lagoon provided to settle sludge draw-off from the clarifiers, waste discharge to the river was obviated. The canal contained an air dispersion system to keep returned waste distributed uniformly in the canal cross section. The plant was equipped to add alum, polymer, and sodium hydroxide as normal treatment chemicals; however, facilities were available to add sodium hypochlorite for prechlorination, carbon and diatomaceous earth for color removal, and floc builder if needed. Mixing was accomplished in the pipe from the intake to the clarifiers. Settling took place in two pulsator–upflow clarifiers where the sludge blanket elevation and compaction were regulated by the intermittent, or pulsating, inflow of coagulated water and periodic sludge withdrawal. The clarified water was introduced onto four filters containing 91 cm (36 in.) of sand. The 3- by 5.5-m (10- by 18-ft) filters operated at a rate of 0.02 m³/s (350 gpm) and were backwashed on a head loss [1.5–1.8 m (5–6 ft)] schedule. From the filters water was conveyed to a 1.9×10^3-m³ (0.5-mil gal) storage tank where sodium hypochlorite, sodium fluoride, and caustic were added. Polyphosphate feed was available for corrosion control but was not used. A 0.15×10^3-m³ (40 000-gal) wet well or pumping well followed the storage tank, from which water was pumped to the distribution system. Auxiliary power was provided. Sodium hypochlorite was produced at the plant by electrolysis of salt and sufficient production was available for both city water plants. Daily plant output averaged about 5.7×10^3 m³/day (1.5 mgd).

The plant had a number of good design features, including flexibility in pumping, chemical feed and application points, no-discharge provisions, sodium hypochlorite production, and auxiliary power. Return of filter backwash and lagoon supernatant to the plant influent appeared reasonable, especially with the low turbidity water normally experienced in the Androscoggin River. Prob-

lems had been experienced with maintaining an acceptable sludge blanket, and the polymer supply was depleted during the day shift on April 23. A problem with floc carryover to the filters was readily observable by deposit formed on the perforated effluent pipes.

Disease in the Community

At the meeting on April 23, the city health officer reported a total of 60 cases of giardiasis in the community, with 20 new cases identified that day. All cases were confirmed by stool examination at the Androscoggin Valley Hospital. A pattern seemed to be developing, in that most of the cases were located in the southwest quadrant of the city, an area served from the Ammonoosuc source; however, there were sporadic cases throughout the community. There were no common foods or events reported among those who were ill. A suggestion of boiling water for home use was made publicly on April 21, but there was general feeling that people did not take the suggestion seriously. The health officer also reported that coliform results from routine sampling were negative.

Plan of Action—1

A plan of action was developed and agreed upon that included immediate installation of *Giardia* samplers, increase in chlorination at both plants, a survey of the Ammonoosuc watershed for beaver activity, contact with the Center for Disease Control for epidemiological assistance, and a need to report information to the public through a single spokesman. Recommendation for a "boil water" order did not receive unanimous approval and was temporarily delayed.

Six locations were selected for installation of *Giardia* samplers. Raw and finished water taps at both treatment plants were selected to identify a single source of *Giardia* intrusion, and two locations in the distribution system were sampled. Chlorine residuals were noted as shown in Table 2.

The samplers remained on line for at least 12 hr and were sent the following day to Cincinnati for analysis for cysts by the Health Effects Research Laboratory. Five samplers were reinstalled at the same sampling locations except for Androscoggin raw water; they were allowed to run for 24 hr. Chlorine residual measurements taken at the start and finish of sampling runs showed increases (Table 2), reflecting higher chlorination doses.

The plan to increase chlorination was controlled by contact time available at the treatment plants and the dose required to destroy *Giardia* cysts. A chlorine contact basin had not been incorporated in the design of the Ammonoosuc facility but there were

approximately 2743 m (9000 ft) of 41-cm (16-in.) transmission main between the plant and the distribution network, which afforded about 1 hr contact time. The Androscoggin plant included a 1.9×10^3-m³ (0.5-mil gal) storage facility and a 0.15×10^3-m³ (0.04-mil gal) pumping well that followed the postchlorine injection point. On the basis of volume displacement calculations and daily plant output, the 2.05×10^3-m³ (0.54-mil gal) combination provided about 7 hr of contact time. The chlorine dose required to destroy *Giardia* cysts is not known; however, work conducted by Chang in 1967* on cysts of *Endamoeba histolytica* yielded "kill curves" as shown in Fig. 2. The resistance of cysts of *E. histolytica* and *Giardia lamblia* to chlorine is not felt to be significantly different,[6] and in the absence of data for *Giardia* disinfection, the dosage requirements for *E. histolytica* were applied. Chlorine feed was increased to 2 mg/l at the Androscoggin plant and to 5–6 mg/l at the Ammonoosuc plant. Because the rather extreme dose at the Ammonoosuc plant would cause immediate consumer complaints about taste and odor, other alternatives were investigated. Prechlorination at Godfrey Dam was selected, in order to increase contact time to 4–5 hr and to decrease the dosage. Since electrical power was not available at the dam, gaseous chlorine feed was considered to be the best solution, but gravity feed of sodium hypochlorite was adopted later because of problems in obtaining gaseous chlorine equipment.

The implication of beavers in transmitting the disease in the Camas, Wash., outbreak[3] and the known presence of the animal in the Ammonoosuc and Androscoggin watersheds prompted a survey, which was restricted to the immediate vicinity of the Godfrey reservoir and was done by the Berlin Water Works and the N.H. State Game Commission.

Assistance from the Center for Disease Control was initiated through contact with the state health officer. The Center was already aware of the outbreak but, by protocol, was awaiting formal invitation by the State Health Dept. which had delayed the invitation until the illness was better defined.

It was agreed that distribution of information to news media would be made through periodic releases from the city manager's office. Meetings among those working on the investigation would be scheduled as needed with the city manager in attendance, and information would be made public as it developed.

Results. *Giardia* cysts were first identified in the Berlin water system by the laboratory at the Androscoggin Valley Hospital. Laboratory personnel fashioned a sampling device that included a

---

CHANG, S.L. Unpublished data

TABLE 3
Giardia Results, Berlin, N.H., 1977

Location	Date On	Date Off	Volume m³	Volume gal	Giardia Results	Coliforms colonies/ 100 ml
First set						
Ammonoosuc raw	4/23	4/24	7.154	1890	+	105 +
Ammonoosuc finished	4/23	4/24	12.490	3300	+	1
Androscoggin raw	4/23	4/24	1.435	379	+	
Androscoggin finished	4/23	4/24	9.890	2613	+	<1
106 Prospect St.	4/23	4/24	9.845	2601	−	<1
City Hall	4/23	4/24	11.892	3142	+	
Second Set						
Ammonoosuc raw	4/24	4/25	10.200	2695	+	
Ammonoosuc finished	4/24	4/25	19.023	5026	−	
Androscoggin finished	4/24	4/25	14.621	3863	−	
106 Prospect St.	4/24	4/25	11.086	2929	−	
City Hall	4/24	4/25	15.541	4106	−	
Third Set						
(repairs completed)						
Ammonoosuc raw	5/29	5/30	6.586	1741	−	
Ammonoosuc finished	5/29	5/30	10.473	2767	−	
Androscoggin raw	5/30	5/31	2.366	625	−	
Androscoggin finished	5/30	5/31	2.381	629	−	
Fourth Set						
Ammonoosuc raw	6/19	6/20	12.358	3265	−	
Ammonoosuc finished	6/19	6/20	19.300	5099	−	
Androscoggin raw	6/18	6/19	5.901	1559	−	
Androscoggin finished	6/18	6/19	13.698	3619	−	

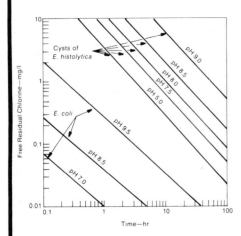

Fig. 2. (Left) Concentration–Time Relations for 99.999 Per Cent Destruction of *E. coli* and Cysts of *E. histolytica* by Free Chlorine at 2–5 C

Source—Chang, 1967, unpublished

TABLE 4
Androscoggin Plant Filter Effluent Test

Filter	Volume m³	Volume gal	Turbidity ntu	pH	Coliforms number/100 ml
Filter 1 (not repaired)	1.079	285	0.23	5	3 (start) <1 (stop)
Filter 2 (repaired)	3.512	928	0.12	4.7	<1 (start) <1 (stop)

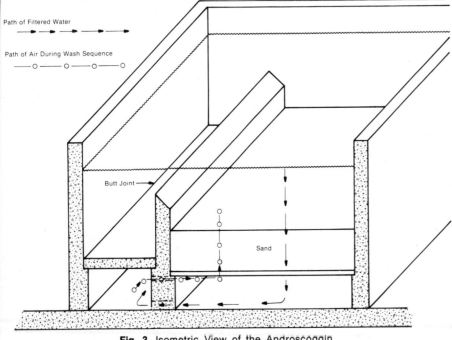

Path of Filtered Water

Path of Air During Wash Sequence

Butt Joint

Sand

Fig. 3. Isometric View of the Androscoggin Plant Filters (Not to Scale)

filter pad attached to the neck of an inverted plastic container with the bottom cut out. When water drawn from a laboratory tap was filtered, the material trapped by the filter pad was examined microscopically and found to be positive for cysts. The identification occurred on Sunday, April 24.

On April 25, a "boil water" order was issued by the city manager through the news media, based on the hospital's finding of cysts in finished water. While the discovery indicated contamination of the water system, the source or extent of the problem was not known. Since the hospital was located in the Ammonoosuc distribution zone, the river was suspected as the source of contamination. This also correlated with the earlier report of a concentration of cases in the southwest quadrant of the distribution system. In addition, survey of Godfrey reservoir detected a beaver lodge on the western bank about 610 m (2000 ft) upstream from the intake. While results and events were confirming an early suspicion that cysts were gaining entry to the system from the Ammonoosuc source, a report was received indicating that raw and finished water samples collected from both treatment plants and a sample collected from the distribution system were positive for *Giardia* cysts (Table 3). The hospital result was now confirmed and contamination of the system was discovered to be more serious than originally believed.

Plan of Action—2

A second plan of action developed after the sampling results indicated wide distribution of cysts in the water system. It was necessary to determine how cysts penetrated the treatment trains and entered the distribution system; desirable to locate the source in the watershed, although this seemed impossible for the 3367 km² (1300 sq mi) Androscoggin drainage area; and obviously important to conduct an epidemiological study to document illness in the community and to determine whether it was related to drinking water.

Treatment. A review of the literature for information on removal of *Giardia* cysts from water produces few answers. The paucity of information stems from the low level of occurrence and historical importance of the disease in waterborne outbreaks, the lack of a culturing technique for testing the efficiency of removal in treatment studies, and the tedious microscopic routine required to identify the cyst on slide preparations. Two studies[7,8] report efficiency of treatment on similar types or sizes of organism including cysts of *E. histolytica* and spores of *Histoplasma capsulatum*. The study on *E. histolytica* showed complete removal of cysts by coagulation and filtration through rapid sand filter beds (similar to the Androscoggin treatment train), and the study on spores of *H. capsulatum* showed that a model rapid sand filter without coagulation did not effect a complete removal of spores (direct filtration without chemical conditioning, similar to the Ammonoosuc plant). Therefore, it was plausible that cysts were penetrating the filters at the Ammonoosuc plant, but how they penetrated the newly built and apparently well-designed Androscoggin Plant was perplexing.

Startup and operational problems, believed to be contributors to the difficulty at the Androscoggin plant, prompted the Water Supply and Pollution Control Commission to request assistance from the process designers. Also, assistance was requested and provided from the Physical and Chemical Contaminant Removal Research Branch, Water Supply Research Div., EPA, Cincinnati, to evaluate plant operation.

Because control of disease in the community was largely dependent upon chlorination, disinfection procedures implemented a few days earlier at the two plants were reviewed. Androscoggin plant records showed that the time between boosting chlorination and detecting a measurable increase in plant effluent was 1¾ hr. This should have been expected when volume-displacement calculations were used, and arrangements should have been made to report the contact time. The same procedures should have been implemented for recording the contact time in the transmission main from the Ammonoosuc plant; however, it had been overlooked.

Because the contact time at the Androscoggin plant was not adequate, an increase in chlorine dosage or in total contact time was needed. Prechlorination feed facilities were available and review of clarifier and filter dimensions indicated an additional theoretical retention time of 2½ hr. By prechlorinating, increased residuals could be avoided. Prechlorination began on April 27 and was halted on April 28 because chlorine caused noticeable disruption in an already weak sludge blanket in the clarifiers. Therefore, postchlorination was adjusted to 5–6 mg/l.

Because of Berlin's somewhat remote location, it was difficult to obtain gaseous chlorine feeding equipment for installation at Godfrey Dam. The intent was to chlorinate sufficiently at the dam to carry a residual to the Ammonoosuc plant, and to postchlorinate to produce a reasonable residual in the distribution system, again avoiding the 5–6 mg/l necessary with postchlorination alone. A plan suggested by the city's consultant was implemented on April 27, by which sodium hypochlorite (available from the Androscoggin plant) was drip fed into the intake at Godfrey Dam. A series of interconnected 0.21-m³ (55-gal) drums was set up on the dam, and the disinfectant was fed through plastic tubing with controlled flow. Although the idea was excellent, it took a great deal of time and patience to control feeding and to produce the desired result.

Representatives of the firm responsible for process design of the Androscoggin plant arrived on April 28 to conduct chemical feed tests and to improve clarifier operation. They concluded that coagulant feed should be changed from nonionic polymer, alum, and caustic soda to anionic polymer, sodium aluminate, and alum. The chemicals had to be shipped in from out of state, so correction of coagulation problems was delayed for a week.

Use of optimum coagulation chemicals to build the sludge blanket became less important, however, when it was noticed that bubbles were escaping from the joints in the slab of the backwash channels during air scour of the filters (Fig. 3). Original design of the joint had specified placement of a sealer gasket in a groove at the joint compressed with a continuous galvanized angle and secured to adjacent walls. The design was changed during construction to a simple butt joint. The escape of air through the joint during backwash and air scour indicated the possibility that applied water was seeping through the joint and contaminating filtered water. The possibility was confirmed by a static hydraulic test of the backwash channel, which showed that daily leakage could attain 45 m³ (12 000 gal) for one of four filters tested. If one test result could be extrapolated, over 3 per cent of the plant output was not filtered. There now was a plausible explanation for cyst detection in finished water from the Androscoggin plant.

There was no question that the joints had to be repaired. A design was adopted that required grinding a groove the length of the joint, filling it with a sealer, covering it with a continuous stainless steel plate, and applying angles 30 cm (12 in.) in length on 61-cm (24-in.) centers for compression. A contractor estimated each filter would require one day to repair; the job was completed in about three weeks.

As work progressed on the filters, the opportunity existed to sample concurrently a repaired filter and one not repaired, to test for removal of *Giardia* cysts. Samplers installed on filter effluent lines operated for about 18 hr. *Giardia* cysts were not identified in either sample; however, the data collected during the sampling run and shown in Table 4 indicate differences in filter operation, the greatest variation being the amount of water filtered by the two samplers. Also, the sampler on filter 1 was dirty after a few minutes of operation and nearly plugged at the end of the sampling run.

After all repairs were completed, raw and finished waters were sampled to determine if cysts were present. The results of two sampling runs indicated that raw water no longer contained cysts and the plant was not being challenged (Table 3). With raw water free of cysts, the removal capability of the plant could not be determined. There was no reason to believe that a properly constructed and operated plant would not remove *Giardia* cysts, although it indeed would have been reassuring to produce sampling results that provided proof.

Activities at the Ammonoosuc plant included inspection of the pressure filters to observe bed condition. This required removal of bolted hatch covers and entry through a pygmy-sized portal. The filters that contained all anthrafilt media (1, 3, 4, 6, and 7) had a fairly uniform bed surface with some mounding. Those that were rebuilt with graded gravel (2, 5, and 8) and 46–61 cm (18–24 in.) of anthrafilt had severely disrupted beds. As shown in the photo (right), the bed surfaces were very irregular in profile and dominated by ridges and pyramids with 30-cm (12-in.) heights not uncommon. The media were separated from the filter walls, with the separations extending 30 cm (12 in.) down at some locations where measurements could be made. The valleys located adjacent to ridges contained only a few inches of anthrafilt on top of the gravel. Filter 8, perhaps, was in the worst condition, with a large conical depression located in the frontal one-third section. Visual inspection during air scour showed that the air discharge was concentrated in the area where the depression was located. Removal of the media to the air manifold uncovered a broken flange that had allowed the full flow of air to escape at one location. The filters were plagued with mud masses that formed clogged areas in the bed and prevented effective backwashing and filtration. This condition is described in water treatment handbooks[9] and is easily detected in filters that can be observed. Detection is not so easy in pressure filters, especially when entry and inspection require a certain amount of torture. There was little doubt why cysts were detected in finished water from the Ammonoosuc plant.

After prolonged backwash and air scour cycles had been used in an unsuccessful attempt to break up the agglomerated mud masses in the filters, the beds were raked with garden rakes and then backwashed to remove the mud. This did improve the bed but did not produce the desired results. Chemically cleaning the beds with chelated sodium hydroxide was suggested by the city's consultant and was used with success, although it

Filters Rebuilt With Graded Gravel, Ammonoosuc Plant

required about two days to treat, flush, and return one filter to operation. With eight filters, the problems in Berlin seemed to be multiplying while solutions were only adding.

When it was recognized that the simple straining mechanism of filtration was not going to remove cysts effectively, arrangements were made to feed alum into the plant influent line, thereby allowing sedimentation and coagulation in the filters. Pore spaces between individual grains of coarse, angular media are reported[9] at 25 to 50 per cent of the grain diameter. For media of 0.5 mm diameter, the pore space would range from 0.12 to 0.25 mm (120 to 250 microns). *Giardia* cyst measurements range from 10–15 microns in length to 7–12 microns in width. Without the aid of a conditioning chemical to enhance formation of larger particles, the filtration mechanisms of straining, sedimentation, and flocculation cannot remove cyst-size particles. The disadvantage of the alum feed arrangement was the gradation in the filter beds, which ran from fine to coarse in the direction of filtration. Thus removal of particles would take place in the top layers, resulting in short filter runs with excessive backwashing. Excessive backwashing in filters that could not be observed was not appealing. Therefore, chemical conditioning was temporarily abandoned, and treatment depended upon the undefined cyst removal provided by filtration and destruction by chlorination.

After all the filters were chemically treated, raw and finished water was sampled for cysts. The results were the same as for the Androscoggin samples—no cysts identified in raw or finished water (Table 3). However, the degree of confidence expressed for cyst removal capability of the Androscoggin

plant could not be applied to the Ammonoosuc plant.

Source of *Giardia*. The discovery of a beaver lodge in Godfrey reservoir presented a likely source of *Giardia* cysts. Additional surveys were conducted to determine the location of other lodges in the watershed, but the prospect of a population count in the large Androscoggin watershed did not attract a tremendous amount of support.

Four beavers were trapped from the Godfrey reservoir lodge by the State Game Commission and delivered to the Androscoggin Valley Hospital for autopsy. Samples were extracted from four locations in the gastrointestinal tract and examined for cysts. Results were initially negative, though persistence of the pathologist in preparing a slide of material from the duodenal mucosa of one beaver resulted in the identification of *Giardia* trophozoites—the reproductive stage.

The sacrifice of four beavers did not receive unanimous public approval. There are alternatives to sacrifice, such as live trapping and keeping the animals in captivity for study of stool specimens over a period of time. If the beaver is infected and actively discharging cysts, the animal can be studied, treated, and returned to a natural habitat. However, a beaver can be infected and not discharging, as was the case with the reservoir beaver. Using resources to support continued study of an animal in captivity, where feeding and shedding habits may change, becomes questionable.

While it would have been desirable to conduct profile sampling of the Androscoggin River for *Giardia* to pinpoint a source, especially a point-source discharge that could be corrected, the logistics and laboratory support did not permit it. The N.H. Water Supply and

Pollution Control Commission, in a sanitary survey conducted upstream of the Androscoggin plant, found numerous sanitary violations. More than 40 cease-and-desist orders were issued to homes and institutions with individual sewage disposal violations.

Epidemiology. Representatives from the USPHS Center for Disease Control began arriving on April 28 to assist state and local health authorities. A physician–epidemiologist and a statistician from the center planned and conducted a survey of the community with support and assistance of local health personnel. A microbiologist experienced in procedures for identification of *Giardia* was also provided to assist the Androscoggin Valley Hospital Laboratory in examination of stool samples.

Health surveys were conducted in Berlin and Gorham to identify magnitude and distribution of illness. Gorham is located 5 mi south of Berlin, has somewhat similar socioeconomic conditions, and was chosen as a control so that illness could be compared between the two communities. A sizable work force commutes from Gorham to Berlin, allowing interaction and opportunity for exposure. When preliminary survey data indicated that interaction between the two communities might influence the results, a third community, Whitefield, was added to the study. Whitefield is located about 40 km (25 mi) west of Gorham and has minimal interaction with Berlin and Gorham.

Results were reported by CDC[10] as follows:

A survey of a group of approximately 700 Berlin residents selected randomly from both water distribution systems was conducted during the first week of May 1977. The survey data indicated that 24% or almost a quarter of this population had experienced the onset of gastrointestinal illness during the preceding 2 months. The attack rate of diarrheal disease was somewhat less, averaging about 17%. There were no significant differences in attack rates between the two Berlin water distribution systems. A case of giardiasis was defined clinically as any diarrheal illness lasting for 7 days or longer. It was found that an average of approximately 5% of the Berlin residents surveyed suffered such a prolonged diarrheal illness consistent with giardiasis during the 2 months preceding the survey and there were no significant differences by water supply. Plotting of the cases of confirmed giardiasis on a map of the City of Berlin by residential address showed that the cases were randomly distributed throughout both water distribution systems; that is the Ammonoosuc and Androscoggin Rivers, respectively.

In order to evaluate the significance of the findings on incidence of giardiasis-like illness in Berlin, a similar survey was conducted among 375 residents of the nearby town of Gorham and 010 residents from the town of Whitefield. There was a significant (over twofold) difference in the attack rate for giardiasis-like illness for the City of Berlin as compared to the two control communities, during the Spring of 1977."

Applying the attack rate of 5 per cent to the population of Berlin, 750 people were affected by a prolonged diarrheal illness consistent with giardiasis. "However," continued the CDC report, "giardiasis may be present as an asymptomatic or only mildly symptomatic infection in man. Illustrating this is the fact that a much larger number (47%) of the population of Berlin was found infected with *Giardia Lamblia* during the epidemic of 1977, the majority of which showed no symptoms."

By the time the epidemic subsided, the Androscoggin Valley Hospital had reported 275 confirmed cases of giardiasis. The temporal distribution of cases (Fig. 4) indicates that the outbreak began about April 15 and subsided on May 10, with the peak occurring around April 25. The success of emergency control measures, i.e., boosting chlorination on April 24 and a recommendation to boil water on April 25, cannot be determined readily mainly because of the variability in the incubation period. Figure 4 shows the number of cases declining after April 25, and with a one- to four-week incubation period for giardiasis,[11] the decline would have been expected about May 5. If the number of cases continued to increase for the length of the early incubation period after April 25 and then declined, the effectiveness of the control measures could possibly be demonstrated. The absence of cysts in finished water samples collected on and after April 25, indicating that contamination had ceased, could be interpreted as support for the variability in incubation period. The decline in cases to May 25 is demonstrated in Fig. 4. The suggestion that people boil their water, issued on April 21, may have accounted for the decrease in cases reported on April 30.

Summary, Conclusions, and Recommendations

An outbreak with a 5 per cent attack rate of prolonged diarrheal illness consistent with giardiasis occurred in Berlin, N.H., in the spring of 1977. *Giardia* cysts were recovered from raw and finished water from the water system and routes of entry were found in both treatment plants. Beaver were implicated in the outbreak as *Giardia* trophozoites were identified in the duodenal mucosa of an animal living in Godfrey reservoir. It is plausible that human contamination, resulting from recreational usage or logging activities, caused infection of the beaver in Godfrey reservoir. One average human stool of 150 grams[12] with a cyst concentration of 15 million per gram[13] would result in a theoretical concentration of 20 cysts/litre in Godfrey reservoir, which is a large enough dose to cause infection in human beings.[14]

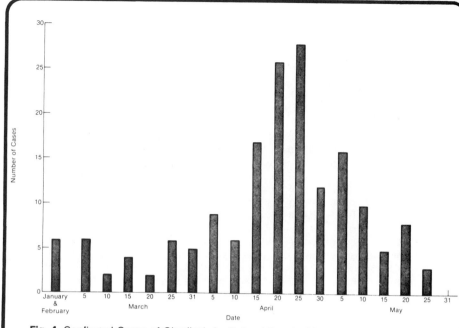

Fig. 4. Confirmed Cases of Giardiasis by Date of Onset of Illness, Berlin, N.H., 1977
Source—Atlanta Center for Disease Control

Berlin Water Works was producing water of acceptable microbiological quality as measured by conventional standards before and during the outbreak; however, the coliform standard is not an acceptable indicator of safety where *Giardia* cysts are present. While chlorination as normally practiced is effective in destroying coliform bacteria, it is not lethal to *Giardia* cysts. Where the threat of contamination exists, effective conventional treatment including coagulation, sedimentation, filtration, and disinfection is the preferred method of protection. Cyst destruction by chlorination is possible, based on studies conducted with *E. histolytica,* but the dosage and contact time requirements for most water systems would require extreme chlorine concentrations that cause and contribute to other quality problems.

Communities relying on surface water sources and chlorination as a method of treatment, and producing water that is acceptable based upon turbidity and coliform requirements, cannot be assured of delivering safe drinking water and protecting public health. This is especially applicable to water systems in the northeast and northwest parts of the country, where *Giardia* outbreaks have been notable.

Giardia cyst penetration of pressure filters has occurred in two outbreaks within the past two years. The filters contribute to an "out of sight, out of mind" attitude because they are difficult to inspect; they cannot be observed during backwash, and obvious indicators of bed stability go unnoticed. Berlin

Water Works operated under a program that required inspection and maintenance every two years. Regular inspection on a six-month schedule to observe bed condition was suggested.

A number of recommendations to improve the water system, included in a summary report to the Berlin Water Works, are summarized here.

Recommendations: Ammonoosuc Plant. *Source.* The watershed that feeds Godfrey reservoir is located in the White Mountain National Forest and is under control of the National Forest Service. Activities promoted by the National Forest Service under a multiple-use-of-land policy have an impact on water quality. Fishing, hunting, logging, snowmobiling, hiking, propagation of wildlife, and so forth, can influence water quality if not properly controlled. Since the Berlin Water Works has no authority to control access and activities, they must attempt to (1) minimize the impact and (2) provide treatment facilities to cope with the water quality conditions that result.

To minimize the impact of public activities in the watershed it is suggested that an agreement be negotiated with the National Forest Service for recreational usage in the immediate area of the reservoir. Access to the dam should be prohibited and limits established in the vicinity of the dam. Patrol and enforcement procedures should be included in the agreement.

Logging operations can adversely affect water quality by increasing the intensity of runoff and contributing to higher turbidity. The sanitary disposal of human waste material from the loggers should be controlled. While the National Forest Service prescribes methods of logging to reduce turbidity, Berlin Water Works should be assured that sanitary facilities are provided to control human waste discharges. While the waste load from loggers would be minimal compared to that resulting from recreation, it could at least be controlled through required toilet facilities in the logging contract. The strategic location of comfort stations may aid in minimizing waste from recreational users, but will not suffice to control it. Therefore, water treatment facilities must act as the barrier of protection against uncontrolled waste materials.

The water supply intake at Godfrey dam is the bottom withdrawal type, which can be influenced by very poor water quality conditions. Because *Giardia* cysts have a specific gravity greater than water, they would be expected to settle to the depths of a reservoir, where a low-level intake would enhance their entry into the water system. It is recommended that Berlin Water Works investigate the feasibility of installing a multi-level intake at Godfrey reservoir to provide flexibility in selecting the best quality of water throughout the depth profile and preclude the use of bottom withdrawal as conditions dictate.

Treatment. The type of treatment currently provided at the Ammonoosuc plant relies heavily on disinfection to produce water that is acceptable from a microbiological standpoint. The unit process of filtration depends on mechanisms of straining, sedimentation, and flocculation within the bed, and each mechanism must operate effectively to reduce the load of silt, organic material, and biological life. For flocculation within the bed to be effective, a conditioning chemical is normally added before filtration to destabilize (reduce the surface charge of) colloids and very small particles so that, by the mechanism of surface attachment, the particles are retained on the sand or coal grains that constitute the filter media. Water treated at the Ammonoosuc plant is not chemically conditioned; therefore it should be expected that part of the applied load is passing through the filters, and that disinfection must destroy the unremoved portion of microbiological life.

Alternatives available at the present time for treatment of cysts include in-line feeding of a conditioning chemical prior to filtration and prechlorination near Godfrey reservoir to provide extended contact time. However, other treatment problems, such as normally low-turbidity water, a high concentration of color (presumably from decomposition products of natural organic matter), low alkalinity and pH, and very low temperatures, hamper flocculation dur-

ing the winter when the threat of cyst discharge seems to predominate.

It would be presumptuous to offer recommendations for improvements at the Ammonoosuc plant without the benefit of more data on water quality, a detailed study of the facilities available, and an economic analysis of several alternatives. Therefore, it is recommended that the Berlin Water Works conduct a study to determine the improvements that are required at the Ammonoosuc Plant. The study should address

1. The treatment requirements necessary to cope with raw water quality conditions in the Ammonoosuc River and Godfrey reservoir, including reduction of color and removal of turbidity, considering low alkalinity, low pH, and low temperature. If prechlorination is instituted as a control measure, the removal of chlorinated organics should be considered.

2. An economic analysis of the costs of renovating existing facilities to obtain desired treatment as compared with the cost of new or additional facilities or expansion of the new facility on the Androscoggin River to serve the entire Berlin system.

During the long-range planning to improve the Ammonoosuc plant, there are several improvements that can be initiated on a short-term basis.

1. The prechlorination facilities currently utilized at Godfrey reservoir are strictly for an emergency and should be replaced with a reliable facility. It is understood that a permanent facility is under consideration by the Board of Water Commissioners and could be installed in a short period of time. Prechlorination is advisable to provide longer contact times and a margin of safety in destruction of cysts. It is suggested that facilities be incorporated into the prechlorination facility for experimental application of conditioning chemicals to aid in filtration and the adjustment of pH.

2. Monitoring for turbidity in the finished water should be done daily in accordance with requirements of the Safe Drinking Water Act. Daily samples should be collected from individual filters as a performance check.

3. The backwash and air scour sequence should be evaluated to determine the reason(s) for excessive turbidity readings when the filters are put on-line after the washing cycle.

4. The filter media should be regularly inspected on a six-month schedule unless individual filter monitoring indicates that more frequent inspection is necessary.

Recommendations: Androscoggin Plant.
Treatment. Entry of cysts into finished water at the Androscoggin plant was apparently through the faulty joints of

the filter channels. It is hoped that these problems will be corrected during acceptance testing, and that the plant will operate in the future as intended. Conventional water treatment of coagulation, sedimentation, and filtration has been shown to be effective in removal of cysts of *E. histolytica,* and it is expected that unit processes properly employed at the Androscoggin plant will be effective in removing *Giardia* cysts. To ensure against leakage through the repaired backwash channels in the future, periodic testing is suggested. Air testing seemed to provide a useful check of the integrity of the joint.

Funds will be provided to Berlin for monitoring raw and finished water for the fall, winter, and spring of 1977–78 to identify temporal distribution of cyst occurrence and to determine whether corrections in treatment are effective. EPA is intensifying its research effort in identification, removal, and destruction of cysts through extramural and intramural studies.

Acknowledgements

The author wishes to acknowledge the support and assistance received from officials and employees of the city of Berlin, N.H. The efforts and many hours of work by James Smith, city manager, Robert Deslisle, city health officer, and Larry Hodgman, superintendent, Paul Roy, Wayne Thompson, and Ben Ouillette of the Berlin Water Works are especially appreciated. The efforts and important contributions of the following are acknowledged and appreciated: Robert W. Christie, pathologist, and Rita St. Armand, laboratory supervisor, Androscoggin Valley Hospital, Berlin, N.H.; Carlos E. Lopez, EIS officer, parasitic diseases div., Bureau of Epidemiology, CDC, Atlanta, Ga.; Bernard Lucey, N.H. Water Supply and Pollution Control Commission, Concord, N.H.; and Charles Larson and Thomas Murphy, Region I, EPA, Boston, Mass.

Also, Gary Logsdon, Municipal Environmental Research Laboratory, EPA, Cincinnati, Ohio; Shih Lu Chang, Health Effects Research Laboratory, EPA, Cincinnati, Ohio; Mr. Vashaw and Mr. Bickford, conservation officers, N.H. State Game Commission; L. M. Pittendreigh and H. Seward, Dufresne–Henry Engrg. Corp., North Springfield, Vt.

References

1. Foodborne and Waterborne Disease Outbreaks. Ctr. for Disease Control, PHS, DHEW, Atlanta, Ga., (1969–1976).
2. SHAW, P.K. ET AL. A Community-Wide Outbreak of Giardiasis with Evidence of Transmission by a Municipal Water Supply. *Ann. Internal Med.* 87 (Oct. 1977). pp. 426–431.
3. KIRNER, J.; LITTLER, J.; & ANGELO, L. A Waterborne Outbreak of Giardiasis in Camas, Washington. *Jour. AWWA,*

70:1:35 (Jan. 1978).

4. Waterborne Giardiasis Outbreaks—Washington, New Hampshire. Morbidity and Mortality Weekly Rprt; Center for Disease Control, PHS, DHEW, Atlanta, Ga. (May 27, 1977).

5. JAKUBOWSKI, WALTER ET AL. Large Volume Sampling of Water Supplies for Microorganisms. HERL, EPA, Cincinnati, Ohio (Submitted for publication).

6. CRAUN, G.F.; McCABE, L.J.; & HUGHES, J.M. Waterborne Disease Outbreaks in the US—1971-1974. *Jour. AWWA*, 68:8:420 (Aug. 1976).

7. METZLER, DWIGHT F.; RITTER, CASSANDRA; & CULP, RUSSELL L. Effect of Standard Water Purification Processes on the Removal of Histoplasma Capsulatum from Water. *Am. Jour. Public Health*, 44:10:1305 (Oct. 1954).

8. SPECTOR, BERTHA KAPLAN; BAYLIS, JOHN R.; & GULLANS, OSCAR. Effectiveness of Filtration in Removing from Water, and of Chlorine in Killing, the Causative Organism of Amoebic Dysentery. *U S Public Health Rpts.*, 49:10:786 (Jul. 1934).

9. *Water Quality and Treatment*. AWWA (3rd ed., 1971). pp. 273-277; second citation, p. 247.

10. LOPEZ, C.E. Community Wide Outbreak of Waterborne Giardiasis Associated with High Rate of Asymptomatic Infection (manuscript in preparation).

11. BENENSON, ABRAM S. Control of Communicable Diseases in Man—An Official Report of the APHA. (12th ed., 1975).

12. ALTMAN, PHILIP L. & DITTMER, DOROTHY S. *Biology Data Book*. Vol. III (second ed., 1974). p. 1489.

13. TSUCHIYA, H. & ANDREWS, JUSTIN. A Report on a Case of Giardiasis. *Am. Jour. Hygiene*, Vol. 12 (1930).

14. RENDTORFF, R.C. The Experimental Transmission of Human Intestinal Protozoan Parasites—Part II. *Giardia Lamblia* Cysts Given in Capsules. *Am. Jour. Hygiene*, Vol. 59 (1954) pp. 209-220.

An article contributed to and selected by the JOURNAL, authored by Edwin C. Lippy (Active Member, AWWA), san. engr., Health Effects Res. Lab., USEPA, Cincinnati, Ohio.

Reprinted from *Jour. AWWA*, 70:9:512 (Sept. 1978).

EPA Project Summary

Giardiasis in Washington State

Floyd Frost, Lucy Harter, Byron Plan, Karen Fukutaki, and Bob Holman

Floyd Frost, Lucy Harter, Byron Plan, Karen Fukutaki, and Bob Holman are with the Department of Social and Health Services, State of Washington, Olympia, WA 98508.

Walter Jakubowski is the EPA Project Officer (see below).

The complete report, entitled "Giardiasis in Washington State," (Order No. PB 83-134 882) will be available only from:

National Technical Information Service
5285 Port Royal Road
Springfield, VA 22161
Telephone: 703-487-4650

The EPA Project Officer can be contacted at:

Health Effects Research Laboratory
U.S. Environmental Protection Agency
Research Triangle Park, NC 27711

This research was initiated to determine the potential for transmission of giardiasis through approved drinking water supplies in Washington State. The project consisted of five separate studies.

The first study, a parasitological stool survey of commercially trapped aquatic mammals, was conducted during each trapping season from 1976 to 1979 and resulted in the examination of 656 beaver stool samples, 172 muskrat stools and 83 other animal stool samples. Positivity for beaver was 10.8%, whereas positivity for muskrat was 51.2%. No Giardia was found in other trapped mammals (nutria, mink, raccoon, river otter, bobcat, coyote, lynx, or mountain beaver).

In the second study, a follow-up of human giardiasis cases identified through medical diagnostic laboratories, 865 Giardia infected Washington State residents were contacted and asked a series of questions designed to identify likely sources or possible risk factors for infection. Two outbreaks were identified which implicated domestic drinking water as the source. Other clusters of cases were linked to day care centers, backpacker excursions or sites for drawing water on outings and foreign travel. No excess of cases was observed for customers of surface drinking water supplies.

The third study was a case-control study to identify risk factors for giardiasis. This study included 349 laboratory-identified cases and 349 controls selected from directory assistance listings. Factors which appeared to place a person at increased risk of giardiasis included consumption of untreated water, foreign travel (for adults) and attendance at a day care center (for children under age 10).

The fourth study examined water filtering techniques for recovery of Giardia cysts from drinking water supplies. Initial application of the technique recovered cysts from several supplies not implicated in giardiasis outbreaks; however, laboratory testing of the technique demonstrated very poor cyst recovery using the recommended filter application and analysis techniques. Changes in the application and analysis techniques (lower water pressure, use of a continuous flow centrifuge, different filter fiber washing techniques, i.e., a 1 micron filter) yielded order of magnitude improvements in cyst recovery. As few as 3000 cysts in 500 gallons of water would be adequate for cyst identification under conditions of low to medium turbidity.

The fifth study was a stool survey of one- to three-year old children in Skagit and Thurston counties. Children were randomly selected from birth certificate listings and parents were paid to submit 2 stool samples for analysis. Overall prevalence of infection was 7.1% for the children surveyed. No differences in the prevalence were found by source of domestic water (surface filtered, surface unfiltered, well or spring).

This report was submitted by the Washington State Department of Social and Health Services, Office of Environmental Health Programs, in fulfillment of Grant No. R-805809 from the U.S. Environmental Protection Agency. This report covers a period from July 1, 1978 to April 1, 1981 and work was completed as of December 31, 1981.

This Project Summary was developed by EPA's Health Effects Research Laboratory, Research Triangle Park, NC, to announce key findings of the research project that is fully documented in a separate report of the same title (see Project Report ordering information at back).

Introduction

Although *Giardia* infections in man have been recognized for centuries, waterborne transmission of this parasite has only recently been recognized as a major mode of dissemination. Drinking water contaminated with human waste was thought to be the likely source of a giardiasis outbreak in Aspen, Colorado in 1966. Contamination of water by aquatic mammal waste was thought to be the likely source of outbreaks in Camas, Washington (1976) and Berlin, New Hampshire (1976). The latter outbreaks were of particular interest to water treatment engineers and public health officials, since the treated water met both coliform and turbidity levels believed to protect against waterborne disease outbreaks. Furthermore, the conditions which resulted in the Camas outbreak were likely to occur commonly throughout Washington State and perhaps throughout much of the West.

Following the Camas outbreak of April and May 1976, the Washington State Department of Social and Health Services (DSHS) together with the U.S. Environmental Protection Agency (EPA) began a series of investigations to determine whether similar outbreaks were occurring elsewhere in Washington State and to estimate the potential for future outbreaks. The Camas outbreak was thought to be related to *Giardia* infected beaver residing in the watershed of the town's surface water supply. Due to problems with the Camas water filter system, cysts (possibly excreted from beaver) passed through the filter. They were probably unaffected by the level of chlorination used at the time of the outbreak. The majority of Washington State residents are served by surface water supplies and many of these supplies use chlorination as the only means of disinfection. Since all of these

watersheds are frequented by beaver, the presence of *Giardia* infected beaver could lead to similar outbreaks.

Information was required on both the potential for human exposure to *Giardia* and the incidence of human illness. To determine the extent of aquatic mammal infection with *Giardia*, stool surveys of commercially trapped animals were initiated in the fall of 1976 and continued through spring 1980. To assess the extent of human illness resulting from giardiasis, a pilot human case follow-up was initiated in 1977 and extended to a statewide human follow-up in July 1978. To identify risk factors for human giardiasis, a case-control study was initiated in March 1979 and continued through March 1980, when case follow-up was also suspended. In July 1978 an investigation was initiated to estimate how frequently *Giardia* cysts could be recovered from drinking water supplies with the use of a large volume water filtration technique developed by the Health Effects Research Laboratory (HERL), EPA. Due to problems with the technique, this aspect of the study was modified so that more effort was placed on evaluating alternative methods for cyst recovery. In September 1980 a human stool survey of one-to-three-year-old children was initiated to determine whether a difference in prevalence of infection existed between areas served by surface water supplies (Skagit county) and areas served by well water supplies (Thurston county).

Conclusions

This project demonstrated a widespread potential for waterborne transmission of giardiasis in Washington State. During the four years of animal surveys, *Giardia* prevalence in beaver ranged from 6% to 19% and in muskrat from 0% to 85%. Infected beaver were found throughout the state in both protected and unprotected watersheds which provide drinking water for Washington State residents.

Statewide human giardiasis surveillance efforts and follow-up substantially increased the number of reported giardiasis cases, identified two outbreaks associated with domestic drinking water supplies, two day care center outbreaks, one outbreak associated with foreign travel and numerous smaller clusters of cases. From the case-control study, foreign travel, consumption of untreated water

and attendance at a day care center (for children) were found to be significantly more common among giardiasis cases than among controls. Among giardiasis cases with foreign travel, only travel to Third World countries was found to be associated with giardiasis.

The human case follow-up revealed that giardiasis follows a bimodal age distribution affecting both young children and young adults. Evidence of secondary transmission was observed, especially in households with young children. No excess of cases was observed among customers of surface water supplies, even after individuals with other likely sources of infection (homosexuals, those who consumed untreated' surface water, day-care center attendees, persons with a history of foreign travel, and case clusters with a likely common exposure) were eliminated.

Results of the stool survey of one- to three-year-old children generally supported the findings of the case-control study and the human case follow-up. No difference in *Giardia* prevalence was observed for children served by deep well water supplies and surface supplies. In both cases one- to three-year-olds were found to have a 7.1% prevalence of *Giardia*. An increased risk of infection was found for children with exposure to untreated surface water and for children with more than two siblings between the ages of three and ten. No increased risk was found for children attending day care centers, contradicting results of the case-control study.

Environmental sampling to recover *Giardia* from natural waters proved to be disappointing. Of the 77 water filter samples examined, only 5 were positive for *Giardia* and three of these were taken in response to a reported outbreak. An examination of recovery efficiency was begun early in the project to test the filter both in the field and in the laboratory. Initial recovery of one cyst out of 30,000 cysts was followed by changes in both the application and analysis procedures. These changes (lower water pressures, more agitation to remove cysts from the filter fibers, and the use of a Foerst centrifuge) resulted in recovery of nearly 10% of the experimentally added cysts. Concentration techniques using sucrose or zinc sulfate were examined but did not provide noticeable improvements when used on filter samples.

The implications of these findings for waterborne transmission of giardiasis in Washington State are as follows: 1) *Giardia* infection among aquatic mammals in Washington is widespread and includes animals in the most remote and protected watersheds. 2) Although recovery of cysts from water implicated in an outbreak has usually been possible, recovery of cysts from other surface water was only occasionally possible. Although animal trapping results suggest that cysts should be commonly found in surface waters, the concentration of cysts required for filter recovery is seldom observed. 3) With the exception of several outbreaks, Washington's surface water supplies were not associated with an increased risk of giardiasis or *Giardia* infection in their customers. The suspected excess level of disease in communities served by surface water supplies was not observed. Consumption of untreated surface water, person-to-person transmission (primarily among children), and travel to Third World countries were the most important risk factors associated with giardiasis.

Recommendations

Waterborne giardiasis does not appear to be a significant public health problem in Washington State, despite the widespread potential for water supply contamination. The waterborne outbreaks detected were associated with operational problems (Leavenworth) and with inadequate design (Boistfort) of treatment plants. No outbreaks were detected in either Tacoma or Seattle, even though infected animals were trapped from the watersheds of the surface water supplies, and the only treatment provided these water supplies is chlorination.

In contrast, untreated surface water does present a significant public health problem. Consumption of untreated water was recognized as a risk factor for giardiasis in all age groups and was also associated with *Giardia* infection among stool survey participants.

Orlon-wound filters proved to be useful in recovering cysts from water supplies implicated in a human giardiasis outbreak but did not yield useful information on water supplies randomly selected. Laboratory evaluation of filter analysis procedures suggests that improvements in recovery and reductions in cost can be achieved by using an algal (Foerst) centrifuge rather than the series of screens recommended in earlier studies.

Results of the stool survey suggest that water contamination may interact with other risk factors by providing an initial infection. The number of children in a household appeared to be a risk factor; however, the risk was only increased among families with a history of untreated water consumption.

ENGINEERING DEFECTS ASSOCIATED WITH COLORADO GIARDIASIS OUTBREAKS JUNE 1980 - JUNE 1982

Richard J. Karlin, P.E.
Chief, Drinking Water Section
Water Quality Control Division
Colorado Department of Health
Denver, Colorado 80220

Richard S. Hopkins, M.D.
Chief, Communicable Disease Control Section
Colorado Department of Health
Denver, Colorado 80220

Introduction

There is widespread belief that a number of waterborne disease outbreaks go unreported due to the lack of active surveillance and investigation programs in this regard. In Colorado, for example, only six outbreaks were reported over the three year period preceding June of 1980.

In June, 1980, the Colorado Department of Health began a 2-year EPA funded project to vigorously pursue and investigate all waterborne disease outbreaks in Colorado. During this period eleven waterborne outbreaks were identified, indicating that intensive surveillance can indeed increase the detection of waterborne disease outbreaks.

Several of these outbreaks were caused by Giardia lamblia, an intestinal parasite common to several cold water areas of the U.S. and the world.

Study Methodology

For the duration of this study, one person was assigned full time to monitor reports of possible waterborne disease in Colorado. Disease cases and clusters of cases reported by citizens, doctors, and local health departments were followed up by epidemiological investigations first to determine whether, in fact, the reported disease was waterborne and second whether it was associated with the consumption of water from a specific source. Outbreaks were confirmed as waterborne only when (1) there was a positive dose - response relationship between water consumption and disease and (2) there was a higher disease prevalance among consumers of a particular water supply than among a control group in another supply, or than among non-water drinkers served by the problem water supply.

Agent identification was based upon appropriate symptom complex, positive stools from cases and/or recovery of an appropriate agent from suspect water supply.

Engineering investigations were conducted either simultaneously with epidemiological studies or immediately following confirmation of an outbreak as waterborne. Such investigation sought to determine what, if any, physical defect(s) led to the outbreak, how to correct such defects, and how to prevent their recurrence in the future.

Investigations were complicated by the fact that in general, and especially in the case of giardiasis, disease symptoms occur after the event which puts the water system at risk for waterborne diesase. Recreation of events at a suspect water treatment facility several days (or even weeks) in the past can be difficult, especially if defects and/or malfunctions have been corrected in the meantime.

Of 30 reported waterborne disease outbreaks investigated, eleven were confirmed as outbreaks actually associated with the consumption of water. Of these, 6 were identified as caused by the flagellated protozoan, Giardia lamblia. It is these six outbreaks upon which the remainder of this paper will focus.

Giardia lamblia

Giardia lamblia is the single agent most commonly identified in association with waterborne disease in the United States. The resulting intestinal disease, giardiasis, is most frequently characterized by nausea, diarrhea and loss of appetite lasting from a few days to several weeks. The disease is transmitted by means of cysts passed via fecal material or the consumption of water contaminated with such material. The six giardiasis outbreaks considered in this paper were all associated with identifiable treatment deficiencies and/or malfunctions. A brief description of each outbreak including date, estimated number of cases, attack rate, brief system description and analytical results follows:

Town of E - August 1981, involved approximately 100 cases and an attack rate of 24%. E utilizes a cold, clear surface water source (raw water turbidity 0.1 N.T.U. to 0.5 N.T.U. year around). The source is disinfected by gas chlorination and distributed to a resident population of approximately 400 persons and a moderate tourist population.

Symptoms and stool samples indicated giardiasis but the organism was not isolated from the water supply.

The system met turbidity and bacteriological requirements prior to and during the outbreak.

Camp A - September, 1981, involved 29 cases with an attack rate of 28%. Camp A drew water from a beaver pond and was equipped to filter (cartridge) and chlorinate. At the time of the outbreak all treatment was being bypassed to the system due to the unavailability of a knowledgeable operator.

Giarida cysts were recovered from water samples collected at the Camp and clinical symptoms were consistent with giardiasis. Bacteriological and turbidity results were not available.

Water and Sanitation District A - November 1981 involved 40 cases and an attack rate of 18.6%.

Water and Sanitation District A obtained water from a clear mountain stream which flowed from an alpine watershed with heavy human useage and beaver dams. Treatment at the A Water and Sanitation plant consisted of pre-chlorination, sedimentation and granular (dual media) filtration without chemical pretreatment (coagulation/flocculation). The filter bed was upset apparently due to wash out by unevenly applied raw water and broken underdrain laterals.

Giardia cysts were recovered from ill persons and from A Water and Sanitation District system finished water. Bacteriological samples concurrent with the outbreak were safe and turbidity never exceeded 0.7 N.T.U.

Water System F - November, 1981, involved 85 persons with an attack rate of 29.2%. System F consisted of a spring supply with simple chlorination. Overflow from the spring box was discharged to a small stream nearby. Beavers constructed a dam on the stream allowing surface water in their reservoir to back up and enter the spring box.

Clinical picture and engineering evaluation of contamination source indicated giardiasis. Bacteriological samples were safe, turbidity was not measured.

Town of P, December, 1981, 135 people with an attack rate of 10%. P utilized a surface source with variable raw water turbidity. Automatic backwash sand filtration and chlorination were practiced with no chemical pretreatment.

Giardia positive stools and clinical symptomology, as well as giardia isolated from finished drinking water, confirmed giardiasis as the disease involved. Bacteriological and turbidity test results were all within standards.

G W and S, January, 1982, involved an undetermined (large number of transient consumers made numerical estimate difficult) number of actual cases and an estimated attack rate of 15.1%.

The water source implicated in this outbreak consisted of a clear mountain stream with only springtime high turbidity in a relatively winter inaccessible mountain valley with some beaver activity. Treatment consisted of unpretreated filtration (dual media) and chlorination.

Clinical picture, stool positive giardiasis cases and giardia in water samples confirmed the nature of this outbreak. No violation of bacteriological or turbidity standards were associated with this outbreak.

Conclusions

The principal engineering conclusion derived from the giardiasis outbreaks observed during the waterborne disease study is that adherence to the multiple barrier concept (The application of redundant steps to assure complete removal of pathogenic organisms [e.g. proper wastewater treatment, coagulation, flocculation, filtration, disinfection] in series to remove micro-biological contaminant) is crucial to the prevention of waterborne disease. All six giardiasis outbreaks resulted from the absence or bypass of one or more of the barriers involved.

A number of common factors were involved in all 6 giardiasis outbreaks.

(1) The consumption of water from a surface water source. The
     beaver have often been implicated as a _Giardia_ carrier,
     however, control (elimination) of the beaver population
     appears to be of limited value in preventing giardiasis
     outbreaks. The prevelance of giardia cysts in canine
     stools and the ubiquitous nature of man and his
     domesticated animals in watersheds places nearly all
     surface water sources at risk for _Giardia_.

(2) The lack of complete conventional treatment (coagulation,
     flocculation, sedimentation, filtration and disinfection)
     of these sources. During this study, no outbreaks occurred
     where full treatment was employed. The importance of the
     multiple barrier concept to waterborne giardiasis
     prevention has previously been mentioned.

(3) No violations of the U. S. Environmental Protection Agency
     and Colorado Department of Health turbidity standards were
     associated with any of these outbreaks. Turbidity is a
     poor measure of the microbiological quality of water in
     areas (like Colorado) where raw water turbidity is low. It
     is estimated that 80% of Colorado's surface supplies could
     meet the 1 N.T.U. USEPA/CDH standard 75% of the time
     without treatment. In fact the existence of a turbidity
     standard (and is often used as a goal) may result in
     systems discontinuing pretreatment during low turbidity
     periods since the standard can be met without such
     pretreatment. Turbidity is a useful operational tool but
     is not sufficient in itself to assure a safe water.

(4) No violations of the USEPA/CDH bacteriological standards
     were associated with the subject giardiasis outbreaks.
     Early in the study all positive coliform bacteriological
     analyses were followed up in an effort to associate these
     results with disease. Not a single waterborne outbreak of
     any kind was uncovered in this way.

Likewise, coliform testing has not been of any value in
predicting potential giardiasis outbreaks. _Giardia's_ inherent
chlorine resistance is part of the reason for this, however, the
historical assumption that the large number of coliform bacteria
relative to the number of giardia cysts would allow coliform to be
predictive of giardia is clearly an error.

These findings are not startling and in fact were predictable
based upon published information. They merely add actual experience
to the theoretical base already established thereby proving the
relevance of the theory to plant design and operation practice.

In summary, compliance with regulated parameters alone is not
sufficient to assure a continuously safe potable water supply. The
proper design application and operation of the multiple barrier
concept is the only reliable tool is this regard.

Reprinted from Proc. AWWA ACE, Las Vegas, Nev. (June 1983).

Causes of a Waterborne Giardiasis Outbreak

Thomas E. Braidech and Richard J. Karlin

In November 1981, a waterborne giardiasis outbreak occurred within the Highlands Water and Sanitation District in Colorado. Although the immediate cause of the outbreak was reduced chlorine contact time because of a pump failure, numerous design and operational deficiencies existed at the plant. Lack of chemical pretreatment, improper backwashing procedures, poor application of raw water to the filter, and failure to monitor plant conditions all contributed to the outbreak. Turbidity and bacteriological records did not indicate any problems, confirming the crucial role of proper design, installation, and operation of the multibarrier technique in preventing waterborne giardiasis outbreaks.

Giardiasis is an intestinal disease most often characterized by nausea, diarrhea, and loss of appetite. In humans, the causative organism is *Giardia lamblia*, a flagellated protozoan that colonizes the upper small intestine of many warm-blooded animals.[1] The organism is transmitted among humans and other animals by means of cysts passed in excreted fecal material. If an infected animal or human defecates in or near a stream, the cysts can then spread via the water. Consumption of water containing cysts of this organism is a principal method of contracting the disease.

During the fall of 1981, a giardiasis outbreak involving approximately 20 of the 165 people served by the system occurred at Highlands Water and Sanitation District (HWSD) in Colorado. The reported cases were evenly distributed throughout the service area. The service area, located approximately 2.5 mi (4 km) northwest of Aspen, Colo., is characterized by rental units, private residences, and restaurants adjacent to a ski area. Surface water from Maroon Creek serves as the raw water source for the system. The stream originates high in the mountains to the south of the district and is fed mainly by snowmelt and groundwater infiltration. Beaver, an important carrier of *Giardia lamblia* cysts,[1] inhabit the watershed.[2] The HWSD has been supplying drinking water to the area for 18 years and never before had reported an outbreak of waterborne disease. The treatment system was the unit installed when the district was formed.

The chronology of the outbreak began on Oct. 31, 1981, when residents of the district contacted the water plant operator to complain that they were out of water. Investigations showed that the district's 150 000-gal (5.7-ML) in-line storage tank was empty. A failure in the pumping system at the treatment plant was determined to be the cause of the water outage. Using an average consumption of 150 gpcd (568 L/d per capita), it was calculated that the pump had failed on approximately October 25th. This malfunction was corrected, and the plant was put back into service to refill the reservoir on that same day.

In late November 1981, the local health department began to receive reports of illness among residents of the district. Symptoms of the illness included diarrhea, bloating, cramping, gas, and vomiting. Subsequently, a team of local and state health department investigators conducted an epidemiologic study to determine the nature and extent of the problem.[2] Symptomatology, coupled with stool examinations, confirmed that the outbreak was, indeed, giardiasis.

Based on a previous arrangement, the HWSD was tied into the city of Aspen distribution system in mid-December 1981. No further cases of confirmed giardiasis were reported after the connection took place. Although both utilities use the same raw water source, the Aspen water treatment plant provides complete treatment, including coagulation, settling, filtration, and postchlorination. This change benefited the HWSD by providing it with a high quality finished water, thus mitigating the chance of future outbreaks of giar-

diasis. Because of this change in supply, the HWSD treatment plant was shut down and left in the condition that existed when the outbreak was identified. This afforded an opportunity for a detailed examination of the system to determine what circumstances led to the giardiasis outbreak.

Findings

On Jan. 20–21, 1982, approximately one month after shutdown, the treatment plant was examined by representatives of the Colorado Department of Health and the US Environmental Protection Agency (USEPA), Region VIII. The following is a discussion of what was learned during this investigation.

The Highlands treatment plant was designed as a 100-gpm (6.3-L/s) package plant with two dual-media filters totaling 50 sq ft (4.6 m²) in area. From bottom to top, the filters were designed with 18 in. (0.46 m) of gravel, 8 in. (0.2 m) of sand, and 30 in. (0.76 m) of anthracite. Treatment consisted of prechlorination, plain sedimentation (detention time of 60 min), and filtration.[3] No provision was made for coagulant addition or postfiltration chlorination (Figure 1).

Investigation of the treatment plant and records disclosed the following:

• Inspection of the treatment unit showed that although it was structurally sound, sand had passed from the filter into the clearwell, indicating possible deficiencies in the media support, the underdrain system, or both.

• Observation of the surface of the filter indicated that the filter media directly below the settling basin overflow had washed out (Figure 2). These washout areas in the filter media, which were at least 10 in. (25 cm) deep in the anthracite, were apparently caused by the practice of restarting when the filter bays were devoid of water. This probably allowed water to short-circuit through the filter in this area.

• Analysis of the filter by probing

97

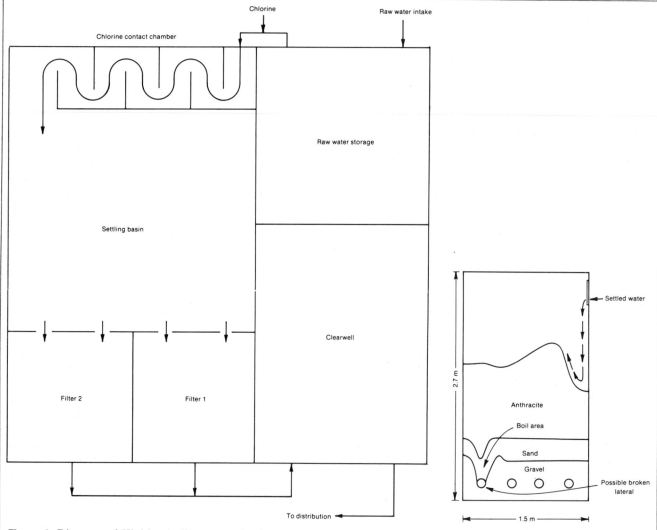

Figure 1. Diagram of Highlands Water and Sanitation District drinking water treatment plant

Figure 2. Cross section of filter bay showing washout of anthracite

with a metal rod[4] to determine the level of the media disclosed that the gravel level in both filters was uneven. A deviation of up to 8 in. (20 cm) below and above the average gravel depth was noted in both filters. This indicated that underdrain problems may have existed in the filters. It appeared that broken laterals were the probable cause of these blowout areas. This, however, could not be confirmed by examination because of the lack of equipment with which to excavate the filter beds.

• A test backwash of both filters resulted in "boil" areas with nonuniform bed expansion during the early phases of the backwash, again indicating underdrain deficiencies. Attempts to backwash the filters were complicated by difficulties with the automatic backwash controls, which were not functioning. Conversations with the operator indicated that he had been manually backwashing for some time and that backwashing was performed at approximately monthly intervals during periods of low turbidity. A review of plant records

showed that the filter had been backwashed approximately one month prior to the failure of the plant.[5]

• Numerous mudcakes were uncovered at the interface between the sand and anthracite, indicating inefficient backwashing procedures. The interface was clearly defined with little mixing of the media despite the apparent underdrain deficiencies.

• Daily turbidity readings for October and November 1981 were below the maximum contaminant level (MCL) of 1 ntu. Records maintained at the treatment plant indicated that the daily turbidity varied from 0.28 to 0.71 ntu during this period. Turbidity on Oct. 31, 1981, the date of the reported water outage, was 0.44 ntu. These levels of turbidity are representative of levels normally found in the ambient water of the Colorado high country during the fall and were accepted as representative.

• No coliform bacteria were detected in the two routine samples collected in October 1981 or in the one sample collected in November, the period associated

with the outbreak. The bacteriological MCL of not more than 1 coliform/100 mL of water was not violated.

• Chlorination was accomplished by using a diluted laundry bleach (sodium hypochlorite) solution, which was applied by a positive displacement metering pump rated at 15 gpd (56.8 L/d) maximum. At the time of inspection it was set to pump 8.1 gpd (30.7 L/d). Based on the reported dilution ratio used at the plant,[5] approximately 0.38 mg total chlorine/L was being applied at the time of the water outage. The actual chlorine residual in the distribution system and the chlorine demand of the raw water during this period are unknown because of the lack of plant data.

Discussion

The multibarrier technique (i.e., coagulation–flocculation, sedimentation, filtration, and disinfection) has been shown to be an effective method of treatment for the removal of cysts and cyst-sized particles from drinking water.[6,7] Indeed, this mode of treatment plant operation has become the most widely used operation procedure in the industry. The method relies on the concept of using several steps in the treatment of raw water. Therefore, if one link in the treatment chain fails or is not operating correctly, safe potable water can still be produced.

The treatment train used at the HWSD plant did not adhere to the multibarrier concept. In fact, the lack of chemical pretreatment was the most significant design deficiency identified in terms of cyst removal. Inadequate pretreatment, i.e., coagulation–flocculation and sedimentation, of water prior to granular filtration has been shown to allow the passage of cysts into the finished water.[8-11] This investigation confirmed that filtration alone, without proper pretreatment, is incapable of continuously removing *Giardia* cysts. Drinking water facilities for treating surface water, or any water suspected of harboring *Giardia*, must include adequate pretreatment if cyst removal is expected.

Jarroll et al[12] showed that *Giardia lamblia* cysts are deactivated at the conventional chlorine dosages and holding times, e.g., 2 mg/L at pH 6 or 7 for 60 min at 5°C, used in the drinking water industry. This indicates that the response of *Giardia* to chlorine is similar to that of other waterborne organisms and, therefore, it would be killed by either a high chlorine dosage and short contact time or a low chlorine dosage and long contact time. After the pump breakdown at the HWSD plant was repaired, the need to supply water to the distribution system allowed water to reach the consumers shortly after treatment. This eliminated the long storage time (estimated to have been 7–10 days) during which the cysts either settled in the reservoir or were deactivated by long contact with low-level chlorine. Cysts could also have been resuspended from the tank sediment upon refilling. Despite possible short-circuiting, apparently settling, chlorine contact time, or both had been effective in preventing previous outbreaks of giardiasis.

The condition of the filters and underdrains was such that cysts could not have been removed reliably even if the multibarrier technique had been properly applied. The probing of the filters, which revealed unevenness in the gravel support media and sand in the clearwell, also indicated breaks in the underdrain laterals. These breaks may have occurred during manual backwashing, which caused sudden surges of water to the underdrain system. Subsequently, water could short-circuit through the filter in the area of the break, giving rise to the formation of the mudcakes found at the sand–anthracite interface. The mudcakes would tend to aggravate the situation by sealing off areas of the filter,

thereby further increasing the velocity of the water and the short-circuiting through the filter. Given the evidence of sand in the clearwell, *Giardia* cysts, if present in the raw water, clearly could have passed through the filter.

The mounds and voids found in the anthracite at the surface of the filter bed were caused by a design that allowed water to cascade onto the empty filter from the settling chamber (Figure 2) after backwashing. The resulting shallow depths of media would exacerbate the short-circuiting problems and enhance the subsequent passage of cysts into the finished water.

Routine maintenance, including probing of the media and observation of clearwell conditions and backwash cycles, should be standard operational practices at all water treatment plants using granular media. Logsdon[11] reports that the practice of using extremely long filter runs with multiple start and stop cycles could result in cyst breakthrough.

Neither coliform nor turbidity tests indicated a potential problem. Although both of these tests are useful operational parameters, these analyses alone do not assure the absence of *Giardia* cysts in the finished water, particularly in granular-media plants where coagulation is not practiced or is inadequate. Indeed it has been the authors' experience[13] that meeting the MCL for these two parameters is of little or no value in determining whether the finished water of a given system may contain cysts.

Conclusions

The most likely cause of the HWSD outbreak was the presence of viable *Giardia* cysts in the potable water system. Three major factors led to the presence of these cysts in the distribution system: (1) a drastic reduction in the chlorine contact time available for disinfection; (2) the absence of chemical pretreatment, i.e., coagulation–flocculation, prior to sedimentation; and (3) the poor design, function, and operation of the filtration system.

It appears that the best alternative for eliminating the potential for outbreaks of giardiasis is to adhere to the multiple-barrier concept in treating water. Specifically, this should include chemical addition for coagulation–flocculation, sedimentation, and filtration followed by disinfection. Experience[13] has shown that systems in which the raw water turbidity is generally less than 1 ntu year-round have tended to eliminate chemical additives for coagulation–flocculation, which may place those served by the system in jeopardy. Further, operation and maintenance programs are commonly inadequate in small systems, particularly with regard to the condition of the filter bed. There must be a concerted effort to upgrade these

programs to detect deficiencies in a drinking water treatment system and to eliminate the inadequacies that may contribute to the outbreak of disease.

Acknowledgments

John R. Blair and Jerry C. Biberstine were of great assistance in the inspection and analysis of the HWSD facility.

References

1. Waterborne Transmission of Giardiasis. Proc. USEPA Sym., Sept. 18-20, 1978. EPA-600.9-79-001 (June 1979).
2. ISTRE, G.R.; DUNLOP, T.S.; & GASPARD, G.B. Waterborne Giardiasis at a Mountain Resort: Evidence for Possible Acquired Immunity. Unpubl., State of Colo. (1982).
3. KOCH, F.O. Operating Manual, Highlands Water and Sanitation District, Water Supply and Distribution System. Pitkin County, Aspen, Colo.
4. American Water Works Association. *Water Quality and Treatment.* McGraw-Hill Book Co., New York (3rd ed., 1971).
5. NELSON, R. Colorado Department of Health Activity Report. Denver, Colo. (Jan. 13, 1982).
6. LOGSDON, G.S. & LIPPY, E.C. The Role of Filtration In Preventing Waterborne Disease. *Jour. AWWA*; 74:12:649 (Dec. 1982).
7. LOGSDON, G.S. & FOX, K.R. Getting Your Money's Worth From Filtration. *Jour. AWWA*, 74:5:249 (May 1982).
8. KIRNER, J.C.; LITTLER, J.D.; & ANGELO, L.A. A Waterborne Outbreak of Giardiasis in Camas, Washington. *Jour. AWWA*, 70:1:35 (Jan. 1978).
9. LIPPY, E.C. Tracing a Giardiasis Outbreak at Berlin, New Hampshire. *Jour. AWWA*, 70:9:512 (Sept. 1978).
10. LOGSDON, G.S. ET AL. Alternative Filtration Methods for Removal of Giardia Cysts and Cyst Models. *Jour. AWWA*, 73:2:111 (Feb. 1981).
11. LOGSDON, G.S.; DE WALLE, F.B.; & HENDRICKS, D.W. Filtration as a Barrier to Passage of Cysts in Drinking Water. *Giardia and Giardiasis* (S.L. Erlandsen and E.A. Meyer, editors.) Plenum Press, New York (1984).
12. JARROLL, E.L.; BINGHAM, A.K.; & ROGER, E.A. Effect of Chlorine on *Giardia lamblia* Cysts Viability. *Appl. & Envir. Microbiol.*, 41:483 (Feb. 1981).
13. HOPKINS, R.S. ET AL. Waterborne Disease in Colorado: Report on Two Years' Surveillance of Eleven Waterborne Outbreaks. Final Rept. EPA Contract 68-03-2927, Colo. Dept. of Health, Denver, Colo. (Jan. 14, 1982).

About the authors: *Thomas E. Braidech is clean lakes coordinator with the US Environmental Protection Agency, Region VIII, 1860 Lincoln St., Denver, CO 80295. Richard J. Karlin is chief of the Drinking Water Section, Colorado Department of Health, 4210 E. 11th Ave., Denver, CO 80220.*

Reprinted from *Jour. AWWA*, 77:2:48 (Feb. 1985).

EPA Project Summary

Occurrence of *Giardia* in Connecticut Water Supplies and Watershed Animals

Henry Adams and Arthur Bruce

Henry Adams and Arthur Bruce are with State Department of Health Services, Hartford, CT 06115.

Theodore H. Ericksen is the EPA Project Officer (see below).

The complete report, entitled "Occurrence of Giardia in Connecticut Water Supplies and Watershed Animals," (Order No. PB 85-151 199/AS) will be available only from:

National Technical Information Service
5285 Port Royal Road
Springfield, VA 22161
Telephone: 703-487-4650

The EPA Project Officer can be contacted at:

Health Effects Research Laboratory
U.S. Environmental Protection Agency
Research Triangle Park, NC 27711

The main objective of this research was to study the occurrence of *Giardia* in selected water supplies and watershed animals in Connecticut.

During the period from October, 1979 to October, 1980, water samples were collected monthly using the U.S. Environmental Protection Agency (EPA) *Giardia* sampling method at selected water utilities and analyzed for *Giardia* cysts. Additionally, samples were analyzed for total coliforms, fecal coliforms, standard plate count, yeast, turbidity, and pH.

Fecal specimens were collected from beaver, deer, squirrel, muskrat, and racoon if these animals were found on the watershed. Descriptions of the watersheds involved in this study included size, type, recreation, human inhabitation, historical water quality data, known sources of contamination, efforts to protect watershed, and species and population estimates of the watershed animals.

Correlations of the presence of *Giardia* cysts with the collected water quality data were not successful because no *Giardia* cysts were detected on the *Giardia* sampling filters.

Six out of 413 fecal animal samples collected on the reservoir watersheds were found to be positive for *Giardia*. Collection of fecal samples from trapped animals yielded a higher percentage of *Giardia* positives than those collected from live animal droppings.

Yeast was found more often in raw water samples than in treated water samples.

This Project Summary was developed by EPA's Health Effects Research Laboratory, Research Triangle Park, NC, to announce key findings of the research project that is fully documented in a separate report of the same title (see Project Report ordering information at back).

Introduction

The purpose of this study was to determine the presence of *Giardia* in selected water supplies and in specific animals found inhabiting the watersheds of these water supplies. Four hundred and thirteen animal fecal samples were collected and tested for *Giardia* during the 12-month study period. One hundred and forty-four sampling filters were collected and examined using the EPA large volume sampling method for *Giardia* cysts.

Water quality parameters including chlorine residual, total coliforms, fecal coliforms, standard plate count, pH, temperature, and turbidity were obtained from samples taken before and after the *Giardia* sampling unit was on-line. Yeast samples were taken after the *Giardia* samples were collected. If *Giardia* was found on the sampling filters, correlations with the other water quality parameters would be determined.

Background

Very little information is known about the presence of *Giardia* in water supplies or in watershed animals in Connecticut. Similar studies to this one have been conducted in the State of Washington where Giardia was found to be a common intestinal parasite of beaver and muskrat. The presence of Giardia would indicate a potential health hazard to the consumers of unfiltered surface water supplies in Connecticut and show Giardia to be potentially a widespread problem in public water supplies. Numerous outbreaks have been documented implicating Giardia in drinking water in such places as Rome, New York; Camas, Washington; and Aspen, Colorado, as well as others.

Methods

This study consisted of two methods to evaluate the presence of Giardia in certain water supplies and the watersheds of these water supplies.

One method used EPA large volume samplers for collecting Giardia cysts. Once a month the selected sites were sampled, with the sampling device on-line for approximately 24 hours. Total coliform, fecal coliform, standard plate count, turbidity, and pH samples were collected for analysis prior to and after the large volume sampler was on-line. Yeast samples were taken after the large volume sampler was on-line. Total chlorine and free chlorine residual were measured at each site prior to and after the filtering unit was installed. The EPA large volume samples were used on both the raw water tap and the treated water tap if the treatment of the water supply included filtration. Where the water supply treatment did not include filtration, samples were taken only from the treated water tap.

The presence of Giardia in fecal specimens from watershed animals that included beaver, deer, squirrel, muskrat, and racoon was determined for those animals found on each watershed. The animal fecal samples were collected by members of the trappers association. The collected fecal samples were placed in formalin containing vials and later analyzed for Giardia. The trappers collected from all the different species of animals so that one species would not account for all the samples to the exclusion of another species more difficult to sample. However, because some animals were not present on the watersheds at the time of collection, all the different animal species were not sampled every month. A maximum of 10 samples per month for each selected watershed was permitted. The fecal specimens were collected by one of two methods. Animals were trapped using the leg-hold trap

during the trapping season when permission was granted by the land owner. After trapping season, all fecal samples were collected from animal droppings. Trappers collecting the fecal droppings from live animals were requested to collect samples from different locations and to take only one sample per animal. This, hopefully, would lessen the possibility of sampling an animal more than once.

Initially, the fecal samples were to be collected by only one method, live-trapping of the animals. This method was found to be more expensive than using the leg-hold trap because the trappers would have to purchase the "live-type" traps. This type of trap required more room when transporting between trapping locations. A method of marking the animals would be necessary when using the live-trapping method to prevent sampling from the same animal more than once.

Comparisons and correlations were made between the different collection methods of obtaining the animal fecal specimens, the site locations, and the physical and bacteriological test data with the presence of Giardia.

Laboratory Methods

The EPA large volume sampling filters were analyzed according to EPA methodology. The animal fecal samples were processed using the formalin-ether sedimentation concentration technique, then examined microscopically. The enumeration and isolations of yeast were based on APHA Standard Methods for the Examination of Water and Waste Water. The bacterial analysis, turbidity, color, odor, and pH determinations were also performed according to Standard Methods for the Examination of Water and Waste Water by the Laboratory Division of the Connecticut State Department of Health Services. The temperature and chlorine residual were determined at each site. The total and free chlorine residual were determined by DPD colorimetric method.

The sites that were sampled for Giardia were selected in two ways: (1) areas with low and high stool positivity rates for Giardia, and (2) watershed with no known human activity (protected) and with human activity and/or human sources of sewage contamination (semi-protected). Giardia-positive stool sample data provided by the State Department of Health were used to calculate human stool positivity rates for each town in Connecticut. Rates varied from .03 per 1,000 to 1.94 per 1,000. Two towns with

low Giardia positivity rates were selected and matched with two towns with high Giardia positivity rates based on population density and population served by a community water supply. The project director selected additional sites based on criterion No. 2. One source that had been selected was changed because it had not been used as a water supply for several years.

Results

A total of 413 fecal specimens were submitted to the laboratory for Giardia determination over the consecutive 12-month study period from November, 1979 through October, 1980. Only six samples (1.4%) were found to be positive for Giardia. Fecal samples submitted during trapping from November, 1979 through February, 1980 were 2.34% positive for Giardia. Fecal samples collected after trapping season were 1.1% positive.

During the fecal specimen examination for Giardia cysts, other parasites were found. The beaver specimens, mainly from one watershed, had the lowest percentage of parasites found of the animal types studied with 16.7%. The racoons studied were found to have the most parasites of the animals examined with 64.8% of the specimens giving a positive result. Most of the other parasites found were helminths.

A total of 144 water sampling filters were collected for Giardia analysis using the EPA large volume sampling method. All of these filters were determined to be negative for the presence of Giardia. A total of 288 water samples were submitted for bacteriological analysis. Of the three types of bacteriological analyses used, total coliform, fecal coliform, and standard plate count, bacteria were found in the majority of the raw water samples. Bacteria were not present in most of the treated water samples.

A total of 74 samples were tested for the presence of yeast. Thirty samples were from raw water sources and 44 were from treated water sources. Yeast was found in 12 of the raw water samples and in 3 of the treated water samples.

Conclusions

1. Giardia cysts were not recovered from any of the 144 Giardia sampling filters examined.

2. Only six, four muskrat (Ondatra zibethicus) and two racoon (Procyon lotor), of 413 animal fecal specimens

examined were found to be positive for *Giardia*.

3. Animal fecal specimens collected from the leg-hold trapped animals showed a slightly higher percent positivity for *Giardia* than specimens collected from the live animal droppings.

4. The racoon fecal specimens were found to have the most parasites of the animal types examined. The beaver specimen had the lowest percentage of total parasites found.

5. Coliform bacteria and yeast were found frequently in raw water samples and less frequently in treated water samples.

6. The bacteriological water quality data collected during the month that *Giardia* positives were found in animal fecal samples showed no relationship to the presence of *Giardia* in animal fecal samples.

7. If the areas selected are representative of Connecticut water supplies and watersheds, then *Giardia* was not prevalent in Connecticut water supplies during the period of the study.

Recommendations

Since the recovery rate for cysts is generally known to be below 10% for the EPA sampling method, it cannot be stated that *Giardia* does not exist in Connecticut water supplies. Gross contamination of the water supplies by *Giardia* appears unlikely in those areas sampled. The presence of *Giardia* in water supplies can be more accurately determined if the recovery rate for this method improves significantly.

Additional collection of animal fecal specimens using the leg-hold trapping method would be the best method for determining the presence of *Giardia* in watershed animals, and thus, the potential for contamination of water supplies.

Further testing of yeast in water supplies might be useful in determining its possible relationship to *Giardia*.

EPA Project Summary

Cross Transmission of *Giardia*

R. B. Davies, K. Kukutaki, and C. P. Hibler

R.B. Davies, K. Kukutaki, and
C.P. Hibler are with Colorado
State University, Fort Collins,
CO 80523.

T.H. Ericksen is the EPA Project
Officer (see below).

The complete report, entitled
"Cross Transmission of Giardia,"
(Order No. PB 83-117 747) will
be available only from:

National Technical Information
 Service
5285 Port Royal Road
Springfield, VA 22161
Telephone: 703-487-4650

The EPA Project Officer can be
contacted at:

Health Effects Research
 Laboratory
U.S. Environmental Protection
 Agency
Research Triangle Park, NC
 27711

Giardia cysts isolated from fecal samples obtained from humans (*Homo sapiens*), beaver (*Castor canadensis*), dogs (*Canis familiaris*), cats (*Felis domesticus*), bighorn x mouflon sheep (*Ovis canadenis x O. musimon*), guinea pig *(Cavis porcellus)*, muskrat *(Ondatra ziethica)*, and mule deer *(Odocoileus hemionus)* were given to a variety of experimental animals. Human source *Giardia* cysts established infections in dogs, cats, beaver, rats *(Rattus norvequicus)*, gerbils *(Gerbillus gerbillus)*, guinea pig, raccoon *(Procyon lotor)*, bighorn x mouflon sheep, and pronghorn antelope *(Antilocapra americana)*. *Giardia* cysts from naturally occurring beaver successfully infected dogs. A dog was infected with *Giardia* cysts from a bighorn x mouflon sheep which had been infected with human source *Giardia*. Human source *Giardia* cysts were used to infect cats and cysts from these cats were used successfully to infect dogs.

Evidence exists that once dogs are treated with metronidazole and then reexposed to *Giardia* cysts they become infected yet do not shed cysts. This most likely occurs in natural cases of giardiasis in dogs where the animal stops shedding cysts yet has a latent infection. Female dogs and cats may start shedding *Giardia* cysts 3-4 weeks after parturition.

This Project Summary was developed by EPA's Health Effects Research Laboratory, Research Triangle Park, NC, to announce key findings of the research project that is fully docu-mented in a separate report of the same title (see Project Report ordering information at back).

Introduction

Giardia (Protozoa: Hexamitidae) has been known as a parasite of humans since Leeuwenhoek found trophozoites of the protozoan in his own feces. A recent report indicates that *Giardia lamblia* is the most common parasite in stool specimens submitted for examination in the United States, with prevalences ranging from 2 to 20% with the average at 3.8%. Within Colorado, giardiasis exists in two forms: endemic and epidemic. Previous parasitological state surveys showed that *Giardia* is the most common parasite identified, with prevalence rates of 5% and 5.6%. A third survey in Colorado reported 3% of the people examined were infected with *Giardia*. Infected persons had diarrhea which lasted an average of 3.8 weeks. A correlation between seasonal distribution of cases and fecal contamination of mountain streams indicates drinking untreated water is an important cause of endemic giardiasis.

Numerous epidemics of giardiasis have occurred in Colorado. The precise source of the *Giardia* in these outbreaks is not known, but *Giardia* cysts have been recovered from samples of public water supplies in various other localities. Cross connections between water and sewage lines were determined to be the cause of one Colorado outbreak, others were associated with incompletely treated surface water. Beaver infected with *Giardia* were found below the water inlets for the water system in a Washington State outbreak.

Sylvatic giardiasis has been described in Colorado with beaver (18%), cattle (10%), domestic cat (25%), and dogs (13%), being positive for *Giardia*. Two of 34 coyotes *(Canis latoans)* from northern New Mexico were also positive for *Giardia*.

Early parasitologists, describing species of *Giardia* from various hosts, named species after the host in which they were found irrespective of morphologic similarities between *Giardia* in the different hosts. In the absence of cross-transmission experiments to determine the validity of speciations, this probably was the safest approach. However, as early as 1952 investigators could not find any morphologic differences between species of *Giardia* described from the laboratory rat and a number of wild rodents. A review of the literature determined that most experimental cross-transmission studies were questionable. This prompted the proposal of two species, *G. muris* in the mouse, rat, and hamster, and *G. duodenalis* in the rabbit, man, dog, cat, cattle, and various rodents.

Although this proposed speciation was based on morphology, the implication that other animals could serve as reservoirs for man was extremely important. Although the author of this proposed speciation did not accept the success obtained by two other investigators in infecting laboratory rats with *Giardia* from man, these early investigators were aware that the various *Giardia* might not be host-specific.

Researchers gave human-source *Giardia* to dogs and reported establishing infections with the prepatent period ranging from 3 to 40 days. However, this experiment was not well controlled. Another experiment, not adequately controlled, infected six dogs with human-source *Giardia* cysts and found the prepatent period was 6-9 days. The dogs used in this experiment were examined for *Giardia* for two weeks, without positive findings, prior to inoculation. All of these results strongly suggest that the premise of only a few species of *Giardia* was probably correct. In another cross-transmission study it was reported that *G. muris* from laboratory mice, *G. simoni* from laboratory rats and *G. peromysci* from deer mice were very host-specific, while *G. microti* and *G. mesocricetus* were not host-specific.

The present cross-transmission studies were stimulated by the increasing number of unexplained epidemics of giardiasis in humans, all apparently waterborne, but not readily traceable to human contamination of the water supplies. They pointed to another possible source of infection, a wild or domestic mammal. This, of course, necessitated more extensive cross-transmission studies involving a multitude of wild animal hosts to determine if, indeed, a wild animal species was responsible for the epidemics.

Results

Animals exposed to *Giardia* cysts from clinically-ill humans produced data which varied both within and among experimental groups. Hamsters, domestic rabbits, laboratory mice, wapiti, mule deer, white-tailed deer, black bear, domestic sheep, and domestic cattle were not infected successfully with human source *Giardia* cysts. Animals which did become infected were laboratory rats, gerbils, guinea pigs, beaver, dog, raccoon, bighorn x mouflon sheep and pronghorn antelope. Cysts from all animals, with the exception of some of those from rats, ranged in size from 9.5 to 11.0 μm x 8.0 to 9.5 μm. After intubated animals became patent, the in-group control often started shedding cysts 8-20 days after exposed animals became patent. This indicated transmission of *Giardia* from the exposed animals to the in-group controls. None of the control groups, held in the same facilities as the exposed animals and the in-group control, became positive for *Giardia*.

Giardia cysts were recovered from the only composite fecal sample from rats on days 22, 25, and 40 postexposure (PE). Cysts were of two sizes, 5 μm long and 10 μm long, but were identical in all other respects. All other experiments were performed with animals in individual cages. In another experiment using rats, cysts were shed for one day at 34 PE. Infected feces from this group were fed to an SPF dog which began shedding *Giardia* cysts eight days PE. Control dogs remained negative.

Gerbils exposed to human source cysts began shedding cysts 8, 13, and 18 days PE. In one test, the in-group control was positive 33 days PE. The cysts were shed in varying numbers and not consistently in all samples. Some exposed animals remained negative until the experiment was terminated in 42 days PE. One exposed guinea pig became positive for *Giardia* two days PE

and continued shedding for 31 days. All other animals remained negative.

Beaver exposed to human *Giardia* cysts from one human source were negative for 40 days preexposure and remained negative for 40 days PE. The control and one exposed beaver were inoculated with human *Giardia* from another source. These beaver started shedding *Giardia* cysts 25 days PE and continued shedding for 22 days, after which they shed cysts intermittently. The beaver used as a control for the second exposure remained negative. The *Giardia* from the first human source apparently were not infective, whereas those from the second source were infective.

SPF beagle puppies exposed to human source *Giardia* cysts began shedding cysts six to eight days PE and the in-group controls began shedding cysts 13 to 15 days after the exposed dogs were inoculated.

A young raccoon in a group of ten shed cysts for one day at eight days PE. A five-month old black bear cub remained negative for cyst shedding.

Bighorn x mouflon sheep exposed to human *Giardia* began shedding cysts nine days PE and shed cysts for four days. Sheep isolated cysts were inoculated to SPF beagles. A pup inoculated with cysts from one sheep started shedding cysts ten days PE, whereas a pup inoculated with cysts from another sheep remained negative for cyst shedding.

A young pronghorn antelope exposed to human *Giardia* cysts started shedding cysts 16-18 days PE and shed cysts for three days.

Wapiti, mule deer, white-tailed deer, and domestic sheep and cattle did not become infected with human *Giardia* cysts. However, naturally-infected mule deer and cattle have been reported in the literature. Infection of muskrats was confounded since all animals including the control were found to be positive 18 hours after being exposed to human source *Giardia* cysts.

Puppies and kittens exposed to human source *Giardia* cysts exhibited prepatent periods of 6-10 days and 6-27 days, respectively. Cysts isolated from one of the exposed kittens were inoculated into puppies and kittens. Most of these inoculated puppies and kittens began shedding cysts. All control animals remained negative throughout these experiments.

Giardia cysts recovered from the feces of naturally-infected, free-ranging beaver were given to mice, rats, hamsters, guinea pigs, and puppies. Only the exposed puppies began shedding cysts eight days PE and the in-group control puppy began shedding cysts 17 days PE of the inoculated animals.

All dogs, including the in-group controls in one group, began shedding cysts at four days PE when exposed to *Giardia* isolated from muskrats. It is assumed that all of these dogs were exposed to *Giardia* before the experiment was begun. A second group of dogs exposed to *Giardia* cysts from muskrats remained negative for 63 days.

Giardia cysts from a naturally-infected mule deer were given to SPF beagle puppies which remained negative for 28 days PE.

Reinfection with human *Giardia* cysts was attempted on positive dogs treated with metronidazole. Following six days of treatment, seven of the puppies were negative for *Giardia* shedding. Seven days after treatment five of the six treated puppies were exposed to human *Giardia* cysts. None of the puppies shed cysts for 40 days PE. The untreated puppy continued to shed; however, cysts were not observed in feces from the puppy which was negative when all of the dogs originally were obtained.

After 40 days PE, intestinal scrapings indicated *Giardia* trophozoites in three of the five exposed dogs. The dog that was treated but not exposed to *Giardia* did not exhibit trophozoites in intestinal scrapings.

Giardia-free dogs and cats were difficult to obtain. Therefore, pregnant dogs and cats were obtained and held in clean rooms until they gave birth and the offspring were weaned. The female dogs began shedding *Giardia* cysts two to four weeks after parturition and all of the offspring were positive seven days after the female started shedding cysts. These female dogs were examined daily and criteria for selection included being *Giardia*-free for seven days prior to acceptance. None of these dogs shed cysts until two or four weeks after parturition. To solve this problem the adult animals were treated with metronidazole for five days and in this manner *Giardia*-free offspring were produced and the adults did not shed cysts after parturition.

Conclusions

1. *Giardia* cysts obtained from human, dog, cat, and beaver sources are not host specific.
2. *Giardia* from human sources will readily infect dogs, cats, and beaver.
3. *Giardia* from beaver sources will readily infect dogs.
4. *Giardia* from dog or cat sources cross-transmit between these species.
5. Therefore, it must be assumed that *Giardia* from dog, cat, and beaver sources will infect humans.

Recommendations

The results of this cross-transmission study showed that the *Giardia* found in humans, dogs, cats, and beaver sources are not host-specific and will readily establish in other animal species. All three species are important as potential sources of *Giardia* for epidemics of waterborne giardiasis, but the beaver probably plays the most important role because of its closer association with water used by communities as their source of domestic supply. Therefore, a study of the host-parasite relationship between beaver and *Giardia*, together with a study of the factors predisposing toward an epidemic of waterborne giardiasis (water pH, hardness, temperature, etc.) is the next logical step necessary to understand waterbrone giardiasis.

Treatment Techniques
for Giardia Removal

REMOVAL OF *GIARDIA LAMBLIA* CYSTS
BY FLOCCULATION AND FILTRATION

Jogier Engeset
Predoctoral Research Associate
Department of Civil Engineering
University of Washington

Foppe B. DeWalle
Assistant Professor
Department of Environmental Health
University of Washington

Giardia lamblia is an intestinal protozoan parasite found in humans and certain animals. It has a two stage life cycle, the pathogenic trophozoite and the dormant cyst stage. The cyst which is the only form found in the environment, ranges in size from 7 μm to 12 μm and can survive for many months in a water distribution system.[1]

The intestinal disease caused by protozoa is called giardiasis. Symptoms of the disease will appear from 2 to 35 days after exposure to the cyst.[2-5]

They include watery, often foul smelling diarrhea, abdominal cramps, marked flatulence, nausea and anorexia. The acute stage of infection may last from a few days to three months.

Waterborne Outbreaks of Giardiasis

In recent years, waterborne outbreaks of giardiasis in the U.S. have become more frequent. Most of the cases reported can be traced back to drinking of water from municipal supplies where the only treatment was chlorination. This indicates that chlorination at the conventional dosages and contact times normally employed in water treatment is inadequate for destruction of *Giardia* cysts.

Some of the recent outbreaks of giardiasis are listed in Table 1. Only two of these outbreaks involved filtered water supplies. They were Camas, WA and Berlin, NH. In both cases ineffective chemical pretreatment or faulty construction of the filtration plant was felt to be the major factor contributing to cysts passing through the treatment plant.

Recovery and Enumeration of Cysts in Water

A study currently underway at the University of Washington in Seattle, WA, is looking at the effectiveness of the conventional water treatment unit processes for removal of *Giardia* cysts, in particular the filtration stage. In order to make an evaluation of the various water treatment processes, it is imperative to have a reliable method for recovering and enumerating cysts from water.

The recovery method which was recently developed in one laboratory utilizes both membrane filtration and centrifugation to concentrate the cyst particles (Figure 1). The 20 l water sample is forced through a 5 μm pore size Nuclepore filter at 10 psi positive pressure using nitrogen gas. The filter is then removed from its holder and the cysts are washed off in 350 ml distilled water using a platform shaker.

Following a final rinse of the filter, the approximately 400 ml wash water is centrifuged at 1500 rpm for 10 min. The supernatant is discarded and the sediment is recentrifuged and concentrated to a final volume of 10 ml. The cysts in the concentrate are enumerated using a Coulter Counter particle counter. In addition, microscopic examination of the concentrated sample is performed for cyst particle verification.

Another method for enumerating cysts is to use a hemacytometer. This microscope counting method is more time consuming than the Coulter technique and the results are less reproducible. This is particularly noticeable at low cyst concentrations. The linearity of the two counting methods is compared in Figure 2.

An example of a recovery test using water from Lake Union in Seattle, is shown in Figure 3. The water sample was spiked with *Giardia* cysts and subsequently concentrated as previously described. As the Coulter Counter plot clearly shows, only a very low number of particles in the size range of *Giardia* cysts is present. Further, recovery tests using this scheme have shown that an estimated 75-80% recovery can be expected. This recovery is similar at several initial cyst concentrations.

Zeta Potential

When determining which coagulant will be most effective for removal of *Giardia lamblia* cysts from water, it is advantageous to know the cysts zeta potential or electric charge. Most natural colloidal particles and microorganisms, including viruses, are negatively charged. Arozarena,[13] working with live *Giardia muris* cysts, found that 93% of the cyst particles had a zeta potential between -30 and -20 mv at pH 5.5. The mean value was -28 mv.

The zeta potential of *Giardia lamblia* cysts was determined using a Zeta-Meter. The cysts were fixed with a 5% buffered Formaline Solution and Lugol's Iodine was added to all suspensions in order to facilitate tracking the cysts under the microscope. At pH 5.5, mean zeta potential value for different cyst suspensions was -25 mv. As pH increased, the zeta potential increased proportionally (Figure 4).

The negative zeta potential of *Giardia lamblia* cysts indicates that the cysts should be removed from water by the addition of coagulants like alum, cationic polymers and nonionic polymers with subsequent flocculation, sedimentation and filtration or flocculation and filtration. The coagulants may be added individually or in combination.

Coagulation and Flocculation

The data available on coagulation and flocculation of *Giardia* cysts are very limited. Working with low turbidity water at pH 6.5, Arozarena[13] found, through jar tests, the optimum alum dosage for turbidity removal to be 25 mg/l for the coagulation, flocculation and sedimentation sequence. When cysts were added to the water, no cysts were detected at alum dosages of 10 mg/l or more, representing a removal efficiency of greater than 99%. At 5 mg/l alum only a 95.7% removal of cysts was achieved.

Nonionic polymers used as primary coagulant give floc with poor settling characteristics and hence low turbidity and cyst removal. This has also been the experience for some cationic polymers, used alone as primary coagulants and at relatively high concentrations. The reason could be that low turbidity waters lack the concentration of particles needed for good floc formation. However, when used in combination with alum and at relatively low dosages, polymers seem to improve the cyst removal efficiency during coagulation, flocculation and sedimentation.

Filtration

Water filtration theory indicates that organisms the size of *Giardia lamblia* cysts should be removed by conventional sand filters provided that proper attention is given to the factors important to particle transport within the filter and attachment of these particles to the filter medium.

Particle transport may include gravitational settling, diffusion, interception and hydrodynamics which are affected by such physical factors as size of the filter medium, filtration rate, fluid temperature and the density and size of the suspended particles. The attachment mechanism may involve electrostatic interaction, chemical bridging or specific adsorption, all of which are affected by the coagulants applied in the pretreatment and the chemical characteristics of the water and the filter medium.

Meeting the turbidity standard for filtered water, less than 1.0 NTU, is no guarantee of low cyst counts in the filter effluent. For example half of the number of cysts present in the filter effluent were found in the effluent when using an inadequate coagulant dosage. Increasing the chemical dosage beyond that necessary for turbidity removal can drastically increase the filter's cyst removal efficiency. As indicated by Figure 5,[14] filtered water turbidity should preferably be kept at 0.3 NTU or less and the filter needs to be backwashed after turbidity increases of 0.1 or 0.2 NTU above the normal value in a filter run. It also shows that even though the cysts were only spiked in the beginning of the run, they are present in the effluent of the filter when the trapped flocs become dislodged during turbidity breakthrough. To achieve low turbidity and good cyst removal, the single most important factor is chemical pretreatment. This is illustrated in Table 2 which summarizes the effect that different coagulant dosages and flow rates have on the filtered water quality.[14]

Summary

The recent outbreaks of giardiasis in the U.S. have demonstrated a need for better understanding of the various factors affecting the removal of *Giardia lamblia* cysts from water. The presently ongoing study is evaluating the removal efficiency during the coagulation-filtration process.

REFERENCES

1. RENDTORFF, R.C. & HOLT, J.C. The Experimental Transmission of Human Intestinal Protozoan Parasites. IV. Attempts to Transmit *Entamoeba coli* and *Giardia Lamblia* Cysts by Water. Amer. Jour. Hyg., 60:3:327 (Nov. 1954).

2. VEAZŪ, L. Epidemic Giardiasis. New Engl. Jour. Med., 281:41:853 (Oct. 1969).

3. WALZER, P.D.; WOLFE, M.S.; & SCHULTZ, M.G. Giardiasis in Travelers. Jour. Infect. Dis., 124:3:235 (Aug. 1971).

4. FORSELL, J.; LANTORP, K.; & STERNER, G. Giardia among Swedish Tourists after a Visit to the Soviet Union. LaCartidringen, 69:1132 (1972).

5. ANDERSSON, T.; FORSAELL, J.; & STERNER, G. Outbreak of Giardiasis: Effect of a new Antiflagulate Drug, Tinidazole. Br. Med. Jour.,2: 21:449 (May 1972).

6. MOORE, G.T.; CROSS, W.M.; MCGUIRE, D.; MOLLAHAN, C.S.; GLEASON, N.N.; HEALY, G.R.; & NEWTON, L.H. Epidemic Giardiasis at a Ski Resort. New Engl. Jour. Med., 281:34:402 (Aug. 1969).

7. SCHULTZ, M.G. Giardiasis. Jour. Amer. Med. Assoc., 233:B:39:1383 (Sep. 1975).

8. BARBOUR, A.G.; NICHOLS, C.R.; & FUKUSHIMA, T. An Outbreak of Giardiasis in a Group of Campers. Amer. Jour. Trop. Med. Hyg., 25:3:384 (May 1976).

9. SHAW, P.K.; BRODSKY, R.E.; LYMAN, D.O.; WOOD, B.T.; HIBLER, C.P.; HEALY, G.R.; MACLEOD, K.I.E.; STAHL, W.; & SCHULTZ, M.G. A community Wide Outbreak of Giardiasis with Evidence of Transmission by a Municipal Water Supply. Ann. Int. Med., 87:4:426 (Apr. 1977).

10. KIRNER, J.C.; LITTLER, J.D.; & ANGELO, L.A. Outbreak of Giardiasis in Camas, WA. Jour. AWWA, 70:1:35 (Jan. 1978).

11. LIPPY, E.C. Tracing a Giardiasis Outbreak at Berlin, New Hampshire. Jour. AWWA, 70:9:512 (Sep. 1978).

12. GEITZAN, T.; HAYNER, N.S.; LANDIS, P.; VERNON, T.M.; & LYMAN, D.O. Giardiasis-Vail, Colorado. Morbid. Mortal Wkly. Rep.,27:19:155 (May 1978).

13. AROZARENA, M.M. Removal of *Giardia muris* Cysts by Granular Media Filtration. M.S. Thesis, Dept. of Environmental Engineering, Univ. of Cincinnati (1979).

14. LOGSDON, G.S. Municipal Environmental Research Laboratory, Water Supply Research Division, U.S. Environmental Protection Agency, Cincinnati, unpublished data (1979).

Reprinted from Proc. AWWA ACE, San Francisco, Calif. (June 1979).

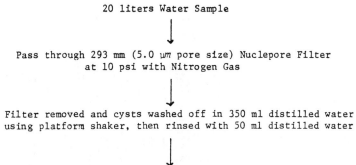

20 liters Water Sample

Pass through 293 mm (5.0 μm pore size) Nuclepore Filter
at 10 psi with Nitrogen Gas

Filter removed and cysts washed off in 350 ml distilled water
using platform shaker, then rinsed with 50 ml distilled water

Centrifuge wash water at 1500 rpm for 10 min. in
50 ml conical tubes

Pipett off top 40 ml from each centrifuge tube
Transfer bottom 10 ml to 2-50 ml conical tubes and recentrifuge

Retain bottom 5 ml from each tube count on
Coulter Counter and verify the particles counted
by microscope examination

FIGURE 1.   Method for Recovery of _Giardia lamblia_ Cysts with
the 293 mm Membrane Filter.

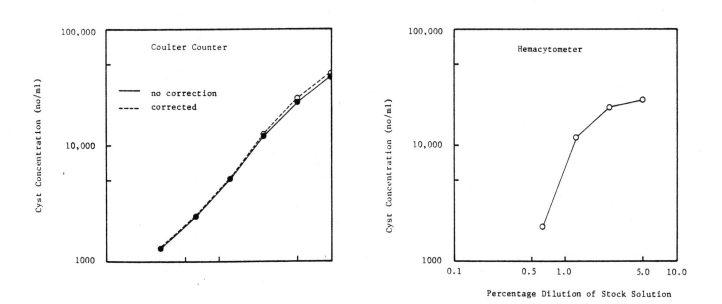

FIGURE 2.   Linearity of Two Counting Methods for Enumeration of
Giardia Cyst in Distilled Water.

113

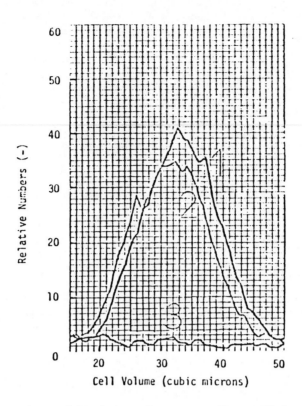

FIGURE 3.   Results of Membrane Recovery Tests Using Lake Union
Water Spiked With .5% Stock Solution of Giardia Cysts
Before (1) and After (2) the Experiment as Compared to
Unspiked Lake Union Water (3).

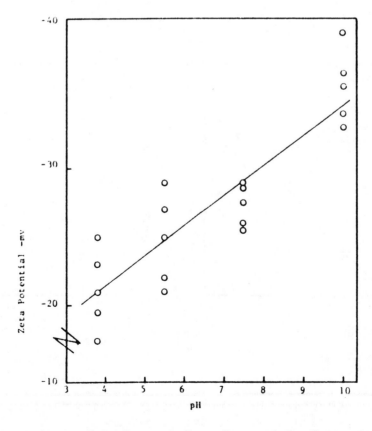

FIGURE 4.   Relationship Between Zeta Potential of Giardia lamblia
Cyst suspensions and pH.

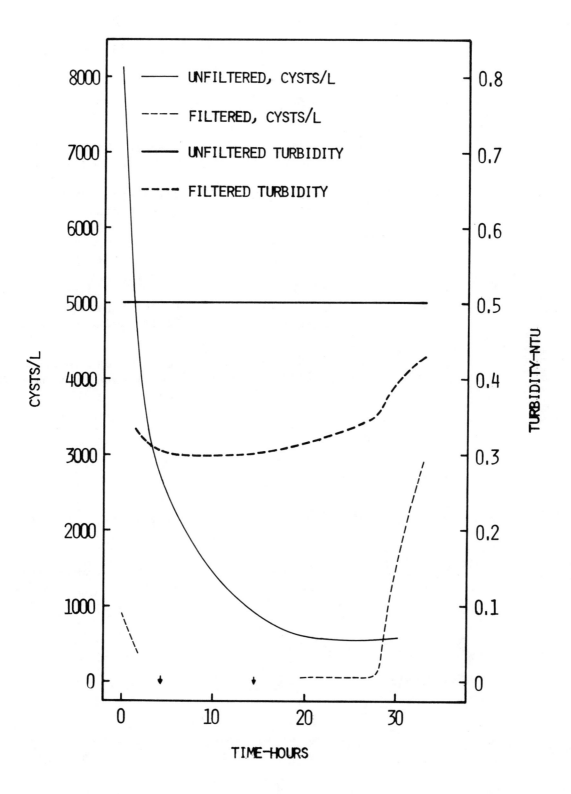

FIGURE 5. Effects of Turbidity Breakthrough, Flowrate 4 GPM/FT^2 with Alum Dosage 7 mg/L and pH 6.5[14]

TABLE 1. Recent Outbreaks of Giardiasis in the United States

Place	Dates	Persons Affected	Attributable Cause
Aspen, CO[6]	1965-66	123/1,094 residents	Sewage leak into wells
Boulder, CO[7]	1972	300 residents	Unknown
Vinta Mts.,UT[8]	9/74	34/54 campers	Suspect stream contamination by beaver and sheep
Rome, NY[9]	11/74-6/75	5300/50,000 residents	Human habitation in watershed contaminated water supply
Camas, WA[10]		600/6,000 residents	Infected beaver in watershed, loss of filter media, faulty alum feeding
Berlin, NH[11]	1/77-5/77	275/5,500 residents	Infected beavers in watershed, faulty joints in filter channel
Vail, CO[12]	3/78-4/78	38 residents	Suspect leak in sewer line contaminated surface water supply

TABLE 2. Effects of Chemical Pretreatment on the Filter Effluent[14]

FLOW RATE GPM/FT^2	COAGULANT DOSAGE	FILTERED TURBIDITY	CYST REMOVAL	INCREASE IN EFFLUENT CYST CONCENTRATION
4	ADEQUATE	≤ 0.3	$\geq 99\%$	
4	INADEQUATE	≥ 0.5	$\leq 50\%$	
4	NO	0.35	POOR	
4	STOP FEED	RAPID INCREASE	Poor	10 - 100 FOLD
4 → 8	STRONG FLOC	CONSTANT 0.3	VERY GOOD	
4 → 10	WEAK FLOC	RAPID INCREASE	Poor	10 - 100 FOLD

Alternative filtration methods for removal of *Giardia* cysts and cyst models

Gary S. Logsdon, James M. Symons, Robert L. Hoye Jr., and Michael M. Arozarena

The recent occurrence of waterborne giardiasis outbreaks has spurred renewed interest in water filtration for cyst removal. Studies with cyst models and *Giardia muris* cysts suggest that properly operated water filtration plants should be able to remove a high percentage of *G. lamblia* cysts. For effective cyst removal, diatomaceous earth (DE) filters need 1.0 kg/m² diatomite precoat plus body feed, and granular media filters require coagulation of raw water and production of filtered turbidity well below 1.0 ntu throughout the entire filter run.

Outbreaks of waterborne giardiasis, a disease caused by a protozoan parasite, have been reported more frequently in recent years. This trend is a cause for concern among public health, environmental, and water utility officials. Because of the need for information on the efficacy of water treatment for *Giardia* cyst removal, a research program on this topic was undertaken by the US Environmental Protection Agency's drinking water research division. Objectives of the study included learning whether or not a relationship existed between cyst removal or cyst concentration in filtered water and the turbidity of filtered water, identifying treatment procedures and operational problems that might allow cysts to pass through filters, and discovering the cyst removal capabilities of well-operated filters.

Literature review

A review of literature revealed no research on water filtration for *Giardia* cyst removal. This is not surprising because until recently the *Giardia* organism was not implicated as a cause of waterborne disease. The hazards of waterborne amoebiasis, however, have been known for decades. Water filtration studies for removal of *Entamoeba histolytica* cysts were carried out in the 1930s and 1940s.

Granular media filtration research. After the Chicago amoebiasis outbreak occurred in 1933, experiments for removal of *E. histolytica* cysts from water were carried out at the experimental filtration plant operated by the Chicago Department of Public Works. In preliminary laboratory experiments, 4-cm diameter filter columns were used. Tests with these small columns showed complete removal of *E. histolytica* cysts from water by coagulation and rapid sand filtration.[1]

Further testing with a pilot plant filter having a surface area of 0.9 m² took place in 1935 and 1936.[2] Cysts were added to previously coagulated water that was then treated by direct filtration at a rate of 5 m/h. Samples were taken at regular intervals during the filter runs, which ranged in duration from 4 to 14 hours. The samples were concentrated in a continuous centrifuge, and the sediment was examined in a counting cell commonly used for algae counting.

On the basis of the total volume of water filtered and the total number of cysts applied during the filtration experiments, removal efficiency exceeded 99.99 percent. Baylis et al[2] concluded that filtration of water through rapid sand filters, in the manner then extensively used in the United States (5 m/h [2 gpm/sq ft]), was effective in removing *E. histolytica* cysts from water.

During World War II the need to provide water free of *E. histolytica* cysts resulted in a cooperative filtration research program involving the Engineer Board of the US Army at Ft. Belvoir, Va., and the National Institute of Health, US Public Health Service. Although a report of this work was prepared by the War Department,[3] it was not widely circulated. This paper presents a summary of that report, subsequently referred to as the Army report.

Granular media filtration tests were performed with a 43-cm inside diameter pressure filter vessel that held sand filter media 46-cm deep, with an effective size of 0.36 mm and a uniformity coefficient of 1.35. Media support was provided by 10 cm of fine gravel.

Filtration tests were performed with and without coagulation, and with and without sedimentation. Results of these tests, summarized in Tables 1 and 2, show the influence of good operating practice on cyst removal by granular media filters. Tests 1, 6A, and 6B involved filtration of uncoagulated water; test 2 was performed with poorly coagulated water. For these tests, the numbers of cysts passing the filter per

TABLE 1

Army portable water purification unit (sand filter).
Summary of the results of E. histolytica experiments under several operating conditions[3]

Experiment Number	Quality of Coagulation	Filtration Rate m/h	Calculated Number of Cysts per Litre		Number of Cysts Passing Filter per Million Applied	Reduction percent
			Influent	Effluent		
1	none	23	550	66	119 000	88.1
2	poor	23	1700	63	38 000	96.2
3	good	23	610	2.4	3900	99.6
4	good	15	1300	1.2	900	99.9
5	good	15	1800	2.6	1400	99.86
6B	none	15	220	44	197 500	80.3

TABLE 2

Army portable water purification unit (sand filter).
Summary of the results of E. histolytica experiments with operation at reduced rate following coagulation and sedimentation[3]

Experiment Number	Pretreatment	Settling Time min.	Filtration Rate m/h	Calculated Number of Cysts per Litre		Number of Cysts Passing Filter per Million Applied	Reduction percent
				Influent	Effluent		
6A	none	90		2600	220	83 800	91.62
6B			15	220	44	197 500	80.25
Combined results						16 500	98.35
7A	alum and soda ash	120		1000	1.4	1400	99.86
7B			15	1.4	0.2	176 000	82.4
Combined results						250	99.975
8A	alum and soda ash	75		1000	15	14 700	98.53
8B			15	15	0.5	35 700	96.43
Combined results						526	99.947

TABLE 3

Diatomaceous earth filters: Summary of operating conditions and E. histolytica cyst data[3]

Test	Precoat kg/m²	Body Feed mg/L	Flow Rate Range		Volume Filtered L	Volume Analyzed L	Cyst Data		
			L/s	m/h			Dosed/L	Total Recovered	Recovered/Applied
1D	0.60	0	0.0063–0.0013	1.9–0.4	4	4	87 500	0	0/350 000
2D	0.50	0	0.088–0.0038		57	23	1565	3	3/36 000
4D	0.50	0				42	2620	1	1/110 000
5D	0.75	0		5,10		38	1580	0	0/60 000
6D	1.8				5.1	5.1	2640–3170	0	0/15 000
7D	0.85	170	1.4–0.8	6.2–3.4	2635	64	1195	7	7/76 500
8D	0.75	120	1.5–1.1	8.8–6.1	1836	23	940	1	1/21 600
9D	0.75	120	1.8–1.5	10.1–8.6	2453	23	505	0	0/11 610
10D	0.75	0	1.5–0.9	8.6–5.0				1	
11D	0.75	0	1.1–0.8	6.5–4.3				0	
12D	0.75	0	1.6–0.8	9.1–4.8				1	
13D	0.75	0	1.8–1.1	10.6–6.5				0	

NOTE: Authors of the Army report did not publish complete data on tests 10D to 13D, but they did make the following comments:
"j. Experiments 10D through 13D inclusive; US Army portable water purification unit, model 1940, converted for use with diatomaceous silica...

The purpose of this series of tests was to determine the effectiveness of various grades of diatomaceous silica in the removal of cysts from water. Attention is directed to the absence of cysts in quantity in the effluent from all of the materials tested. Since cysts were removed by all of the materials tested, the significant feature of these tests is that no cysts passed the unit in experiment 13D. The filter aid used in the experiment was Johns-Manville Celite 545, which is too coarse for water filtration. Notwithstanding the passage of turbidity in excess of 10 ppm no cysts were found in the effluent samples... It appears, therefore, that any currently manufactured filter-aid in the range of the materials tested will produce satisfactory results."

million applied were much higher than for other tests with good coagulation and comparable experimental conditions. Filtration rates were 15 or 23 m/h for the Army tests.

Diatomite filtration. In the United States the Army took the lead in developing diatomaceous earth (DE) filtration for drinking water with a research and development program during World War II.[3] A discussion of DE filtration principles and a summary of bacteriological data developed in the project were also published by Black and Spaulding,[4] but the paper merely mentioned a need to remove cysts of E. histolytica, without giving actual test results. Experimental data are given in the Army report.

Experiments on diatomaceous earth filtration for cyst removal were carried out with a variety of DE filter systems, ranging from devices providing batch filtration of as little as 4 L of water to equipment capable of filtering approximately 75 L/min. Table 3 summarizes operating conditions and E. histolytica data for DE filters.

Newly calculated data include all data for the ratios of cysts recovered from filtered effluent to cysts in the same volume of filter influent, as well as some data on filter aid usage and flow rates. Some data were not available in the Army report so no values are given in certain instances in Table 3.

Data in Table 3 support the Army's conclusion that DE filtration could remove essentially all cysts. The largest number of cysts recovered, seven, resulted from using contaminated water (225 000 cysts in 130 L) to precoat the filter. Clean, filtered water should always be used for filter precoating.

Giardia-related research methods

Past research on cyst removal was related to treatment for E. histolytica cysts. Research results demonstrated that these cysts could be removed and suggested that filtration should also be effective for reducing the concentration of Giardia cysts in water. To demonstrate this specifically, a program of filtration research with Giardia cysts and cyst models was conducted at EPA's environmental research center. Both granular media and DE filters were studied. Certain aspects of the work, such as preparation of cyst suspensions and microscopic examination of samples, were common to both kinds of filtration research. They are presented first.

Use of cyst models. When the DE filtration research was begun in 1977, the difficulty of detecting and counting Giardia cysts in water necessitated a model for the cysts. Radioactive, 9-μm diameter microspheres* were used because they were very similar in size to Giardia cysts and were easy to trace. The resin bead microspheres were coated with cerium-141 to prevent the isotope from leaching. The radioactivity of a single microsphere could be differentiated from background radiation in the low-level scintillation counter that was used to measure gamma radioactivity on the 5-μm pore diameter membrane sampling filters.

The first phase of diatomaceous earth filtration research was performed with the microspheres. Because DE filters remove particles in the Giardia size range by a straining process, the radioactive particles were useful models for the cysts. The ease and sensitivity with which the microspheres were detected facilitated performing a large number of tests related to the effects of various aspects of filter operation on passage of

---

*NEN-TRAC, New England Nuclear, Boston, Mass.

cyst-sized particles through the DE filter. Subsequent studies with *G. muris* cysts confirmed the efficacy of DE filtration for cyst removal.

Use of *Giardia* cysts. Because it was difficult to obtain a constant source of viable *G. lamblia* cysts for research, mice were infected with *Giardia muris* and used as a source of cysts. Nearly all of the filtration research was done with the radioactive beads or with *G. muris* cysts, rather than with *G. lamblia* cysts.

G. muris should be a suitable surrogate for *G. lamblia* in filtration studies because both species are similar in size, shape, and surface electrical charge. EPA analysis* showed that *G. muris* cysts in distilled water had a zeta potential of −26 mV at pH 5.5. This figure agrees well with research conducted at the University of Washington,[5] where different equipment† gave a *G. lamblia* zeta potential of about −25 mV at pH 5.5. The *G. lamblia* cysts showed increased electronegativity at pH 8. Some physical characteristics of *E. histolytica* cysts and *Giardia* cysts and beads are given in Table 4.

Cyst preparation and counting. Certain laboratory techniques for handling the cysts, such as removal and concentration from fecal samples and cyst counting, were common to all of the research. Others, such as dosing the raw water and separating the cysts from filtered water, varied from time to time.

Cysts were obtained from mouse feces, and the suspensions were cleaned by a sucrose discontinuous gradient technique.[8,9] The washed cyst suspensions were refrigerated in "normal" saline solution until used.

Cyst enumeration by the hemocytometer method. The sample in the hemocytometer was examined under a light microscope, and cysts were identified on the basis of their morphology. Because examination of even 1 or 2 mL of water in the hemocytometer would be very time-consuming, only a fraction of the filtered water could be examined. As a result, a variable lower limit of detection always existed, and the finding of no cysts in the hemocytometer never could be interpreted as representing a cyst concentration of 0 cysts/L.

Cyst recovery from samples. Cysts were separated from the filtered water by use of 5-μm pore size polycarbonate filters. Effluents from the DE filtration research tests were passed through a 142-mm diameter membrane. After termination of the filter run the cysts were removed from the membrane by repeated washings with 0.85 percent saline. The washings were concentrated by centrifugation, the supernatant liquor was aspirated off, the concentrate was vortex-mixed, and a small, measured

*Laser zee meter, PenKem, Inc., Croton-on-Hudson, N.Y.
†Zetameter, Zeta Meter, Inc., New York, N.Y.

amount was counted in the hemocytometer.

In granular media filtration tests, water samples were periodically collected in 1-L cubitainers for refrigeration until they could be filtered through 47-mm diameter polycarbonate filters having 5-μm diameter pores. The polycarbonate filters were repeatedly washed with 0.85 percent saline, and the analysis continued as described above.

In order to evaluate the loss of cysts to DE filter apparatus and the failure of cysts to be recovered from the 142-mm membrane filter, tests were performed in which 7–21 × 10[6] cysts were instantaneously dosed to the filter when it was being operated with an uncoated septum. When the filter was operated at 3.8 L/min, four recovery tests averaged 10.3 percent recovery with a standard deviation of 2.7 percent. Therefore, during DE filtration experiments at 3.8 L/min the number of cysts recovered was corrected by dividing by 0.10. Later, two tests at 5.3 and 6.6 L/min showed 40 and 45 percent recovery, considerably higher than 10 percent. The higher velocities caused by greater flow rates may have reduced the tendency of the cysts to adhere to the DE filter apparatus. Nevertheless, to provide a conservative interpretation of the results of the higher rate tests, cysts recovered were divided by 0.10 rather than by 0.40 to 0.45 to estimate cyst removal percentages. This would tend to underestimate the percentage somewhat and yield conservative recovery rates.

The recovery efficiency of the 47-mm diameter polycarbonate filters used in the granular media study was determined by measuring cyst recovery from six separate 1000-mL samples, each of which had been dosed with about 3.5 × 10[6] cysts. Cyst recovery ranged from 55 to 65 percent for an average of 61 percent. For granular media filtration research, recovered cysts were divided by 0.61 to account for the efficiency of cyst recovery.

The water used for this research was obtained from an inactive gravel pit. This surface water was selected because of its clarity (less than 5 ntu) year-round. With respect to turbidity, this water resembles those where giardiasis outbreaks have occurred.

Diatomaceous earth filtration studies

The drinking water research division's pilot plant facilities included a 0.1-m² DE filter,[10] which was used in treatment studies with *Giardia* cysts and radioactive bead cyst models. The test results, presented more fully elsewhere,[11] are briefly summarized here.

Experimental methods. The DE test filter was set up for operation as a pressure filter with a slurry feeder for body feed. The 0.1-m² filter was operated at 2.4 or

3.5-m/h rates. The filter system is shown in Figure 1. Operation of the DE filter consisted of three steps—precoating, filtration, and backwashing.

A diatomaceous earth slurry was prepared by adding the desired weight of DE to tap water in the precoat tank and mixing the contents of the tank during precoating. The weight of diatomite added to the precoat tank was used to calculate precoat thickness expressed as kg/m². The slurry was recirculated until the water in the precoat tank appeared clear and free of diatomite.

During the filtration process a small amount of diatomaceous earth body feed was continuously added to the raw water in order to maintain good hydraulic characteristics of the precoat filter cake. Without the body feed, clay, bacteria, and other particulates found in natural waters and removed by the filter would soon coat the surface of the filter cake. The pores would become plugged, and a pressure drop would occur in the filter.

When body feed is added, the depth of the filter cake increases during the run and the major constituent of the filter cake is DE rather than removed particulates. The result is a porous filter media with good flow characteristics. In order to properly adjust body feed dose, some DE filter plant operators add body feed in proportion to raw water particulates. Turbidity is used as an indicator of the particulate content of water, so a body feed to turbidity (mg/L to ntu) ratio is sometimes used as a guideline for the needed dose of body feed. An economical body feed to turbidity ratio would vary for different raw waters. Body feed to turbidity ratios in this work generally ranged from 2:1 to 35:1.

A diatomaceous earth filter run generally is terminated because of head loss considerations. Because the filter cake increases in thickness during the run, filtered water turbidity should not increase rapidly (turbidity breakthrough) as with granular media filters. Turbidity breakthrough can occur, however, if the flow of water through the filter is interrupted, because interrupting the flow can cause the diatomaceous earth cake to fall off the filter septum. This should not be allowed to occur.

When a DE filter run is completed, the cake is removed from the septum, spent diatomite is discharged, the septum is cleaned, and the filter is made ready for precoating again.

An important feature of the filter system was the membrane sampling filter located downstream from the DE filter. The purpose of this membrane filter was to sample the entire effluent from the test filter and thus attain more reliable data on removal of cysts and cyst models.

When radioactive microspheres were used, the total number of microspheres passing the DE filter in a run could be determined directly by counting the membrane filter in a low-level gamma counter. The ability to identify a single bead in the total effluent for a filter run meant that the passage of one 9-μm radioactive bead could be detected. The microspheres were dosed on a continuous basis throughout the runs. Filter runs with microspheres ranged in duration from 2 to 27 hours with a typical run time of 6 to 8 hours.

When *G. muris* cysts were used in DE filtration tests, they were dosed in one massive slug, generally of 20–40 × 10⁶ cysts, just before the raw water entered the pressure filter vessel. Filter operation continued long enough for at least 99 percent of the cysts to pass through the filter, as calculated by a first order differential equation for complete mixing within the filter vessel. Usually this operation ran 20 or 30 min. After the filter run, the 142-mm diameter membrane filter was carefully removed from its holder. The filter was washed and cysts were concentrated and counted as described in the section on cyst enumeration.

Results and discussion. Results of tests with radioactive beads (Figure 2) indicated that in most cases reduction of microsphere concentrations exceeded 99.9 percent. Filter performance was not related to turbidity reduction, a finding very similar to that in the Army report.[3] A probable explanation for this is that the cysts were removed even though clays and other particulates that caused turbidity in the gravel pit water were small enough to pass through the filter. The smallest cyst dimension would be about 7 μm, whereas clays and bacteria would be as small as 1 μm and thus more likely to pass through the pores of the filter cake. Therefore, performance is related to good operating procedures and not necessarily to turbidity removal.

Data obtained from the filter runs in which microspheres were used are plotted to show the differences in filter performance that are related to filter operating procedures. Specifically, reduction of microsphere concentration met or exceeded 99.9 percent in 21 tests when the precoat applied was at least 1.0 kg of diatomite per m² of filter surface and when body feed was utilized. Reduction of the microsphere concentration was less than 99.9 percent in only two tests with these operating conditions.

Filter operation with a precoat thickness of less than 1.0 kg/m² (two runs), with no body feed (three runs), or with less than 1.0 kg/m² of precoat as well as no body feed (three runs) resulted in removal of less than 99.9 percent of the microspheres. The test results show the importance of properly establishing the filter cake by using an adequate amount of precoat diatomite and of maintaining the integrity of the cake during a filter run by applying body feed.

TABLE 4
Physical characteristics of beads and cysts

Parameter	E. histolytica[7]	G. lamblia[6]	G. muris[6]	Microsphere
Shape	usually spherical	ovoid to ellipsoidal	ovoid to ellipsoidal	spherical
Dimensions	10–20 μm	8–12 × 7–10 μm	7–13 × 5–10 μm	9 μm
Zeta potential at pH 5 to pH 8	unknown	electronegative	electronegative	not measured

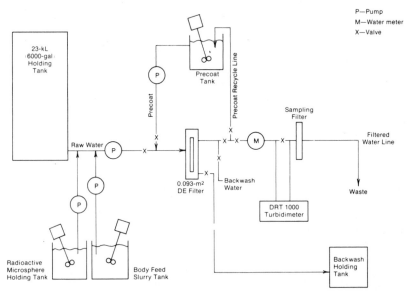

Figure 1. Schematic diagram of a diatomaceous earth (DE) pilot plant filter system

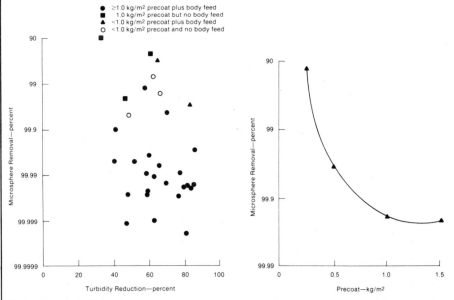

Figure 2. Relationship of water quality and DE filter operation

Figure 3. Effect of precoat amount on microsphere removal

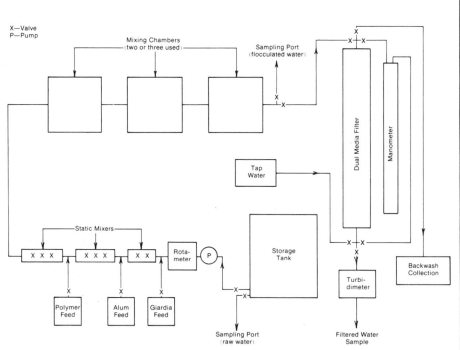

Figure 4. Schematic of granular media filtration

X—Valve
P—Pump

Mixing Chambers (two or three used)

Sampling Port (flocculated water)

Dual Media Filter

Manometer

Tap Water

Static Mixers

Rota-meter

Storage Tank

Backwash Collection

Turbi-dimeter

Polymer Feed

Alum Feed

Giardia Feed

Sampling Port (raw water)

Filtered Water Sample

TABLE 5

Diatomaceous earth filtration for G. muris cyst removal*

Filtration Rate m/h	Turbidity Raw Water ntu	Turbidity Filtered Water ntu	Body Feed mg/L	Cysts Dosed	Cysts Recovered	Reduction percent
2.3	2.5	0.58	31	22.0×10^6	2.8×10^4	99.87
2.4	2.5	0.56	30	18.7×10^6	3.3×10^4	99.82
2.2		0.76	33	24.4×10^6	1.56×10^5	99.36
2.4	1.7	0.49	15	19.2×10^6	1.02×10^4	99.947
2.4	1.7	0.49	29	36.4×10^6	4.2×10^3	99.988
3.5	0.24	0.31	31	22.5×10^6	3.9×10^4	99.83
3.5	0.24	0.32	31	20.8×10^6	1.4×10^3	99.993
3.5	0.95	0.42	31	26.0×10^6	2.8×10^4	99.89
3.5	0.95	0.41	31	27.1×10^6	1.1×10^4	99.959
3.5	0.45	0.38	1.5	39.0×10^6	5.6×10^3	99.986
3.5	0.45	0.40	1.5	34.5×10^6	6.4×10^4	99.81

*All tests used 1.0 kg/m² J-M 535 precoat and body feed, Johns-Manville, Denver, Colo.

TABLE 6

Granular media filtration with low coagulant doses

Length of Filter Run h	Alum Dose mg/L	Filtered Water Turbidity ntu	G. muris cysts/L Raw Water	G. muris cysts/L Filtered Water	Cysts Reduced percent	Test Conditions
Run 18						
0	2.2					raw water
0			8500			turbidity—4.5 ntu
0.5–1.5		0.58–0.51		1000	88	pH 6.5
2.5	2.1					temperature—20° C
3.0–4.0		0.63–0.46		1900	78–63	filtration rate—10 m/h
4.0	1.9					
4.8–5.7		0.79–0.77		4000	23	
5.7			5200			
Run 20						
0	1.4					raw water turbidity—4.4 ntu
0			5900			pH 6.5
0.9–1.4		1.0–0.95		1500	75	temperature—22° C
1.5	1.8					
2.3–2.7		0.65–0.60		420	94	filtration rate—10 m/h
2.8			7100			

Because data indicated that filter efficiency was related to precoat thickness, a series of tests was performed in which the filter was operated with precoat only. The tests were made with a single batch of water so that turbidity differences would not influence the results (Figure 3). Efficiency of microsphere removal increased with increasing precoat thickness up to 1.0 kg/m². Additional precoat did not improve filter performance.

Results from the membrane sampling filters showed that the microsphere removal capability of the DE filter usually improved during the run. The sampling filter was usually changed once during a filter run because of head loss buildup in the sampling filter. Counting data showed that the proportion of radioactive beads trapped during the first portion of the run almost always exceeded the proportion of the DE effluent that had passed through the filter. Conversely, the fraction of beads caught on the filter for the latter part of the run was almost always less than the proportion of effluent passing the sample filter in the latter part of the run. These data suggested that addition of body feed and removal of particulates produce a filter cake of increasing depth that is progressively more effective for particulate removal as the filter run continues.

Results of DE filtration experiments performed at 2.4 and 3.5 m/h show that *G. muris* cysts were very effectively removed at either filtration rate (Table 5). After the test with the lowest reduction of cyst concentration (99.36 percent) an above-normal amount of diatomite appeared in the concentrate from the centrifuge. Larger amounts of both diatomaceous earth and cysts passed through in this test. Although the septum had been cleaned before this test, it was cleaned again very carefully and examined for physical defects. None being found, the filter was reassembled and performed well thereafter.

Tests with cysts confirmed the findings with microspheres that turbidity removal is not a good indicator of cyst removal. Even when turbidity removal was low, cyst removal was high. This finding may be attributable to very small particles that could pass through the filter cake and cause a turbidity reading even though the larger cysts were strained out by the diatomite. High quality operation is the key to effective operation of DE filters.

Granular media filtration studies

Granular media filtration research emphasized assessment of the causes of poor and effective filter performance. Operating conditions were usually selected to achieve good cyst removal, but in some tests coagulant chemical doses were deliberately selected to give poor cyst removal. Filter operation was intentionally disturbed during some runs to show how this would affect cyst removal. One experimental goal was to determine whether changes in the cyst concentration of filtered water could be related to changes in filtered water turbidity.

Experimental methods. Jar tests were first performed in order to select appropriate doses of treatment chemicals and to make a preliminary assessment of the efficacy of coagulation and sedimentation for *Giardia muris* cyst removal. Combinations of treatment chemicals that were found effective were later tested in a pilot filter column. Alum, either alone or in combination with cationic polymer, was effective for coagulation of the cysts because of their negative zeta potential.

The filtration system (Figure 4) generally operated at 0.2 L/min with low-turbidity, gravel-pit water pumped through three inline mixers where cysts and coagulation chemicals were added, then through two or three 2-L enclosed jars with a 30-rpm stirrer in each for flocculation, and finally to a 3.8-cm diameter filter column with 46 cm of 1.27-mm effective size (e.s.) coal on top of 15 cm of 0.36-mm e.s. sand. Head loss through the filter column was measured with a mercury manometer. Filtered water was passed through an inline laboratory turbidimeter before discharge to waste or collection for analysis.

Samples to be analyzed for cysts were collected in 1-L cubitainers, refrigerated, concentrated, and counted by the method previously described.

Results. Test conditions included a broad range of possible field situations so that data on the ability of granular media filters to remove *G. muris* cysts could be developed. Alum coagulant doses ranged from none to adequate; cationic and nonionic polymers were also used. The effects of filter rate changes and loss of chemical feed were evaluated.

One filtration test was performed to determine the ability of a clean filter bed to remove uncoagulated cysts. Cysts were mixed into 67 L of unchlorinated, filtered well water at pH 8.2. The water was mixed and sampled for cysts, and the contaminated water containing 68 000 cysts/L was then pumped directly to the filter column for four hours. Cyst concentrations in filtered water ranged from 4100/L to 28 000/L in four samples, for removals of 94 to 59 percent.

Since those regions of the United States that have experienced *Giardia* problems tend to have clean, clear surface waters, filtration with minimal alum dose was also studied. Low coagulant doses are not unusual in locations where surface waters frequently have turbidities near or less than 5 ntu. The objective of the tests with low alum doses was to meet the 1-ntu maximum contaminant

level (MCL) for turbidity, but not to produce the lowest possible turbidity. In these experiments the alum feed was changed during the filter run to vary the filtered water turbidity. Cyst samples were taken beginning about one half hour or longer after the change of coagulant dose, when filtered water turbidity was stable or declining. Results of these experiments are shown in Table 6. Higher filtered water turbidity was clearly associated with lower cyst removal efficiency, although the level of efficiency varied in the two runs.

Effective coagulation was attained in nine filter runs, resulting in filtered water turbidities ranging from 0.05 ntu to 0.30 ntu (generally greater than 0.2 ntu) during the portions of the runs that preceded turbidity increases or breakthrough. Cyst removal was good when filtered water turbidity was stable.

In six of the filter runs a typical ripening period occurred, and turbidity declined from above-normal to equilibrium values in approximately the first half hour of operation. In three such runs during the ripening phase, when turbidity was above normal, cysts were detected in concentrations substantially greater than those found later during equilibrium operating conditions. Initial cyst concentrations of 990, 690, and 910 cysts/L exceeded later values by factors of $\geq$ 25, 10, and 21, respectively, even though the turbidity changes observed during filter ripening were all decreases of less than 0.10 ntu.

Filtration rate increases during some filter runs caused effluent turbidity and cyst concentration to increase. Rate increases of 50, 100, or 150 percent in less than 10 s were studied to show the effect of sudden increases in flow at a treatment plant. A strong alum floc conditioned with nonionic polymer withstood the added shear at 20°C, but alum floc at 20°C and alum–nonionic polymer floc at 10°C broke through when flow rates increased.

During one filter run the rate was increased from 11 to 27 m/h for a duration of 2 min and then decreased again to 11 m/h. Turbidity rose sharply during this brief period of high rate operation, but it rapidly declined after the filtration rate was lowered. Cyst concentrations followed the trends in turbidity (Figure 5). When the rate of filtration was increased abruptly, a fourfold increase in turbidity was accompanied by a 25-fold increase in cyst concentration in filtered water.

Turbidity increases associated with loss of coagulant chemical feed also were associated with increases in the cyst counts. When chemical feed was interrupted during the run with 22 mg/L alum, cyst concentration rose from less than 80 to 790 and then to 1760 cysts/L as turbidity increased from 0.27 to 0.37 and

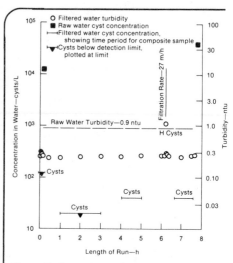

Figure 5. Operating data for filter run 11 *Filtration rate 10 m/h except during evaluation of rate increase effects; pH 6.4; 9.8 mg/L alum; temperature 20° C.*

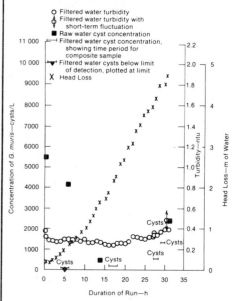

Figure 6. Operating data for filter run 15

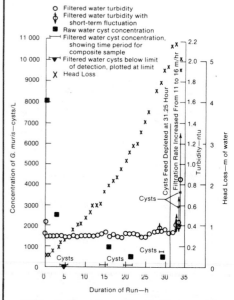

Figure 7. Operating data for filter run 16

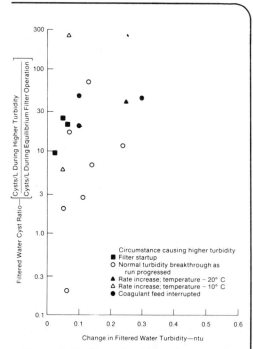

Figure 8. Relationship of cyst concentration changes to filtered water turbidity changes

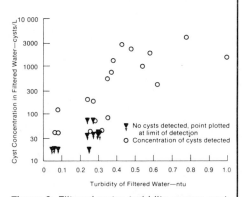

Figure 9. Filtered water turbidity versus cyst concentration for filtration of coagulated water at 10 m/h (constant rate)

TABLE 7

Summary of G. muris cyst removal during stable granular media filter operating conditions

| Filtered Water Turbidity ntu | Cyst Concentration cysts/L | | Removal percent |
	Raw Water	Filtered Water	
0.27	14 000	< 40	> 99.71
0.27	16 000	< 80	> 99.50
0.27	45 000	180	99.60
0.28	61 000	70	99.88
0.24	5700	< 40	> 99.30
0.24	29 600	< 80	> 99.73
0.25	12 200	< 20	> 99.84
0.26	12 200	40	99.67
0.26	36 200	40	99.89
0.30	3200	< 40	> 98.7
0.24	490	200	59.2
0.30	2500	< 41	> 98.3
0.32	950	< 43	> 95.5
0.32	470	43	90.8
0.08	180 000	120	99.93
0.08	190 000	40	99.98
0.08	16 000	40	99.75
0.08	30 000	< 20	> 99.93
0.08	72 000	< 20	> 99.97
0.08	50 000	< 20	> 99.96

then to 0.57 ntu. When alum and cationic polymer feed were halted in another test, cyst concentration increased from 70 to 3300 cysts/L as turbidity rose from 0.28 to 0.38 ntu.

In order to evaluate the effects of turbidity breakthrough at the end of a longer filter run, two filter runs were continued for more than 24 hours. Graphs of head loss, turbidity, and cyst concentration are shown in Figures 6 and 7. Because the pilot column was installed as a pressure filter, the head loss in these runs exceeded the normal terminal head losses at which gravity filtration plants operate. Near the end of the two longer runs, slight turbidity increases were accompanied by large rises in the *G. muris* cyst concentrations.

In run 16 (Figure 7) filter operation was continued after the stock suspension of cysts being fed into the raw water was exhausted. Before breakthrough, turbidity was 0.30 ntu and 40 cysts/L were detected in the effluent. As the run continued, turbidity fluctuated between 0.35 and 0.4 ntu, and cyst count rose to 740 cysts/L. Two hours after the addition of cysts had ceased, filtered water turbidity varied between 0.36 and 0.7 ntu, and the cyst count of a sample taken at this time was 2900 cysts/L. This suggests that cysts removed in the filter migrated down through the media during the filter operation and were discharged along with floc sheared off the filter media.

Most of the research was done at water temperatures near 20°C but three of the runs were carried out at the low temperatures more typical of mountain streams or winter and spring conditions that have been associated with waterborne outbreaks of giardiasis. Raw water was chilled to 1°C. The flocculator jars were kept in an ice bath, but because the laboratory was at room temperature the water in the second (last) flocculator jar rose to 5° or 6°C. Even though the filter was insulated, water temperature rose to about 15°C at the filter outlet.

When the low temperature runs were made, raw water turbidity was very low, ranging from 0.15 to 0.20 ntu. Filtered water turbidity during the runs was 0.05 to 0.08 ntu. When abrupt rate increases occurred during these experiments, slight turbidity rises (to 0.10–0.17 ntu) and increases in cyst concentrations were observed.

Discussion of granular media results. The results of granular media filtration research indicate that a dual media filter operating in an equilibrium condition does remove a high percentage of *Giardia muris* cysts. Changes in the operation of a filter, such as a rate increase or a loss of chemical feed, can adversely affect effluent quality, causing pronounced deterioration. Cyst concentration increased when turbidity increased, but cyst counts rose more than turbidity.

Figure 8 shows the relationship between turbidity changes and cyst concentration changes. To emphasize the need for very careful control of filter effluent turbidity, the actual turbidity changes are shown. Because cyst counts varied from run to run, ratios of high to normal cyst counts are plotted. The ratios are minimum values because when no cysts were detected in filtered samples the detection limit was used as the denominator for the ratio. Most of the data represent increases of turbidity caused by breakthrough, rate increase, or coagulant feed loss plotted against the corresponding ratio of cysts at high turbidity to cysts at low turbidity. Three data points show the turbidity decrease during filter ripening plotted versus the cyst concentration ratio. In the filter ripening stage of treatment both turbidity and cyst concentration decreased.

The data showed one exception. During one of the longer runs the lowest observed turbidity of 0.24 ntu corresponded to a sample with 200 cysts/L, whereas the lowest cyst concentration, less than 40/L, occurred when turbidity averaged 0.30 ntu. Calculations for cyst and turbidity changes were based on the lowest observed turbidity in this run. This was the only observation of a decrease in turbidity accompanied by a rise in cysts.

The most significant aspect of the data in Figure 8 is that the equilibrium values of turbidity during the filter runs ranged from 0.05 ntu to 0.30 ntu, and in no case did the change in turbidity cause the filtered water turbidity to exceed the 1-ntu MCL.

The need for attaining filtered water turbidity at a level considerably below 1 ntu can also be seen in Figure 9, which shows cysts in filtered water plotted versus filtered water turbidity. Cyst counts as high as 4000 cysts/L were observed even though the 1-ntu MCL was being met. The data in Figure 9 were obtained from runs made with coagulated water at a rate of 10 to 11 m/h. These data suggest that turbidity should perhaps be as low as 0.3 ntu for effective cyst removal. Merely attaining the 1-ntu limit is not adequate for removal of *G. muris* cysts.

Because *G. lamblia* has caused a number of waterborne disease outbreaks, assessing the efficacy of granular media filtration for cyst removal was an important aspect of this research. In order to do this, Table 7 was prepared. It presents data taken from tests in which filter operation at 10 m/h with adequate coagulation produced water with a turbidity of 0.30 ntu or less.

Under the conditions described, cyst concentrations in filtered water ranged from less than 20 to 200 cysts/L. Only three of twenty samples had more than 100 cysts/L. As would be expected, the highest removal percentage, 99.98 percent, was attained when the influent cyst concentration was highest, 190 000 cysts/L. During incidents of contamination when the cyst concentration in raw water is very high, properly operated granular media filters should remove a substantial portion of the cysts. In this research, when raw water cyst concentration exceeded 10 000 cysts/L, the filter always removed more than 99.5 percent of the cysts when operated properly.

On the other hand, two samples in Table 7 show that when cyst concentrations in flocculated water were lower (5700 cysts/L or less), some cysts passed through the filter. Cysts were not detected in four other samples. The continuous passage of a small number of cysts through the filter cannot be ruled out. Because of the detection limit problem inherent in microscopic enumeration of the cysts, in no case was the entire volume of concentrated filter effluent examined. Thus no data exist to show that all cysts were removed; i.e., the detection limit is always above zero.

At this time granular media filters cannot be considered a totally effective barrier against passage of *Giardia* cysts into drinking water. Another barrier, disinfection, must also be used. Disinfection research is still under way and definitive data are not available, but preliminary results suggest that substantially higher concentrations of chlorine are required to inactivate *Giardia* cysts as compared to coliforms. Chlorination practice that is barely adequate for inactivation of coliforms should not be expected to inactivate *Giardia* cysts.

Water utilities that are located in regions where giardiasis outbreaks have occurred and that rely on disinfection alone to treat low turbidity, colored surface waters face a particularly difficult situation. Use of very high doses of chlorine to control *Giardia* cysts could result in violation of the 0.10-mg/L MCL for trihalomethanes, a limit that is applicable to utilities serving 10 000 or more persons. Use of very low chlorine doses to control trihalomethanes might leave the water supply susceptible to a giardiasis outbreak. Water utility managers who find themselves in this situation should discuss the problem with regulatory officials and obtain advice on how to deal with it in the short term. The long-range solution would be installation and proper operation of a filtration plant.

Summary and conclusions

Diatomaceous earth research. Diatomaceous earth filtration research conducted since the early 1940s has shown that DE filters can effectively remove cysts and cyst-sized particles. Removal of particles in the 7 to 20-μm size range is not necessarily related to turbidity removal, as shown by both the US Army research

and this work. Therefore, strict adherence to recommended operating technique is suggested for successful cyst removal. The following conclusions can be drawn:

● DE filtration for *Giardia* cyst removal is practical and effective as long as these filters are properly maintained and operated.

● Thickness of DE precoat had a greater effect on microsphere removal than did DE grade (particle size).

● A precoat of 1.0 kg/m² provided an effective barrier to cyst passage.

● DE filter efficiency for microsphere removal generally increased as the filter runs progressed.

● DE filters used in the production of potable water should not be operated without body feed addition.

● Because effluent turbidity is not an effective indicator of a DE filter's ability to remove *Giardia* cysts, reliance must be placed instead on use of proper treatment techniques with an adequate amount of DE precoat and body feed.

Granular media filtration tests. This work represents initial efforts in the investigation of filtration techniques for removal of *Giardia* cysts from drinking water. As only limited data are available at this time, it would be premature to suggest a filtered water turbidity goal that would assure successful cyst removal. Although pilot plant work done at the University of Washington through 1979 generally confirmed earlier EPA results, further research is needed before a turbidity goal can be recommended. On the basis of this early work, however, some conclusions can be formed.

● Meeting the 1-ntu MCL for turbidity did not result in effective cyst removal.

● Increases in filtered water turbidity, even as low as 0.05 to 0.10 ntu, were usually associated with large increases in cyst concentrations.

● Cyst concentrations during the filter ripening period are likely to be higher than they are after turbidity has stabilized. One way to avoid problems of higher cyst concentrations at this time would be to filter to waste for perhaps the first half hour of a filter run.

● After a stable effluent turbidity has been achieved, increases of turbidity should be avoided in order to minimize passage of *Giardia* cysts through the filter.

● Although a granular media filter should not be expected to remove all cysts from the raw water, a filter treating water dosed with an adequate amount of coagulant chemical and operated in a manner that prevents filtered water turbidity increases should remove a very substantial portion of the *Giardia* cysts.

● The multiple barrier concept should be employed to treat water likely to contain *Giardia* cysts. Use of both filtration and disinfection is necessary.

Acknowledgment

R. L. Hoye Jr. and M. M. Arozarena partially fulfilled academic requirements of the University of Cincinnati by participation in this research when they were employed by the USEPA.

References

1. SPECTOR, B.K.; BAYLIS, J.R.; GULLANS, O. Effectiveness of Filtration in Removing from Water, and of Chlorine in Killing, the Causative Organism of Amoebic Dysentery. *Public Health Repts.*, 49:27:786 (Jul. 6, 1934).
2. BAYLIS, J.R.; GULLANS, O.; & SPECTOR, B.K. The Efficiency of Rapid Sand Filters in Removing the Cysts of Amoebic Dysentery Organisms from Water. *Public Health Repts.*, 51:46:1567 (Nov. 13, 1936).
3. Efficiency of Standard Army Purification Equipment and Diatomite Filters in Removing Cysts of *Endamoeba histolytica* from Water. War Dept. Rept. 834. Submitted to the Engineer Board, Ft. Belvoir, Va. and the Chief of Engineers, US Army, Washington, D.C., by Water Supply Equipment Branch, Technical Division III, The Engineer Board, Ft. Belvoir, Va. and the National Institute of Health, US Public Health Service. (Jul. 3, 1944).
4. BLACK, H.H. & SPAULDING, C.H. Diatomite Water Filtration Developed for Field Troops. *Jour. AWWA*, 36:11:1208 (Nov. 1944).
5. ENGESET, J. & DeWALLE, F.B. Removal of *Giardia lamblia* cysts by Flocculation and Filtration. Proc. AWWA Annual Conf. San Francisco, Calif. (1979).
6. LEVINE, N.D. *Protozoan Parasites of Domestic Animals and of Man.* Burgess Publishing Co., Minneapolis, Minn. (1st ed., 1961).
7. BROOKE, M.M. Intestinal Protozoa. *Manual of Clinical Microbiology* (J.E. Blair, E.H. Lennette, and J.P. Truant, editors). American Society for Microbiology, Bethesda, Md. (First ed., 1970).
8. ROBERTS-THOMPSON, I.C., ET AL. Giardiasis in the Mouse: An Animal Model. *Gastroenterology,* 71:1:57 (Jan. 1976).
9. SHEFFIELD, H.G. & BJORVATN, B. Ultrastructure of the Cyst of *Giardia lamblia.* *Am. Jour. Tropical Med. Hygiene,* 26:1:23 (Jan. 1977).
10. Johns-Manville One Square Foot Test Filter Unit Operating Instructions, FF-113A. Celite Division, Johns-Manville. New York, N.Y. (1967).
11. LOGSDON, G.S.; SYMONS, J.M.; & HOYE, R.L. Water Filtration Techniques for Removal of Cysts and Cyst Models. Proc. Symp. on Waterborne Transmission of Giardiasis, Cincinnati, Ohio. EPA-600/9-79-001 (Jun. 1979).

Gary S. Logsdon (Active Member, AWWA) is a research sanitary engineer at the Drinking Water Research Division of the US Environmental Protection Agency, Cincinnati, Ohio 45268. James M. Symons (Active Member, AWWA) is chief of EPA's physical and chemical contaminant removal branch, and Robert L. Hoye Jr. and Michael M. Arozarena are environmental scientists with PEDCO Environmental, Inc., Cincinnati, Ohio.

Reprinted from *Jour. AWWA*, 73:2:111 (Feb. 1981).

SURROGATE INDICATORS FOR ASSESSING
REMOVAL OF GIARDIA CYSTS

David W. Hendricks
Professor
Department of Civil Engineering
Colorado State University
Fort Collins, CO 80523

Mohammed Y. Al-Ani
Research Associate
Department of Civil Engineering
Colorado State University
Fort Collins, CO 80523

William D. Bellamy
Environmental Engineer
CH2M-Hill Consulting Engineers
1301 Dove Street, Suite 800
Newport Beach, CA 92660

Charles P. Hibler
Professor
Department of Pathology
Colorado State University
Fort Collins, CO 80523

John M. McElroy
Environmental Engineer
CH2M-Hill Consulting Engineers
P.O. Box 9150
Bellevue, WA 98009

BACKGROUND

Measurement of Giardia cyst removals by filtration is a task not likely to be adopted for routine water treatment practice. Cysts may not be present at the time of sampling, or they may be too low to recover. Also, the sampling effort required is considerable. Further, analysis of the samples requires appreciable labor and skill. Thus there is interest in having a surrogate indicator of cyst removal by filtration.

This paper reviews experimental work in which several surrogate indicators of Giardia cyst removal by filtration were investigated. The surrogates included: turbidity, particles, total coliform bacteria, and standard plate count bacteria. The investigation was one facet of an EPA sponsored project to ascertain removals of Giardia cysts by the filtration technologies of rapid rate, slow sand, and diatomaceous earth, reported by Al-Ani, et al. (1984), Bellamy, et al. (1984), and Bellamy, et al. (1984).

EXPERIMENTS

The experiments to ascertain the influences of process variables were comprised of many simultaneous measurements of the four parameters in both raw water and filtered water. The data were generated using pilot plants in which raw water was spiked with Giardia cysts. Since the cysts were obtained from dog feces the spiking procedures usually resulted in high levels of particles and bacteria in the spiked, now filtered water. The pilot plants used were: (1) a laboratory scale rapid rate pilot plant using 5 cm and 10 cm diameter filters, (2) nine 30 cm diameter slow sand filters, and (3) a 730 cm^2 (1 ft^2) diatomaceous earth pilot filter. The rapid rate pilot plant used only rapid mix and filtration as the operating mode.

RESULTS - RAPID RATE FILTRATION

Relationships using data generated by the laboratory-scale pilot plant were ascertained using two types of plots: three-dimensional histograms and probability. Conventional plots were used for data generated from the field-scale pilot plant. These results are reviewed in the following paragraphs. The emphasis was on the use of low turbidity waters, e.g. less than 1 NTU.

Turbidity. Figure 1 shows all of the percent removal data points for cysts and turbidity. The medians for groups of eight data points from left to right are indicated by the black dots, which are connected by lines to show the trend. The associated data points show that more scatter occurs at the lower percent removals of turbidity, and less scatter toward higher percent removal of Giardia cysts, which corresponds to higher percent removal of turbidity. This indicates that if a relatively high turbidity reduction occurs for the low turbidity raw water, e.g. greater than 80 percent, then a relatively high cyst removal can be expected.

Figure 2 is a three-dimensional histogram plot which utilizes data from the upper right portion of Figure 1, showing again percent removals of Giardia cysts for corresponding percent removals of turbidity. Figure 2 permits easier visualization of relationships and also illustration of quantitative relationships. For example, it shows that if turbidity removal is high, e.g. greater than 70 percent, then removal of Giardia cysts is likely to be high also. To be more specific, the plot shows 44 observations when turbidity removal exceeds 70 percent. Of these, 37 have Giardia cyst removals of 99 percent or more. In other words, if turbidity removal exceeds 70 percent the probability is 0.85 (e.g. 37/44) that removal of Giardia cysts will equal or exceed 99 percent.

Figure 3 is another kind of plot utilizing the data in Figure 1. The plot shows the probability of removal of Giardia cysts due to a given level of turidity removal. For example, if turbidity removal is 80 percent, Figure 3 shows that the probability of achieving 95 percent removal of Giardia cysts is 0.93. The message in the plot is clear: with higher percent removals of turbidity, higher percent removals of Giardia cysts can be expected, with corresponding high probabilities that the latter will occur. If turbidity removals are low the probability is low that Giardia removals will be significant.

To explain how Figure 3 is constructed, consider the locus of points having turbidity removals of 80 percent or greater. There are 35 data points having this level of turbidity removal. The Giardia data are arranged first in ascending order in one column; the associated turbidity data are shown in the other column. For the sample shown in Figure 2 there are 35 data points, and the first data point, showing 85 percent Giardia removal, has a probability of 1.0 of occurrence (e.g.

35/35). The last point, showing 99.99 percent Giardia removal, has a 0.029 probability of occurrence (e.g. 1/35). These calculated probability of occurrence data are arranged in another column. From these data the 80 percent turbidity removal locus was plotted.

Total Coliform Bacteria. Figure 4 shows the data points for cysts and total coliform bacteria (T.C.), and the medians for groups of nine data points. Figure 5 is a histogram for a portion of the plot. It shows that if percent removals of T.C. is high, percent removal of Giardia cysts will be high. For example, counting observations in the range 90 to 100 percent removal of T.C., there is 0.85 probability that removal of Giardia cysts will exceed 96 percent, and 0.70 probability that cyst removal will exceed 99 percent. Figure 6 is the corresponding probability of occurrence plot.

It should be noted that the probability of occurrence plots indicate the trends in the data, but are not absolute. There are not enough data at low percent removals of turbidity and T.C. to develop accurate relationships. Nevertheless, the trends indicated are clear.

Standard Plate Count Bacteria. The sequence of plots shown for turbidity and T.C. as surrogates are not shown for standard plate count bacteria (S.P.C.). They have about the same in trends, but the plots are omitted for the sake of brevity.

Particles. Particle counts for the 2.52 to 50.8 micrometer size range were obtained during 66 test runs in which Giardia cysts were measured. A histogram plot of the data, similar to Figure 2 for turbidity, showed a similar trend, but more scatter. For example, the plot showed that if removal of particles was greater than 70 percent, the probability that removal of Giardia is 99 percent or higher was only 0.60 (32/53), which compares with 0.85 probability of removal for turbidity.

It is of interest also, that in plotting 100 percent removal points for turbidity vs particles, the points were distributed widely over the grid. Measurements of particles were discontinued after 124 tests (out of 178). This was done because the trends were not strong, and particles were less essential and caused diminuation of effort.

Field Scale Testing. Figure 7 shows Giardia cyst and total coliform bacteria removals obtained using a field scale pilot plant (a trailer mounted Neptune Microfloc Waterboy ® on loan from EPA). The water used was low temperature, e.g. less than $1^{o}C$, and low turbidity, e.g. less than 1 NTU, obtained from the Cache La Poudre River during the period November 1983 to January 1984. The low percent removals were a part of the testing plan in which "no chemical pretreatment" and "ineffective chemical pretreatment" used. The one high percent removal point was for "optimum" chemical coagulation. Turbidity removal was 42 percent for this point, while it was zero percent for the other points.

Some tests were conducted using raw water from Horsetooth Reservoir having 7 NTU turbidity and $10^{o}C$ temperature. This water was easier to coagulate than the low-temperature low-turbidity water. Nine of the ten points were in the vicinity of 95 to 100 percent Giardia removal and 90 to greater than 99 percent T.C. removal. Turbidity removals ranged from 76 to 97 percent. One point was at 40 percent Giardia removal, 25 percent T.C. removal, and 3 percent turbidity removal; no coagulants were used for this test.

RESULTS - SLOW SAND AND DIATOMACEOUS EARTH

Slow Sand. Three 30 cm diameter slow sand pilot filters were operated in parallel at hydraulic loading rates of 0.04, 0.12, and 0.40 m/hr over a 16 month period to determine removals of Giardia cysts. At the same time removals were determined for turbidity, T.C., and S.P.C., particles and fecal coliform bacteria. The latter two parameters were measured during five months of this period. Water from Horsetooth Reservoir, just above the Engineering Research Center, was used for all experiments. The raw water had very low concentrations of all bacteria, e.g. 0 to 1 coliforms per 100 ml, and so it was spiked with Giardia cysts obtained from dog feces, and settled raw sewage provided the spike for bacteria and particles. Table 1 shows the ranges of these parameters in the raw water, as spiked. The turbidity was not changed signficantly by the spiking, e.g. the changes were always $\leq$ 0.2 NTU.

Table 1 summarizes results of all testing, showing the number of analyses, and average percent removal data from all tests. The cyst removal percent generally exceeded 90 percent nominally for a new filter and 99.9 percent for a biologically "mature" filter. Possible surrogates for removal of Giardia cysts can be evaluated based upon this fact.

The sand particles comprising the sand bed act as a substrate for attachments of biological growths. The filter sheds these growths and they appear in the effluent as S.P.C. and as particles. Thus neither could be a suitable surrogate. This was borne out by measurements, which showed wider variation than the T.C. or fecal coliform bacteria. Neither was turbidity removal suitable; the particles comprising the turbidity were nominally one micrometer or smaller in size and most passed through regardless of filter conditions.

T.C. was the most suitable surrogate. This was borne out by data in which there was the closest correspondence between removals of T.C. and Giardia cysts. Since the coliform bacteria is foreign to the column, and apparently is removed by the same mechanisms that remove Giardia cysts, such correlation could be expected. Percent removals of T.C. showed improvement with time after the filter was started up, leveling off to greater than 99 percent removal after some time (measured in weeks). When this occurs the filter was felt to be "biologically mature." At this time Giardia cysts could not be detected in the filter effluent. Thus percent removal of T.C. was a surrogate in the sense that if removals exceed 99 percent, the filter is biologically mature, and thus removal of Giardia cysts will be 100 percent (qualified by detection limits).

Diatomaceous Earth. Removals of Giardia cysts were determined over a wide range of operating conditions, including graded of diatomaceous earth and hydraulic loading rates. Removals were 100 percent for all test runs except one (in which influent cyst concentration was 33,000 cysts/liter). Removals of turbidity and T.C. were affected by grade of diatomaceous earth. Particle counts in the 6.35 to 12.67 micrometer size range were discontinued after 35 measurements since removals were high with no relationship to operating conditions. Removals were generally 90 to 95 percent. It is believed that attrition from the filter comprised a portion of the particles in the effluent stream. The conclusion of these results must be that there is no useful surrogate for cyst removal. On the other hand, since removals were essentially 100 percent for all condition qualified by detection limits, there was no evidence of need.

DISCUSSION

The data show that each of the four parameters (e.g. percent removals of turbidity, T.C., S.P.C., particles) examined can serve as a surrogate for percent removal of Giardia cysts for rapid rate filtration. The student t-distributions for each showed positive associations at 99.5 percent confidence level, indicating it is virtually certain that functional relationships exist. In slow sand filtration, T.C. removal can be used as a surrogate for Giardia cyst removals because the former is an index of the biological maturity of the filter. A surrogate is not needed for diatomaceous earth filtration since removals were always 100 percent. The following paragraphs apply to rapid rate filtration.

Turbidity. The most appealing surrogate is turbidity removal, both because it is relatively fast, easy, and inexpensive to measure, and because it showed a strong correlation with cyst removal. This is especially notable since the strong correlation occurred for testing with low turbidity waters having raw water turbidities in the range 0.5 to 1 NTU. In other words, if turbidity removal is about 70 percent or more, e.g. from 0.5 to 0.15, Giardia cyst removal can be expected to be 98 percent or more.

Total Coliform Bacteria. Removal of T.C. could be an effective surrogate also, because it is a routine measurement in water treatment practice. One problem with its use in low turbidity waters is the low T.C. counts, typically 10 to 100 per 100 ml. In the experiments conducted, the raw water usually was spiked. Thus to use T.C., a pilot column would be advisable, run side by side with the full scale filter. The measurement requires 24 hours to complete.

Standard Plate Count Bacteria. Removal of S.P.C. could be an effective surrogate, for the same reasons as T.C., and it has the same problems, e.g. low counts are prevalent in low turbidity water. But, in addition, it is not used quite so frequently. The measurement requires 48 hours.

Particles. There is logic to the use of particle removal in the 2.52 to 52.8 micrometer size range, since this range encompasses the 10 micrometer nominal size of Giardia cysts. But the relationships were not strong as compared with the other parameters. In addition, particle counting is not routine.

SUMMARY AND CONCLUSIONS

For rapid rate filtration, exploration of relationships between the dependent variables, e.g. turbidity, standard plate count bacteria, total coliform bacteria, particles, and Giardia cysts, were done by means of plots and statistical analyses. All of the above dependent variables were examined to determine if a relation existed between percent removal of the given parameter and pecent removal of Giardia cysts. Histogram plots showed definite relationships, e.g. high percent removals of turbidity were associated with high percent removals of Giardia cysts. Statistical tests, e.g. the student t-distribution, showed high (99.5) confidence levels, indicating that functional relationships exist between removals of the above parameters and removals of Giardia cysts.

All parameters examined were found to be suitable as indicators of percent removal of Giardia cysts for rapid rate filtration. Turbidity is recommended because it is easy to use. In general, if 70 percent turbidity removal is achieved, then there is 0.85 probability that removals of Giardia cysts exceed 95 percent. Similar relationships exist for the other parameters investigated, but there was less confidence in the use of particle counting as a surrogate indicator.

Turbidity was found to be a good indicator of percent removals of other parameters as well, e.g. standard plate count bacteria, and total coliform bacteria. High removals of turbidity, e.g. from 0.5 NTU to 0.1 NTU, is evidence that filtration has occurred with effective coagulation. If effective coagulation-filtration occurs, then removals of 99 percent or greater can be expected for all substances, e.g. bacteria cysts, etc.

Coliform bacteria would be an excellent indicator, except in low turbidity water situations the ambient water's total coliform bacteria concentrations may be 0-100 org/100 ml, which is for use in detecting removals with confidence. A pilot filter if constructed adjacent to the full scale filter could be used to evaluate the filtration process; however, the influent water would probably be needed to be spiked using a pure culture of T.C. Percent removal attained would be an indication of whether effective chemical pretreatment has occurred. Chemical pretreatment does not include adding chlorine for prechlorination or any other disinfectant.

For the slow sand filtration, T.C. removal can be a useful surrogate of cyst removal because the former is an index that the filter is "biologically mature." In both rapid rate filtration and slow sand filtration, the surrogates are useful really because they provide an index as to whether the filters are operating under conditions intended. This includes effective chemical pretreatment in rapid rate filtration and having a biologically mature filter in slow sand filtration. For diatomaceous earth filtration, Giardia cyst removal was uniformly 100 percent qualified by detection limits. Cyst breakthrough may occur only if the system is improperly operated, e.g. a hole in the septum or leaky-connection. Abberations in behavior of other parameters during monitoring could detect this.

ACKNOWLEDGMENTS

The plots used in this paper were obtained from reports listed in References. The project generating these reports was funded by the Environmental Protection Agency, Drinking Water Research Division, Project Number CR808650-02. Dr. Gary S. Logsdon was project officer.

REFERENCES

1. Bellamy, W. D., Lange, K. P., and Hendricks, D. W., "Filtration of Giardia Cysts and Other Substances, Volume 1: Diatomaceous Earth Filtration," Environmental Engineering Technical Report 5847-84-1, Colorado State University, Fort Collins, March 1984.

2. Bellamy, W. D., Silverman, G., and Hendricks, D.W., "Filtration of _Giardia_ Cysts and Other Substances, Volume 2: Slow Sand Filtration," Environmental Engineering Technical Report 5847-84-3, Colorado State University, Fort Collins, December 1984.

3. Al-Ani, M. Y., McElroy, J. M., and Hendricks, D. W., "Filtration of _Giardia_ Cysts and Other Substances, Volume 3: Rapid Rate Filtration," Environmental Engineering Technical Report 5847-84-4, Colorado State University, Fort Collins, December 1984.

Reprinted from Proc. AWWA WQTC, Denver, Colo. (Dec. 1984).

Table 1. Average percent removals for dependent variables in slow sand filter columns. Columns were 30.5 cm diameter. Data shown are for all operating conditions.

Dependent Variable	Total Number of Analyses	Range of Variable in Raw Water	Percent Removal of Parameter		
			Filter 1 $v=0.04$ m/hr	Filter 2 $v=0.12$ m/hr	Filter 3 $v=0.40$ m/hr
Giardia cysts	222	50–5,075 cysts liter	>99.991	>99.994	>99.981
Total Coliform	243	0–290,000 colif. 100ml	99.96	99.67	98.98
Fecal Coliform	81	0–35,000 colif. 100ml	99.84	98.45	98.65
Standard Plate Count	351	10–1,010,000 org. ml	91.40	89.47	87.99
Turbidity	891	2.7–11 NTU	39.18	32.14	27.24
Particle (6.35–12.7um)	39	62–40,506 part. 10ml	96.81	98.50	98.02

131

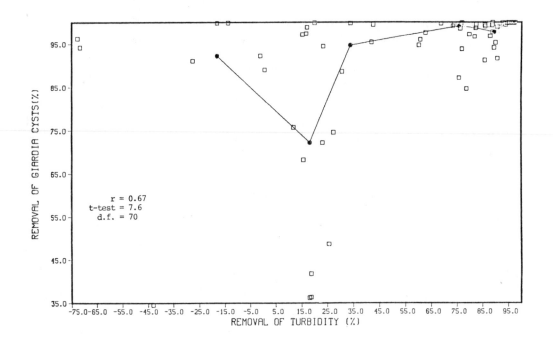

Figure 1. Observations of percent removal of <u>Giardia</u> cysts with corresponding percent removal of turbidity. The solid points show the median for each consecutive 8 data points. Data points are for water having turbidity of 1 NTU or less, and temperatures between 3° and 5°C. Data were obtained using 5 cm diameter rapid rate laboratory scale pilot filters packed with anthracite and sand.

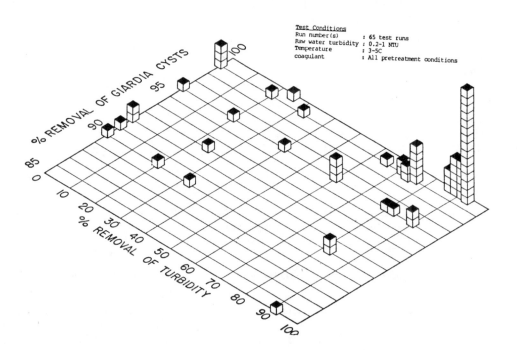

Figure 2. Histogram showing percent removal of <u>Giardia</u> cysts with corresponding percent removal of turbidity. Each block represents one measurement set. Data points are for water having turbidity of 1 NTU or less, and temperatures between 3° and 5°C. Data were obtained using 5 cm diameter rapid rate laboratory scale pilot filters packed with anthracite and sand.

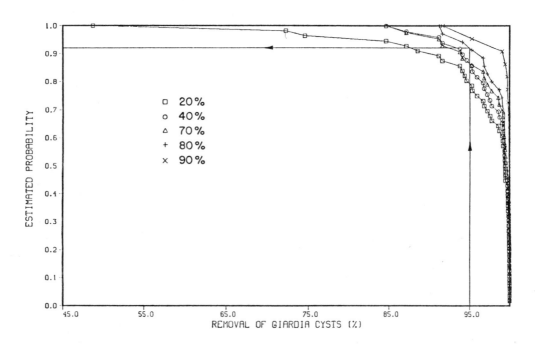

Figure 3. Probability of a given percent removal of <u>Giardia</u> cysts for specified percent removal of turbidity. Data points are for water having turbidity of 1 NTU or less, and temperatures between 3° and 5°C. Data were obtained using 5 cm diameter rapid rate laboratory scale pilot filters packed with anthracite and sand.

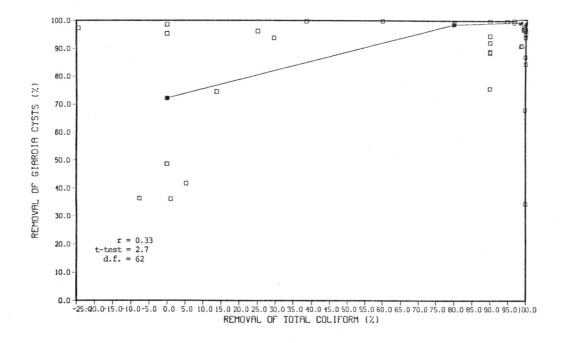

Figure 4. Observations of percent removal of <u>Giardia</u> cysts with corresponding percent removal of total coliform bacteria. The solid points show the median for each consecutive 9 points. Data were obtained using 5 cm diameter rapid rate laboratory scale pilot filters packed with anthracite and sand. Raw water turbidities were 1 NTU or less with temperatures between 3° and 5°C.

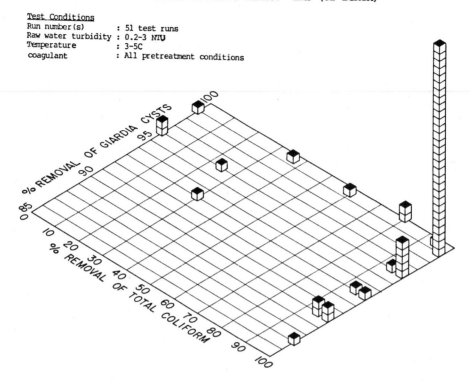

Type of Filtration: Rapid Rate
Scale of Pilot Plant: Lab (or Bench)

Test Conditions
Run number(s)   : 51 test runs
Raw water turbidity : 0.2-3 NTU
Temperature   : 3-5C
coagulant   : All pretreatment conditions

Figure 5. Histogram showing percent removal of <u>Giardia</u> cysts with corresponding percent removal of total coliform bacteria. Each block represents one measurement set. Data were obtained using 5 cm diameter rapid rate laboratory scale pilot filters packed with anthracite and sand. Raw water turbidities were 1 NTU or less with temperatures between 3° and 5°C.

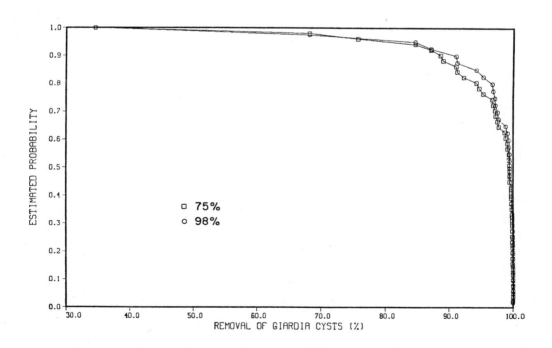

Figure 6. Probability of a given percent removal of <u>Giardia</u> cysts for specified percent removal of total coliform bacteria. Data were obtained using 5 cm diameter rapid rate laboratory scale pilot filters packed with anthracite and sand. Raw water turbidities were 1 NTU or less with temperatures between 3° and 5°C.

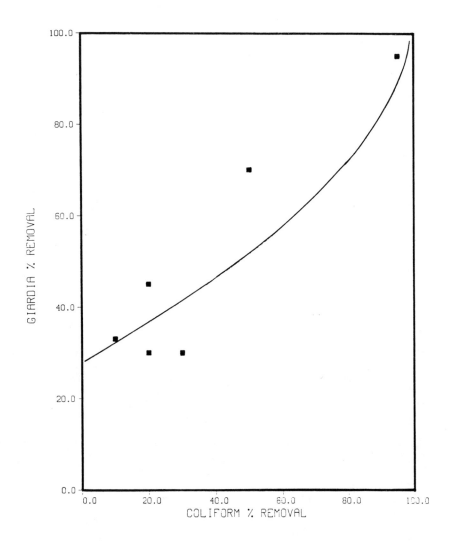

Figure 7. Percent removal of Giardia cysts vs percent removal of coliform bacteria for low-turbidity water. Raw water characteristics were: <1°C and <1 NTU. Data were obtained using a field scale pilot plant having a 61 x 61 cm filter bed packed with anthracite and sand. Plant was operated using only rapid mix and filtration mode.

Evaluating Sedimentation and Various Filter Media for Removal of Giardia Cysts

Gary S. Logsdon, V. Carol Thurman, Edward S. Frindt, and
John G. Stoecker

The primary purpose of the pilot study described in this article was to evaluate the efficiency of sedimentation and of coarse anthracite media for removing *Giardia* cysts. Removal of cysts by sedimentation was observed to range from 65 to 93 percent and was generally similar to removal of turbidity. The coarse anthracite media produced water of inferior quality when alum was the only coagulant, but its performance was found to be improved by the use of a high-molecular-weight, slightly anionic polymer. Cyst concentrations appeared to be higher than usual during the initial phase of a filter run, indicating that a filter-to-waste period may be desirable. The authors conclude that effective control of *Giardia* cysts depends on informed application of the multiple-barrier concept.

During the winter of 1983–84, an outbreak of waterborne giardiasis (342 confirmed cases) occurred in McKeesport, Pa., a town located south of Pittsburgh at the confluence of the Monongahela and Youghiogheny rivers.

In late December 1983, McKeesport experienced an unusually high water demand that diminished stored water and prevented effective backwash of the filters. Because an elevated backwash tank was out of service at the time, all backwash water was being supplied from the distribution system. Filters were run for several days without backwashing, and large-scale breakthrough led to significant deterioration of the finished water turbidity. Data entered in the plant report from Dec. 25, 1983, through Jan. 14, 1984, indicate that composite turbidity from the plant rose to 5 nephelometric turbidity units (ntu) on December 29. After Jan. 11, 1984, filtered water turbidity was generally 1.0 ntu or lower. During this time, the free chlorine residual at the plant was 0.7–1.9 mg/L and was below 1.0 mg/L only twice in 21 days. Total chlorine residual was generally about 0.3 mg/L higher than the free residual at the plant. Positive coliform results were found in one of five 10-mL tubes one day (confirmed) and in two of five 10-mL tubes another day (one of two confirmed). Coliforms were not alarmingly high.

The high turbidity episode was followed by an unusually high incidence of giardiasis that became apparent the second week of February 1984. Microscopic analysis of sediment from large volumes of finished drinking water sampled by cartridge samplers during the week of Feb. 27, 1984, by the US Environmental Protecton Agency's (USEPA's) Health Effects Research Laboratory later confirmed the presence of *Giardia* cysts in the raw water, the finished water at the plant, and the water in the distribution system.[1] These results showed that the treatment plant was not performing as expected.

On Feb. 22, 1984, a "boil water" notice was issued by the Allegheny County Health Department to all consumers in the affected communities, based on epidemiological evaluations of 15 cases of giardiasis.[2] The conclusion that the drinking water was the source of the giardiasis in McKeesport was strengthened because the time interval between the turbidity problem in late December and early January and the subsequent disease outbreak was similar to the incubation time for giardiasis.

Background

Plant design. The McKeesport Municipal Water Authority operates a 9-mgd (34.1-ML/d) conventional treatment plant that was constructed in 1907 and serves about 51 000 residents in the communities of McKeesport, Versailles, Port Vue, and White Oak. Raw water is obtained from the Youghiogheny River about 0.5 mi (0.8 km) above its confluence with the Monongahela River. Raw water typically has a turbidity of 2.5–200 ntu, a pH of 6.9–7.8, an alkalinity of 20–53 mg as calcium carbonate ($CaCO_3$)/L, and a hardness of 90–228 mg as $CaCO_3$/L. Clarification processes include hydraulic mixing (chemical solution poured on upwelling raw water), baffled flocculation (no direct power input), sedimentation, and filtration.

In 1960 the sand filters at the plant were converted to monomedia anthracite filters. A core sample of the media was obtained in the spring of 1984, when the filters contained about 30 in. (0.75 m) of media. The USEPA's Drinking Water Research Division (DWRD) performed three separate sieve analyses on the media and determined that it had a mean effective size of 0.92 mm (range 0.89–0.93 mm) and a mean uniformity coefficient of 1.64 (range 1.61–1.70). This effective size is considerably larger than the size usually used for sand filters (0.5 mm) and is also larger than the fine media used in dual-or mixed-media filters. Thus, although the bed depth was typical, the grain size was not.

Previous research. Previous USEPA-sponsored research (DWRD studies, and projects at the University of Washington, Colorado State University, and McIndoe Falls, Vt.) had focused on direct filtration, slow sand filtration, and diatomaceous earth filtration. No thorough evaluation of a conventional treatment train had been performed.

The earlier filtration research[3-6] had demonstrated a relationship between filtered water turbidity and cyst concentration, with higher turbidity waters generally having much higher cyst concentrations. Much of the previous work was done on raw water with turbidities ranging from 1 to 5 ntu; settling was seldom employed in the treatment train.

Little information is available on *Giardia* cyst removal by sedimentation. Arozarena[7] reported that cyst removal in jar tests involving alum coagulation and sedimentation ranged from 58 to 99 percent. That work was done with a clear (usually <5 ntu) gravel-pit water to simulate the water quality in places where *Giardia* cysts are found. The results suggested that sedimentation could remove a portion of the cysts present in raw water.

Research objectives

A primary objective of the DWRD pilot study was to evaluate the efficacy of the McKeesport anthracite media for *Giardia* cyst removal and to compare it with other media types. This was done by operating either three or four filters simultaneously and applying the same pretreated water to all filters, thereby allowing direct comparison of granular activated carbon (GAC), sand, dual media, and anthracite monomedia.

Another objective of the research was to evaluate a conventional treatment train for *Giardia* cyst removal. Before the McKeesport giardiasis outbreak, most outbreaks had occurred in locations having low-turbidity raw waters, for which direct filtration might be appropriate. An evaluation of cyst removal by sedimentation was needed.

An additional objective was to study cyst removal from somewhat turbid

TABLE 1
Characteristics of pilot-plant treatment processes

Plant Equipment	Dimensions						Theoretical Detention Time at 1.9 gpm (0.12 L/s)
	Length		Width		Depth		
	in.	m	in.	m	in.	m	
Rapid mix tank 1	8.4	0.21	5.4	0.14	8	0.20	0.8 min
Rapid mix tank 2	8.4	0.21	5.6	0.14	7.6	0.19	0.8 min
Flocculation basin	48	1.22	18	0.46	22	0.56	43 min
Sedimentation basin	120	3.05	48	1.22	24	0.61	5.2 h

TABLE 2
Media used in filters

Media	Depth		Effective Size mm	Uniformity Coefficient
	in.	m		
Dual media				
Anthracite	20	0.51	0.99	1.13
Sand	10	0.25	0.46	1.24
Sand	30	0.76	0.42	1.55
McKeesport anthracite	30	0.76	0.92	1.64
12 × 40 GAC*	32	0.81	0.55–0.65	≤1.9

*Filtrasorb F-400, Calgon Corp., Pittsburgh, Pa.

TABLE 3
Removal of Giardia cysts by sedimentation

Raw Water Turbidity ntu	Chemical Dosage—mg/L		Turbidity Reduction percent	Cyst Removal percent
	Alum	Polymer		
22–25	27.5	none	81	
22–25	27.5	none	79	
22–25	27.5	none	79	
22–25	27.5	none	77	
11–15	25.4	0.048	77	79
11–15	25.4	0.048	82	93
11–15	25.4	0.048	76	80
11–15	25.4	0.048	71	70
7.5–9.5	24.8	0.095	81	81
7.5–9.5	24.8	0.095	80	86
7.5–9.5	24.8	0.095	78	87
7.5–9.5	24.8	0.095	75	83
27–32	13.7	none	72	71
27–32	13.7	none	67	68
27–32	13.7	none	69	83
27–32	13.7	none	66	65

TABLE 4
Effect of chemical pretreatment on length of filter run

Test Series	Alum mg/L	Polymer mg/L	pH of Settled Water	Estimated Time to 8-ft (2.4-m) Head Loss—h			
				GAC Filter	Sand Filter	Anthracite Filter	Dual-Media Filter
1	27.5	none	6.9–7.3		30	140	210
2	25.4	0.048	7.4–7.6	91	28	100	110
3	24.8	0.095	7.2–7.4	35	13	45	93
4	13.7	none	7.2–7.3	100	34	290	290

(7–30 ntu) waters and to learn whether the cyst concentrations in the filtered water were related to the turbidity of the filtered water.

This pilot-plant study allowed for the first time an evaluation of sedimentation and a simultaneous comparison of four types of filter media. Apparently, no previous research used anthracite as a single medium or granular activated carbon for Giardia cyst removal.

Procedures and equipment

Pilot plant. The pilot plant used in this research was DWRD's 2-gpm (0.13-L/s) conventional plant, constructed of plastic and fiberglass. A complete description has been given elsewhere.[8] The plant was operated at 1.9 gpm (0.12 L/s); this rate of flow through the rapid mixers, flocculator, and settling basin provided enough settled water for four filters operating at 0.3 gpm (0.02 L/s) each, with surplus water available for sampling. Excess settled water was wasted.

The water source was the Ohio River, and raw water was trucked from the Cincinnati Water Works pumping sta-

tion to the pilot plant and stored in a 7000-gal (26 250-L) stainless steel tank. Turbidity in the river declined to about 7 ntu during the summer, so a slurry of finely blended topsoil and clay subsoil was added for the last series of tests. The soil particles readily remained in suspension during the entire run and kept the turbidity at the 30-ntu level.

The pilot-plant treatment train consisted of two rapid-mix basins in series followed by flocculation, sedimentation, and filtration. Because the research was done to evaluate cyst removal by sedimentation and filtration, no disinfectant was added. A solution of 3 percent alum was added to the first rapid-mix basin. The flocculation basin had three sets of vertical paddles, with baffles between each set. When polymer was added, it was fed into the pipe that conveyed coagulated water to the flocculator. The polymer was a high-molecular-weight, slightly anionic polyacrylamide,* which was intended to strengthen the floc and thus improve the performance of the settling basin and the filters. A single settling basin was used. Dimensions of all basins, along with theoretical detention times for operation at 1.9 gpm (0.12 L/s), are given in Table 1.

Four filters were used, each with a diameter of 4.25 in. (108 mm) and a surface area of 0.0985 sq ft (0.0092 m²). The filters were equipped with piezometers so that head loss data could be collected every 4 in. (102 mm) through the media. Filter effluent was piped to a progressing cavity pump, which was used as a rate controller. The filters were intended to be operated at 3.0 gpm/sq ft (2.0 mm/s); measured rates ranged from 2.4 to 3.0 gpm/sq ft (1.6 to 2.0 mm/s). The rate typically used at McKeesport is 2.0 gpm/sq ft (1.4 mm/s). Table 2 describes the media used in the filters.

Cyst spiking. When the pilot plant was operated to study cyst removal, it was run continuously for about two days before cysts were added. This was done so that the raw water could be spiked with Giardia cysts when the plant was in a condition of equilibrium. After spiking was started, the plant was operated for one half day or one and one half days, depending on the technique. Each three- to four-day experimental period is referred to as a test series; four test series were conducted in this research.

The initial series of tests was conducted with G. canis obtained from an infected dog. When further attempts to find Giardia cysts in dog feces proved futile, G. muris cysts were obtained from infected mice.

In the first two test series, cysts were added to the raw water in a 5-min slug

*Separan NP10, Dow Chemical Co., Midland, Mich. (Mention of commercial products does not constitute endorsement by USEPA)

dose (6 mL/min). The fraction of cysts expected to pass over the weir of the settling basin at a given elapsed time after dosing (time zero) was assumed to be the same as the fraction of added total dissolved solids that had passed over the settling basin weir at the same elapsed time during an earlier salt dosing experiment. The number of cysts expected to pass over the weir during a given time interval would be the fraction expected at the end of the interval less the fraction expected at the beginning, with this difference multiplied by the total number of cysts dosed. Cyst concentration was then calculated based on the volume of raw water that flowed over the weir during the same time interval.

In the last two series, the cysts were kept iced in a 2000-mL stirred container and were fed into the raw water continuously at 1 mL/min. Cyst concentration was based on the volume of raw water, the volume of cyst suspension fed, and the calculated concentration of cysts in the stock suspension.

Calculations of alum and polymer concentration were based on volume of raw water treated, volume of chemical solution fed, and concentration of chemical in the feed solution. Chemical solutions and cyst suspensions were fed by means of peristaltic pumps. Data on cyst and chemical concentrations are given with treated water data for each test series.

Sampling and analysis. Water samples to be analyzed for *Giardia* cysts were filtered through polycarbonate membrane filters having a 5-μm pore size. For filtered water samples, the entire flow of water from the deep-bed granular media filter was passed through the membrane during the sampling period. Back pressure on membranes was measured, so that sampling could be discontinued if pressure exceeded 10 psi (69 kPa). Generally, pressure rose slowly and was not a problem. Pressure rose rapidly and exceeded 10 psi (69 kPa) for a few turbid filtered waters. For filtered waters, 142-mm-diameter membrane holders were used for the dual-media, anthracite, and sand filters. Because only three of these filter holders were available, the GAC filter was sampled with a 293-mm-diameter membrane sampler. The volume of the filtered water sample varied from 3 to 154 gal (10 to 583 L). Small volumes were collected during filter ripening or turbidity breakthrough; large volumes were collected after the filters had ripened. Settled water samples (5 gal [20 L]) were pumped through a 293-mm-diameter membrane sampler four times during the second, third, and fourth test series. The 16 gal (60-L) settled water samples collected during series 1 contained excessive debris and could not be analyzed satisfactorily.

After collection, samples were taken to the laboratory where cysts were washed off the membranes and holders with a 0.01 percent solution of polysorbate 20 followed by distilled water. Samples of filter concentrate were refrigerated until the time of counting. Sampling equipment was washed in hot, soapy water, rinsed in hot water, and reused. After being concentrated by centrifuge, the cysts were counted in a hemacytometer. "Dirty" samples were centrifuged in 50-mL centrifuge tubes containing 25 mL of $1M$ sucrose solution overlaid with 25 mL of distilled water containing the cyst suspension. Centrifugation at $800 \times G$ for 10 min concentrated the cysts at the sucrose–water interface and separated the cysts from the debris. On the basis of three experiments in which known numbers of cysts were diluted, filtered, concentrated, and counted, a correction factor of 0.76 was applied to account for cysts lost.

Grab samples of raw, settled, and filtered water were analyzed for turbidity, pH, and alkalinity. Sampling frequency was once each hour; filtered water was tested more frequently when turbidity was rising or immediately after backwash. Alkalinity was determined by titration to the methyl orange endpoint, pH was measured electrometrically, and turbidity was measured with a turbidimeter.*

Pilot-plant flows were measured with nutating-disk, plastic domestic water meters; head loss was indicated by piezometer tubes. The chemical solutions or cyst suspensions remaining in the graduated cylinders were recorded periodically. Chemical feed rates, applied cyst concentrations, and filtration rates were calculated from these data.

Settled water samples were collected and analyzed to obtain information on the cyst removal capabilities of sedimentation. About one quarter of the samples were of settled water, and about three fourths were of filtered water. In order to express the performance of the complete conventional treatment train, percentage removals of cysts for filtered water samples were based on the cyst concentrations of the raw water.

Results

The results obtained for a continuously flowing sedimentation basin are shown in Figure 1, which indicates that as turbidity removal in the settling basin improved, so did cyst removal. The results in Table 3 show that both cyst removal and turbidity removal were enhanced by use of a polymer and a higher dosage of alum.

Both the rate of head loss development during filtration and filtered water quality influence the choice of chemical

---

*Hach 2100 A, Hach Chemical Co., Loveland, Colo.

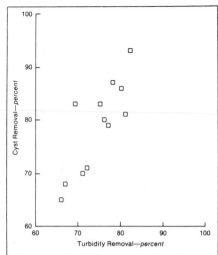

Figure 1. Turbidity removal and cyst removal by sedimentation

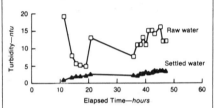

Figure 2A. Turbidity of raw and settled water (test series 2)

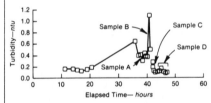

Figure 2B. Turbidity of effluent from anthracite filter (test series 2)

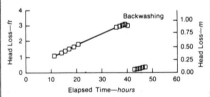

Figure 2C. Head loss in anthracite filter (test series 2)

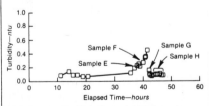

Figure 2D. Turbidity of effluent from dual-media filter (test series 2)

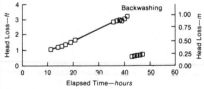

Figure 2E. Head loss in dual-media filter (test series 2)

TABLE 5
Removal of G. canis *cysts during test series 1**

Media	Condition	Head Loss ft	Head Loss m	Rate gpm/sq ft	Rate mm/s	Turbidity—ntu Raw Water	Turbidity—ntu Filtered Water	Cysts/L Applied†	Cysts/L Filtered Water	Cyst Removal percent
Sand	Ripened	1.4–2.4	0.43–0.73	2.74	1.86	22–25	0.09–0.28	426	4.6	98.9
Sand	Ripened	2.4–4.0	0.73–1.22	2.74	1.86	22–24	0.08–0.16	141	<1.3	>99.0
Anthracite	Ripened	0.9–1.2	0.27–0.37	2.79	1.89	23–25	0.12–0.45	514	11	97.8
Anthracite	Ripened	1.2–1.3	0.37–0.40	2.79	1.89	22–23	0.14–0.51	337	26	92.3
Dual media	Ripened	1.2–1.4	0.37–0.43	2.87	1.95	22–25	0.09–0.15	336	3.7	98.9
Dual media	Ripened	1.4–1.7	0.43–0.52	2.87	1.95	22	0.08–0.16	128	3.4	97.3

*pH of settled water—6.9–7.3; alum dosage—27.5 mg/L
†1.65×10^6 cysts dosed in 5 min; concentration calculated

TABLE 6
Removal of G. muris *cysts during test series 2**

Media	Condition	Head Loss ft	Head Loss m	Rate gpm/sq ft	Rate mm/s	Turbidity—ntu Raw Water	Turbidity—ntu Filtered Water	Cysts/L Applied†	Cysts/L Filtered Water	Cyst Removal percent	Sample Code in Figure 2
GAC	Ripened	2.0–2.1	0.61–0.64	2.55	1.73	11–15	0.06–0.15	22 000	2.9	99.987	
GAC	Ripened	2.3–2.6	0.70–0.79	2.55	1.73	12–16	0.07–0.14	8 600	<2.0	>99.977	
Sand	Ripened	5.7–6.6	1.74–2.01	2.68	1.82	11–15	0.06–0.08	22 000	4.1	99.981	
Sand	Ripened	7.0–7.9	2.13–2.41	2.68	1.82	12–16	0.07–0.13	8 600	<1.9	>99.978	
Anthracite	Turbidity fluctuating	3.0	0.91	2.82	1.91	12	0.31	11 200	630	94.4	A
Anthracite	Turbidity breakthrough	3.0–3.1	0.91–0.94	2.82	1.91	11–13	0.38–1.1	25 000	340	98.6	B
Anthracite	Backwashed, ripening	0.2	0.06	2.95	2.00	14–15	0.28–0.13	17 000	150	99.1	C
Anthracite	Backwashed, ripened	0.2–0.3	0.06–0.09	2.95	2.00	12–16	0.09–0.17	8 600	17	99.80	D
Dual media	Turbidity rising	2.9	0.88	2.81	1.91	13	0.21–0.25	22 000	44	99.80	E
Dual media	Turbidity rising	3.1	0.94	2.81	1.91	13	0.26	25 000	210	99.16	F
Dual media	Backwashed, ripening	0.6	0.18	2.92	1.98	15	0.15–0.08	17 000	<8.9	>99.94	G
Dual media	Backwashed, ripened	0.6–0.7	0.18–0.21	2.92	1.98	12–16	0.08–0.10	8 600	5.1	99.941	H

*pH of settled water—7.4–7.6; alum dosage—25.4 mg/L; polymer—0.048 mg/L
†82×10^6 cysts dosed in 5 min; concentration calculated

TABLE 7
Removal of G. muris *cysts during test series 3**

Media	Condition	Head Loss ft	Head Loss m	Rate gpm/sq ft	Rate mm/s	Turbidity—ntu Raw Water	Turbidity—ntu Filtered Water	Cysts/L Applied†	Cysts/L Filtered Water	Cyst Removal percent
GAC	Ripened	4.6–5.2	1.40–1.58	2.42	1.64	8.0–9.5	0.06–0.08	31 000	17	99.94
GAC	Ripening after wash	0.3	0.09	3.04	2.06	7.7	0.17–0.08	31 000	42	99.86
GAC	Backwashed, ripened	0.6–1.4	0.18–0.43	3.04	2.06	7.5–8.5	0.06–0.09	31 000	13	99.958
Sand	Ripening after wash	1.3	0.40	2.86	1.94	8.1	0.14–0.13	31 000	8.3	99.973
Sand	Backwashed, ripened	1.8–5.6	0.55–1.71	2.86	1.94	7.5–8.5	0.07–0.09	31 000	5.2	99.983
Anthracite	Ripened	3.4–4.0	1.04–1.22	2.92	1.98	8.0–9.5	0.10–0.14	31 000	19	99.94
Anthracite	Ripening after wash	0.2	0.06	2.90	1.97	7.7	0.35–0.13	31 000	35	99.89
Anthracite	Backwashed, ripened	0.3–0.5	0.09–0.15	2.90	1.97	7.5–8.5	0.10–0.16	31 000	11	99.96
Dual media	Ripened	2.0–2.8	0.61–0.85	2.90	1.97	7.5–9.5	0.06–0.09	31 000	12	99.96

*pH of settled water—7.2–7.4; alum dosage—24.8 mg/L; polymer dosage—0.095 mg/L
†Cysts dosed to raw water continuously; concentration calculated

TABLE 8
Removal of G. muris *cysts during test series 4**

Media	Condition	Head Loss ft	Head Loss m	Rate gpm/sq ft	Rate mm/s	Turbidity—ntu Raw Water	Turbidity—ntu Filtered Water	Cysts/L Applied†	Cysts/L Filtered Water	Cyst Removal percent	Sample Code in Figure 3
GAC	Ripened	2.0–2.1	0.61–0.64	2.96	2.01	29–30	0.08–0.12	11 400	5.3	99.95	
GAC	Ripened	2.1–2.8	0.64–0.85	2.95	2.00	27–29	0.08–0.11	11 400	5.5	99.95	
Sand	Ripened	5.6–6.6	1.71–2.01	2.68	1.82	29–30	0.11–0.13	11 400	3.5	99.97	I
Sand	Ripened, turbidity rising	7.7–7.9	2.35–2.41	2.68	1.82	27–32	0.17–0.23	11 400	16.7	99.85	J
Anthracite	Turbidity breakthrough	0.8	0.24	2.92	1.98	30	0.80–0.84	11 400	440	96.1	K
Anthracite	Turbidity breakthrough	0.9	0.27	2.92	1.98	30	0.88–1.2	11 400	240	97.9	L
Anthracite	Ripening after wash	0.2	0.06	2.89	1.96	29	0.78–0.44	11 400	8.7	99.92	M
Anthracite	Backwashed, ripened	0.2	0.06	2.89	1.96	29	0.26–0.40	11 400	14.5	99.87	N
Dual media	Ripened	1.2–1.4	0.37–0.43	2.91	1.97	29	0.08–0.18	11 400	7.9	99.93	
Dual media	Ripening after wash	0.6	0.18	2.92	1.98	28	0.37–0.17	11 400	61	99.46	

*pH of settled water—7.2–7.3; alum dosage—13.7 mg/L; no polymer
†Cysts dosed to raw water continuously; concentration calculated

dosages at water filtration plants. In this research the primary emphasis was on filtered water quality, and some dosages used would not be considered appropriate for use with certain media types, because of either inferior water quality or short filter runs. The estimated times to reach 8 ft (2.4 m) of head loss under various conditions are given in Table 4. For most of the runs, terminal head loss was not developed, so the run length was based on a linear extrapolation of head loss to the 8-ft (2.4-m) value. Head loss would probably have increased exponentially, so the estimated times exceeding 35 h may be somewhat long. Nevertheless, they show that filter runs longer than 24 h could be obtained with all chemical combinations used, except for the sand filter with the combination of 24.8 mg alum/L and 0.095 mg polymer/L.

The first series of filter runs was conducted with sand, dual media, and anthracite media, but not with GAC. During this test series, all filters operated in a "ripened" or physically and chemically matured condition. Data in Table 5 clearly show that both turbidity and cyst concentration were lower in sand and dual-media effluent than in the effluent from the anthracite media.

During the first test series, all filters were shut off in the evening when *Giardia* sampling was completed, and the entire 1.9-gpm (0.12-L/s) flow of settled water was wasted overnight. The next day, the dual-media and anthracite filters were restarted without being backwashed. Within a few minutes, the sampling membrane filter located downstream from the anthracite filter had plugged and backpressure was close to 50 psi (345 kPa). Head loss after restart was 0.65 ft (0.20 m) of water, compared with 1.70 ft (0.52 m) at the time of shutdown the previous evening. The membrane was so filled with debris that the sample could not be analyzed with confidence for cysts. This episode points out the need to restart only clean coarse-media filters, especially when they are to be brought up to the normal operating rate very quickly.

All four types of media were compared in series 2. When alum and 0.048 mg/L of polymer were used, the quality of the effluent from the ripened anthracite filter was inferior to that from the other media types tested (Table 6 and Figures 2A–E). When water samples were collected from anthracite and dual-media filters during conditions of rising turbidity (Figures 2B and 2D), cyst concentrations were much higher than during operation of the lower turbidity, matured filter. Cyst concentrations in filtered water rose by factors of 20 to 40 even though turbidity rose by factors of only 3 to 10. When the anthracite filter was backwashed and restarted, cyst concentration just after backwashing was considerably higher than it was after the filter had matured.

Pilot plant series 3, in which the polymer dose was 0.095 mg/L, showed that the coarse anthracite media could produce water similar in quality to that from dual media or GAC when the floc in the settled water was tough enough to adhere to the larger pores of the anthracite bed (Table 7). With regard to cysts, the quality of the effluent from the sand filter was slightly better than from the other filters. Turbidity was comparable from all matured filters. Some quality deterioration was observed when the GAC filter and the anthracite filter were maturing after backwashing. In both cases, cyst concentration during filter ripening was about three times that after the filter had matured.

In an attempt to study turbidity breakthrough conditions, the last series of filter runs (Table 8 and Figures 3A–E) was conducted with a lower alum dosage and higher raw water turbidity. Breakthrough did not occur with the GAC filter, but when head loss exceeded 7 ft (2.1 m) of water in the sand filter, turbidity rose to 0.23 ntu and the cyst concentration was somewhat higher (Figure 3D). When turbidity rose to the 0.8–1.2 ntu level in the anthracite filter, cyst concentrations were 20 to 30 times those observed for a backwashed, matured filter (Figure 3B). Cyst concentration was higher just after backwashing in the dual-media filter effluent but not in the anthracite filter effluent. This appears to be an anomaly.

Discussion

Continuous-flow sedimentation basins are capable of removing a significant fraction of *Giardia* cysts. Cyst removal appears to improve as turbidity removal increases. The performance of the continuous-flow sedimentation basin in the pilot plant, however, was inferior to the performance of batch settling (jar test results of Arozarena[7]) for cyst removal. Although short-circuiting existed in the pilot-plant basin, the pilot-plant results are considered to be a better indication of the capability of settling basins to remove *Giardia* cysts than are jar test results.

The results of sampling for the first 30 to 35 min after backwashing indicate that *Giardia* cysts can be expected to pass the filter in higher concentrations immediately after backwashing than after a filter has ripened. Similar data were obtained earlier for *Giardia* cysts[3] and several decades ago for bacteria.[9] Studies of filter behavior indicate that if *Giardia* cysts are expected to be in raw water, water quality could be improved by filtering to waste for perhaps 15 to 30 min after backwashing.

The data obtained during times of

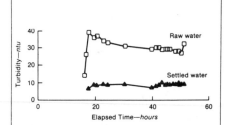

Figure 3A. Turbidity of raw and settled water (test series 4)

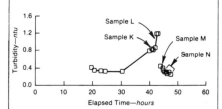

Figure 3B. Turbidity of effluent from anthracite filter (test series 4)

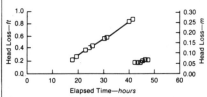

Figure 3C. Head loss in anthracite filter (test series 4)

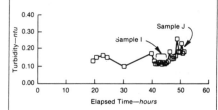

Figure 3D. Turbidity of effluent from sand filter (test series 4)

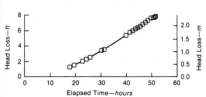

Figure 3E. Head loss in sand filter (test series 4)

rising turbidity and turbidity breakthrough have important implications for filter operation. During sampling, none of the filtered water turbidities exceeded 1.4 ntu, the highest value that can be rounded down to 1 ntu, which is the maximum contaminant level for turbidity. Nevertheless, cyst concentrations increased by factors as high as 20 to 40 when turbidity rose from the lower values attained during mature operating conditions. This observation held true even when the raw water turbidity was about 30 ntu and turbidity removal in the treatment train ranged from 93 to

>99 percent and is similar to an earlier finding for a pilot filter operated in a direct filtration mode.[3] These results lead to the conclusion that in order to keep the cyst concentration in filtered water as low as possible, treatment plant operators must keep the filtered water turbidity as low as possible.

Results of simultaneous filter runs clearly showed that the 0.92-mm-effective-size monomedia anthracite filter produced a water of inferior quality when alum was the only coagulant. This situation was remedied by the use of the polymer, which produced a tougher floc. The tougher floc resisted penetration and breakthrough in the coarse anthracite media, but backwashing was made more difficult because the polymer–alum floc trapped in the top inch (25 mm) of the anthracite media resisted breaking up when the filter was washed. During the initial period of the surface wash, the floc and media broke up into pieces about 1 cm^3 in volume, with shapes resembling broken concrete fragments. Eventually these floc–media fragments were broken apart by the surface wash.

At a plant like McKeesport's, where surface wash facilities are not available, failure to break up a tough floc and attain cleanly washed media could result in the growth of mudballs. This could eventually require removal of the media and perhaps the support gravel and rebuilding of the filters. Thus, the use of high-molecular-weight nonionic or slightly anionic polymer to improve the quality of the effluent from monomedia anthracite filters may not be advisable if filter backwashing capabilities are poor.

As an alternative to the use of such polymers, older plants that lack surface wash facilities may choose to optimize the pH of coagulation for the iron or aluminum salt being used. This might promote formation of a better floc. Careful control of filtered water quality is essential. If possible, the turbidity of the water from each filter should be monitored continuously. Small rises in turbidity (0.2 or 0.3 ntu) should be considered an appropriate reason for backwashing filters. A goal of 0.1 ntu (one of AWWA's quality goals for potable water[10]) might be set, even if it cannot be attained at all times, so that plant operators would strive to improve water quality.

In order to produce a filtered water with turbidity in the range of 0.1–0.2 ntu, some water filtration plant operators may need to accept filter run lengths in the 24–48-h range, rather than the 50–100-h range that can be attained with a weaker floc. The trade-offs involved in this decision would need to be evaluated on a plant-to-plant basis.

The importance of effective multiple barriers must be emphasized. For rivers like the Youghiogheny River, which do not have controlled watersheds, reliance has to be placed on both the disinfection barrier and the filtration barrier. Tables 5–8 show that even though substantial reductions in cyst concentration were achieved during these studies, cysts were usually detected in the filtered water. Although cyst concentrations were frequently only 3–10 cysts/L when the filters were operated properly, this concentration may be sufficient to cause infection. Rendtorff found that one *Giardia* cyst was insufficient to cause infection, but 10 cysts caused infection in two volunteers.[11]

Care should be taken to achieve adequate disinfection by operating at an effective pH and by using an adequate free chlorine residual and a sufficient contact time. Contact time should be based partly on the fact that when water is being filtered and added to the clearwell and water is being withdrawn from the clearwell at the same time, some short-circuiting is almost inevitable. Thus, a portion of the water leaving the plant receives less than the theoretical contact time. At most plants, operators probably have better knowledge of chlorine concentration than of contact time, and caution should be exercised when the product of chlorine dosage times contact times is being estimated.

Conclusions

Removal of *Giardia* cysts by sedimentation ranged from 65 to 93 percent and was generally similar to percentage removals of turbidity.

The concentration of *Giardia* cysts was influenced by relatively small turbidity changes (0.2 to 0.3 ntu), increasing greatly as turbidity increased.

The coarse anthracite media generally produced water of inferior quality when the bed depths of all filters were similar.

Use of high-molecular-weight, slightly anionic polymer improved the performance of the anthracite filter but made backwashing more difficult.

Because cyst concentrations may be higher than usual during the first portion of a filter run (just after backwashing), operators should consider a filter-to-waste period when a new run is begun.

Effective control of *Giardia* cysts depends on informed application of the multiple-barrier concept. Adequate disinfection is absolutely essential.

Acknowledgments

The authors thank John Ashton, McKeesport Municipal Water Authority; Steven Steranchak, Pennsylvania Department of Environmental Resources; Wilder Bancroft, Allegheny County Health Department; Herbert Braxton, USEPA Drinking Water Research Division; and Frank Schaefer III, USEPA Health Effects Research Laboratory.

References

1. AKIN, E.W. & HOFF, J.C. Microbiological Risks Associated with Changes in Drinking Water Disinfection Practices. *Water Chlorination: Environmental Impact and Health Effects*, Vol. 5 (in press).
2. RICHARDS, N.M. Allegheny County Health Dept. Personal communication (Mar. 15, 1984).
3. LOGSDON, G.S. ET AL. Alternative Filtration Methods for Removal of *Giardia* Cysts and Cyst Models. *Jour. AWWA*, 73:2:111 (Feb. 1981).
4. DE WALLE, F.B.; ENGESET, J.; & LAWRENCE, W. Removal of *Giardia lamblia* Cysts by Drinking Water Treatment Plants. EPA-600/2-84-069 (Mar. 1984).
5. AL-ANI, M. & HENDRICKS, D.W. Rapid Sand Filtration of *Giardia* Cysts. Proc. 1983 ASCE Natl. Conf. Envir. Engrg. ASCE, New York (1983).
6. HENDRICKS, D.W.; BELLAMY, W.D.; & AL-ANI, M. Removal of *Giardia* Cysts by Filtration. Proc. 1984 Natl. Conf. Envir. Engrg. ASCE, New York (1984).
7. AROZARENA, M.M. Removal of *Giardia muris* Cysts by Granular Media Filtration. Master's thesis, Univ. of Cincinnati, Ohio (Apr. 1979).
8. SORG, T.J. & LOGSDON, G.S. Treatment Technology to Meet the Interim Primary Drinking Water Regulations for Inorganics: Part 2. *Jour. AWWA*, 70:7:379 (July 1978).
9. Report of the Filtration Commission of the City of Pittsburgh, Pennsylvania (Jan. 1899). Data in *Jour AWWA*, 74:12:653 (Dec. 1982).
10. AWWA Statement of Policy: Quality Goals for Potable Water. *Jour. AWWA*, 60:12:1317 (Dec. 1968).
11. RENDTORFF, R.C. The Experimental Transmission of *Giardia lamblia* Among Volunteer Subjects. Proc. Sym. Waterborne Transmission of Giardiasis. EPA-600/9-79-001 (June 1979).

About the authors: *Gary S. Logsdon is a research sanitary engineer with the Drinking Water Research Division, US Environmental Protection Agency, 26 W. Saint Clair St., Cincinnati, OH 45268, where he has conducted research on water filtration for the past 10 years. A graduate of the University of Missouri, Columbia (BS civil engineering, MS sanitary engineering) and of Washington University, St. Louis, Mo. (DSc), Logsdon is a member of AWWA and ASCE. His work has been published previously in such journals as* JOURNAL AWWA *and* Journal Environmental Engineering Division—ASCE. *V. Carol Thurman and Edward S. Frindt are both cooperative education students at the College of Engineering, University of Cincinnati, Cincinnati, OH 45221. John G. Stoecker is an environmental scientist, Region III Water Supply Branch, USEPA.*

Reprinted from *Jour. AWWA*, 77:2:61 (Feb. 1985).

CONTROL OF GIARDIA CYSTS BY FILTRATION: THE LABORATORY'S ROLE

Gary S. Logsdon
U. S. Environmental Protection Agency
Cincinnati, Ohio 45268

David W. Hendricks
Colorado State University
Fort Collins, Colorado 80523

Gordon R. Pyper
Dufresne-Henry, Inc.
North Springfield, Vermont 05150

Charles P. Hibler
Colorado State University
Fort Collins, Colorado 80523

Robert Sjogren
University of Vermont
Burlington, Vermont 05405

INTRODUCTION

Dramatic improvements have been made in the quality of drinking water in the USA since the turn of the century. Waterborne epidemics of cholera and typhoid fever are rare events now. Nevertheless, the need for vigilance continues, as in recent decades we have become aware that other biological agents are causing waterborne disease.

One notable example is giardiasis, a disease that received very little attention before the 1960's. According to Rendtorff (1), "Giardia was thought in the 1950's to cause occasional problems of diarrhea in children but its appearance was so common and, in adults so lacking in clinical symptomology, that most considered it a non-pathogen." Recently Craun and McCabe (2) reported that for the 1971-1979 time period, giardiasis was the cause of 11 percent of the waterborne disease outbreaks in the USA. Waterborne giardiasis outbreaks have continued to occur in the 1980's. The need to improve water treatment practice continues in some locations, as proven by the giardiasis outbreaks.

For several years, the Drinking Water Research Division has been conducting and sponsoring research on water filtration in order to develop information needed by water utilities that treat waters that might be contaminated by Giardia cysts. This paper presents information on some of the laboratory methods used in this work and briefly relates preliminary findings of recent research.

CYST SOURCES

One of the significant obstacles to conducting research on water filtration for Giardia cyst removal has been and continues to be the lack of a readily available and easily maintained source of the cysts. Giardia cysts can not be cultured and grown, as can E. coli, for example, so they must be obtained from warm blooded animals that are infected and shedding cysts. Donors have included mice, dogs, and humans.

In early cyst removal research performed by the Drinking Water Research Division (DWRD), cyst models (9 µm radioactive beads) and G. muris cysts from mice were used (3). Giardiasis is not a common disease in Cincinnati, where the DWRD laboratory is located, so obtaining cysts from human donors has been very difficult. In order to obtain research results in a timely fashion, DWRD performed exploratory studies with cysts other than G. lamblia.

Studies later funded by DWRD have used Giardia cysts from human donors or from dogs. Granular media filtration research at the University of Washington (4) was done with G. lamblia from humans. The quantity of cysts was limited, and the supply varied from time to time; thus much of the work was based on counting of cyst-sized (8 to 12 µm) particles.

Research at Colorado State University (CSU) has been conducted with Giardia cysts obtained from dogs. Cysts from dogs have been reported to have caused human infection (5), and G. lamblia cysts from human stool specimens and G. lamblia trophozoites (Portland-1 strain) have caused infections in dogs (6). The physical characteristics of Giardia cysts from humans and from dogs seem to be very similar or identical. For purposes of water filtration studies, use of Giardia cysts from dogs should be as appropriate as use of Giardia cysts from human donors.

Filtration research at McIndoe Falls, Vermont has been conducted with cysts from human donors. G. lamblia cysts were obtained from stool samples through the cooperation of the Vermont Health Department and the Mary Hitchcock Hospital in Hanover, New Hampshire.

CYST DOSING AND SAMPLING PROCEDURES

CSU Dosing

To obtain cysts for pilot scale filtration experiments conducted at the Environmental Research Center (ERC) at Colorado State University, cysts were separated from canine fecal samples by mixing the feces in water, allowing the sediment to settle, and saving the supernatant (7). If the sample was quite dirty it was filtered through cheesecloth. The processed cyst samples were refrigerated until use.

When a filtration experiment was to begin, cysts were added to a temperature controlled milk cooler filled with Horsetooth Reservoir water and kept at 5° or 15° C. The cyst concentrations in the feces suspensions were determined by three to five analyses of each suspension prior to adding it to the milk cooler. The concentration of cysts in the milk cooler was then calculated, and designated as the "concentration added." The cyst concentration in the milk cooler was determined also by analyzing a sample from the cooler. This sample was obtained by pumping water from the milk cooler through the polycarbonate membrane sampling system, and the measured cyst concentration was designated as the influent "detected" Giardia concentration.

The size of the water sample taken from the milk cooler ranged from 2 to 15 L. The sample volume was limited by the amount of water that could be filtered through the membrane filter before a pressure of 10 psi was reached. A 200 mL grab sample was also collected from the cooler and used for standard plate count and coliform analyses.

For field testing at CSU using a 1 gpm diatomaceous earth filter, cysts were added to a 700 L raw water feed tank. Raw water was sampled by the membrane filter method to determine the concentration of cysts in the raw water. When the 20 gpm mobile rapid sand pilot plant was used, cysts were dosed on a continuous basis from a suspension of known concentration. The cysts were added to the raw water and mixed before treatment chemicals were added.

Vermont Dosing

In the Vermont research, cysts were stored in a 5% formalin solution at 0° to 4° C until processed for concentration by sucrose separation. The concentrate after the separation step was stored at 4° C until used.

In the research at McIndoe Falls, cyst concentrates were counted at the University of Vermont. Three subsamples were used and counted with a hemacytometer to determine the number of cysts. The cyst concentrate was then shipped on ice to McIndoe Falls. Because the filters were operated at 10 to 15 gpm flows (characteristic of very small systems) the cysts were added as an instantaneous slug, in order to present the maximum challenge to the filters.

CSU Sampling

Polycarbonate membrane filters were used to sample for Giardia cysts at CSU for both influent feed tank water and filter effluent. For most experiments, filters having 142 mm diameter and 5 µm pores were used. Larger filters (293 mm) were used to sample the effluent from the 20 gpm mobile filtration plant.

Sampling protocol for the feed tank was described by Silverman (8): The filter holder was washed and the membrane was installed, tap water was introduced into the filter device from the effluent side, the filter was then attached to the sampling pump, the remaining air was bled off, and water was pumped through the filter. This was done very carefully in order to avoid creating pressure disturbances when filter effluent samples were taken. The volume of water passed through the membrane was measured and recorded so that the cyst concentration could be calculated. When sampling was complete, the flow of water to the membrane filter was shut off, excess water in the filter chamber was removed by vacuum, and the apparatus was opened. The top and bottom of the filter holder and the membrane were all thoroughly washed into a Pyrex® dish to recover cysts. The washings were sealed in a mason jar, labeled, and refrigerated until they could be concentrated and counted. Sampling protocol for the effluent side was identical, except the pump was not needed for pilot filters operating under pressure, which included the laboratory scale rapid sand filter, three slow sand filters, and the DE filter. Evaluation of the membrane sampling procedure gave 45% average recovery, with a range of 36% to 54%.

Vermont Sampling

Sampling of the filter effluent for cysts at Vermont consisted of pumping water at 1 gpm (about 8% of sand filter effluent) from the test filters to a membrane filter holder equipped with a water meter and pressure gauge. Polycarbonate membrane filters having a 293 mm diameter and 5 µm pores were used. The meter reading was recorded and timed at the start and at the end of the sampling period in order to determine the volume of water sampled for cysts. Sampling was continued for 24 hours per sample, in the slow sand filter runs, and pressure on the influent side of the filter never exceeded 10 psi.

When sampling was completed, all the water in the filter holder was drained by an aspirator. The filter membrane was removed from its holder by means of tweezers, folded to the center to retain any moisture, and placed in a one quart wide mouth mason jar. Adding 50 mL of distilled water to the jar containing the filter membrane was found to be advantageous when reduction of interference from background material was needed. Formalin-fixed cysts did not disintegrate, but some naturally occurring biological materials did. The jars were tightly sealed and shipped in an iced condition to the laboratory for counting.

The cyst sampling methods used in these two research projects differed from the sampling technique described in Standard Methods (9). The method described in that publication has been used to sample raw and finished waters during outbreak investigations. Previous filtration studies (3, 4) used membrane filtration for cyst sampling because small (1 liter) and medium volumes (about 100 liters) could be sampled by membrane filters, in contrast to the 1900 L volume recommended for the yarn-wound orlon filters. Thus small pilot filters could be used and the volume of filtered water produced would be sufficient for analysis. Membrane filters have continued to be used for research on filtration for Giardia cyst removal even when larger systems are tested. This provides some continuity in procedures and is helpful when comparisons are made between various studies. Another factor favoring use of membrane filters for these studies is the high variability of cyst recovery inherent in the orlon yarn-wound filter sampler. Cyst recovery by membrane filter is higher and less variable, by comparison.

CYST COUNTING METHODS

CSU Counting

At CSU the cyst enumeration procedures were done under direction of Dr. C. P. Hibler at the Pathology Laboratory, and depended on the nature of the sample. The zinc sulfate flotation technique was used in 1981 at the start of the project. After investigation of other techniques the micro-pipette technique was adopted in July, 1982, as the latter method was found to be more accurate.

Fecal samples of cyst suspensions were analyzed according to the estimated concentration of cysts present. If the initial zinc flotation procedure indicated a relatively large number of cysts in the fecal sample, the actual number of cysts in the sample could be determined by the Stoll dilution technique. Samples having low cyst concentrations were less desirable and had to be concentrated and then counted by either the zinc flotation or the micro-pipette method.

The Stoll dilution technique consisted of adding 3 mL of Lugol's Iodine to a Stoll flask which was then filled to the 56 mL mark with cool distilled water. Four mL of thoroughly mixed fecal suspension were placed in the Stoll flask, and the flask was shaken to mix the contents. A 0.075 mL aliquot was removed from the flask via micropipette and placed in a vaseline well on a glass slide. A coverslip was affixed to the slide, and the slide was examined for Giardia cysts at 400 x magnification. The total number of cysts seen on one slide was multiplied by 200 to give the total number per mL in the sample. A minimum of 2 slides was prepared as described, read, and averaged.

Cyst suspension containers storing processed cysts ready for filtration experiments were labelled with the date and the number of cysts per mL. The samples were counted at least every third day, or before a portion was used, to ascertain that the cysts had not begun to degrade.

Samples obtained during operation of the pilot filters were processed by either the zinc flotation method or the micropipette technique. In the zinc flotation method, a fecal sample the size of a pea was placed in a centrifuge tube, 5 to 6 drops of Lugol's Iodine were added, and the sample was mixed well. The tube was filled half way with zinc sulfate ($ZnSO_4$, S.G. = 1.18) solution and mixed. The tube was then filled with zinc sulfate solution until the meniscus bulged at the top. A coverslip was affixed to the tube, and the coverslip was tapped with the end of a pencil to firm a secure bond. Next the tube was centrifuged for 5 minutes at 1500 rpm. The coverslip was removed from the tube, placed on a glass slide, and examined for Giardia cysts at 100 x magnification.

Samples obtained by the membrane filter sampling technique were brought to the CSU Pathology Laboratory and stored in the refrigerator overnight to settle the cysts and debris. The following day the samples were reduced in volume by pipette to less than 50 mL without disturbing the sediment that had formed. This remaining volume was remixed and poured into a 50 mL conical centrifuge tube. The tube was centrifuged for 5 minutes at 1500 rpm. The supernatant was then pipetted off until about 5 mL remained, and the procedure was repeated until finally all the sample had been concentrated to 1 mL and the sample jar had been rinsed well with distilled water.

Once the sample had been concentrated to 1 mL volume, either of the two counting methods were used to determine the cyst concentration. The first method used was zinc flotation. Later, the micropipette technique was used. This method entailed adding 5 to 6 drops of Lugol's Iodine to the 1 mL concentrated sample, mixing thoroughly, and withdrawing an aliquot of 0.05 mL. This volume was then placed in a vaseline well on a glass slide, a coverslip was affixed to the slide, and the slide was examined at 400 x magnification. If cysts were seen, a minimum of two aliquots were counted and averaged. The total number of cysts in the sample was found by multiplying by a factor of 20. Counts of replicate samples counted by micropipette varied from the mean by $\pm$ 20%.

Vermont Counting

In the McIndoe Falls study, cyst enumeration procedures were done under the direction of Dr. Robert Sjogren at the University of Vermont. Polycarbonate membrane filters were transported on ice from the field in one quart wide mouth glass mason jars. Jars and folded membrane filters were stored at 0° to 4° C with distilled water or 5% buffered formaldehyde at the University of Vermont.

Filters were processed by cutting them into 5 cm pieces and placing the pieces in a 40 oz. glass Waring blender cup. Any excess fluid in the sample jar was poured into the blender cup, and the jar was washed twice with about 20 mL of distilled water. This was added to the blender cup, and the volume of fluid for blending was made up to 200 mL. A few drops of liquid dishwashing detergent (Ivory®) previously diluted to 1:100 were added. The cut membrane was blended for 30 seconds at 20,000 rpm using a single speed blender base.

The contents of the blender were strained through two thin layers of cheese cloth held in a large glass funnel supported by a 1 L glass beaker. . The blender cup was washed two or three times with about 30 mL of distilled water from a spray bottle. The cut pieces of membrane were stirred briefly with a wooden applicator stick during this process. The cheese cloth was then carefully squeezed to remove excess water. All fluid was collected in the 1 L beaker.

After the filtrate was collected, it was distributed among several 50 mL glass conical centrifuge tubes which were spun at 450 rpm for 15 minutes using a swinging bucket rotor centrifuge. The supernatant was then carefully aspirated from each tube, leaving the undisturbed pellet in the bottom along with about 2 mL of liquid. The pellets were again suspended and the contents of the 50 mL tubes were combined in a 15 mL conical glass centrifuge tube. All of the 50 mL tubes were rinsed in serial fashion with distilled water that was also added to the 15 mL tube. The Pastuer pipette used to transfer material was also rinsed and that liquid was added to the 15 mL tube. The 15 mL tube was then centrifuged in a clinical table top centrifuge for 15 minutes at 450 rpm.

Upon completion of the second centrifugation, the supernatant in the 15 mL tube was carefully removed using a Pastuer pipette until a final volume of 0.3 mL of pellet and fluid was left. To this was added 0.1 mL of fresh 5% Lugol's Iodine solution. The tube contents were mixed, and a portion of the cyst suspension was examined using a counting chamber and binocular microscope. Experiments demonstrated that a volume of 0.4 mL gave the best counting reproducibility.

Laboratory tests were performed to evaluate several aspects of the University of Vermont procedure. The blender method using detergent gave satisfactory results in determinations of cyst recovery when diluted suspensions were prepared as a method check. A study to determine if blending destroyed cysts revealed that the 30 seconds of blending did not result in loss of cysts. Four separate experiments were conducted to determine average cyst recovery from the 293 mm membrane filters. Recovery ranged from 50% to 60%. Formalin treated cysts, when stored in distilled water, did not decrease in numbers or show signs of deterioration during two months of storage. Of several different treatments of the cysts after they were in pellet form, staining with Lugol's Iodine was preferred, although use of acridine orange showed some promise for easier cyst identification.

Laboratory methods for other water quality parameters of interest in this study have been developed and standardized for a number of years. Coliform measurements in the CSU and McIndoe Falls studies were done according to Standard Methods (9). Analyses were performed at CSU and at Spectrum Environmental Research Laboratory in Montpelier, Vermont. Samples collected at sites away from the analytical laboratories were shipped on ice to the laboratories. Turbidity was measured on site in both projects with bench top turbidimeters or flow through turbidimeters or both.

FILTRATION EQUIPMENT USED

Filtration research at CSU has been conducted with pilot scale rapid rate granular media filters, slow sand filters, and a diatomaceous earth (DE) filter. Full scale, small system sized filters were also tested in these research projects. CSU used a 10 to 30 gpm rapid rate granular media package filtration plant. A 13 gpm slow sand filter and a similarly sized DE filter have been studied at McIndoe Falls.

CSU Pilot Filters

The CSU pilot scale rapid rate filters were operated in an "in-line" mode, using two stage rapid mixing followed by filtration (10). Rate of flow was from 0.4 to 0.7 L/min, and filtration rates were in the 5 m/hr to 12.5 m/hr range. Filter columns of 5 and 10 cm diameter were used. Sand filters had 76 cm of 0.45 mm e.s. (effective size) sand with a uniformity coefficient (u.c.) of 1.5. The dual media filters had 46 cm of 0.9 mm e.s. anthracite (u.c. = 1.7) over 30 cm of 0.45 mm e.s. sand (u.c. = 1.5). Coagulation was accomplished with alum or polymer or with combinations of alum and polymer. Some filter runs were made with uncoagulated water to show the detrimental effects of failing to coagulate clear waters before they are filtered.

Three pilot scale slow sand filters were assembled at CSU for studies of this process. The filters, described by Bellamy et al. (11), were 30 cm diameter columns filled with 96 cm of 0.28 mm e.s. sand (u.c. = 1.5) supported by 46 cm of coarse sand and gravel. The filters operated at rates of 0.04, 0.12, and 0.40 m/hr (1, 3, and 10 million gallons per acre per day, or mgad) with raw water fed from a common tank so the effects of filtration rate could be evaluated. No coagulation chemicals were used, because water generally is not coagulated before it is applied to slow sand filters. Water temperatures for these experiments were approximately 5° C or 15° C. The three filters were operated to conventional slow sand filter head losses (about one meter) but because of limited ceiling height they were operated in a pressure filter mode. Later seven more filter columns were constructed. These have been operated in a gravity mode because they were located in facilities with adequate head room for columns 3.3 meters high. The seven columns included some with different sand sizes and different media depths, so other design variables could be evaluated.

The diatomaceous earth pilot filter used at CSU was a 0.093 m^2 DE filter similar to the one described by Logsdon et al. (3). The filter was operated at filtration rates ranging from 2.4 to 9.8 m/hr (4 to 15 L/min or 1 to 4 gpm flow rate). Operation of the DE filter consisted of pre-coating, filtration, and backwashing. DE grades used for precoating and for body feed included Celite 545®, Celite 535®, Celite 503®, and Hyflo Super Cel®. Although normal DE filtration practice is to terminate a run because terminal head loss has been reached, in this research most of the runs were terminated because of operating time considerations rather than because terminal head loss had been reached.

CSU Small System Filter

The small scale granular media filter used at CSU was a Neptune-Microfloc WB-27 package plant that previously had been mounted on a trailer so it could be transported to locations where water filtration studies were to be carried out. It had been used earlier at Leavenworth and Hoquiam, Washington (4) and at Canton and Oneida, New York (12). This plant has a filter area of 0.37 m^2 (4 ft^2) and can operate in the 38 to 114 L/min (10 to 30 gpm) flow range. It was operated in an in-line or direct filtration mode by CSU.

Vermont Small System Filters

The slow sand filter at McIndoe Falls was one of two filter beds in a plant built in 1976 when the springs that had previously served as the community's water source were replaced. Later, ground water was developed by the community, so the slow sand filter was removed from service. Because it was not being used to produce potable water, it could be used for EPA-sponsored research. The filter bed used in this study was 5.9 x 6.7 m (19.5 x 22 ft). Flow at 51 L/min (13.6 gpm) resulted in a rate of filtration of 0.08 m/hr (2 mgad). The filter was designed to have 1.07 m (42 in) of 0.33 mm e.s. sand (u.c. = 2.7).

The DE filters used during the McIndoe Falls research were pressure filters. The first set of experiments was performed with a rented unit that had a septum area of 0.74 m^2 (8 ft^2). Later tests were made with a filter having a septum area of 0.93 m^2 (10 ft^2). Both units used slurry pumps to add body feed to the raw water. The grade of DE studied at McIndoe Falls was Celite 503®.

PILOT SCALE RESULTS

CSU Research Results

Preliminary results of rapid rate granular media filtration studies have been presented by Al-Ani and Hendricks (10). They reported that coliform removal by filtration (without disinfection) could be as high as 99.9% or more, when raw water spiked with sewage was filtered after being adequately coagulated. Previously presented results plus more recent data show that high removals of turbidity relate well with coliform removal when clear surface waters of less than 1 NTU are coagulated and filtered. Table 1 shows that when filtered water turbidities were 0.10 NTU or lower coliform counts were equal to or less than 1 per 100 mL and coliform removal efficiencies met or exceed 99.0% for 70% of all samples taken.

Granular media filtration research results from CSU also showed that Giardia cysts could be removed effectively when properly coagulated water was filtered. Cyst removal percentage exceeded 99% in numerous filter runs. The relationship between cyst breakthrough and turbidity in coagulated, filtered water is shown in Table 2. The frequency of occurrence of effluent samples having 5 or fewer cysts per liter decreased as filtered water turbidity increased. The same trend was seen for samples showing cyst removals of 99% or greater. A similar trend could be seen in the data analyzed if the occurrence of finding no cysts in the filtered water had been reported. The detection limit varied from run to run, however, depending on the volume of filtered water passed through the membrane filter sampling device, so those results would be more difficult to interpret.

The effect of using no chemical pretreatment is shown clearly in Table 3. Percent removals of Giardia cysts ranged from 36% to >99.99%, while percent removals of total coliform bacteria were -7% to 99.9%, and turbidity decreased as much as 77% in some runs but increased up to 72% in others. These results contrast sharply with removals of cysts and total coliforms when effective chemical pretreatment is used. The message is clear: effective chemical pretreatment is imperative if cysts and bacteria are to be removed adequately. It is clear also that if turbidity removal is 80% or greater, when raw water turbidity is less than about 1 NTU, removals of total coliform bacteria and Giardia cysts will exceed 99%.

Results of the CSU slow sand filtration studies, shown in Table 4, indicate that the filters were able to remove a variety of microorganisms. Total coliform removal ranged from >99.99% to >99.29%. Standard plate count removals were lower, ranging from 92.7% to 95.8%. Fecal coliform removal ranged from >98.4% to >99.8%. Removals were lowest at the highest filtration rates for all three bacterial parameters studied. An interesting aspect of this research was the lack of a relationship between removal of biological particles (total coliform, standard plate count, and fecal coliform) and removal of turbidity when Horsetooth Reservoir water was the raw water source. Average filtered water turbidity from all filters exceeded 1 NTU, ranging from 3.5 to 4.1 NTU. Raw water turbidity ranged from 5 to 10 NTU. The failure of the slow sand filters to produce filtered water meeting the 1 NTU MCL was not anticipated. When the filtered water turbidity exceeded expected values, samples of raw water were subjected to additional testing to determine the size characteristics of the particles. The results of turbidity analysis after 5.6 NTU water was passed through various sizes of membrane filters ranging from 0.22 to 8 μm indicate that 64% of the turbidity remained after filtration through the 1.2 μm membrane and 27% remained after filtration through the 0.45 μm membrane. Later the fine particles passing through the slow sand filter were reported to be kaolinite and montmorillonite clay particles.

The capability of slow sand filters to remove Giardia cysts was demonstrated in the CSU research. Results from Bellamy et al. (11) have been analyzed and are presented in Table 5. Cyst removal is influenced strongly by filter operating conditions. Removal of cysts was least effective when a new filter was placed in operation, before a biological population (a variety of organisms including protozoa, rotifers, and bacteria) could become established in the filter sand and in the coarse sand and gravel support media.

When the slow sand filters had a well developed biological population (minimum sand bed exposure was 29 weeks), Giardia cyst removals were nearly 100 percent, even though in some cases the filter had been scraped and the age of the schmutzdecke was as low as three weeks. The lowest percentages of cyst removal attained with the developed sand bed was 99.77%, and cyst removal was 99.90% or greater in 22 of 24 observations. In 12 of 24 observations, cysts were not found when filtered water volumes ranging from 81 L to 1,440 L were passed through 5 um pore diameter membrane filters to test for presence of Giardia cysts.

An evaluation of more severe operating circumstances showed that cyst passage could occur when no schmutzdecke was present on the sand surface and when new sand and gravel support media had been installed in the slow sand filter. Slow sand filtration often was effective for Giardia cyst removal, even in the absence of a schmutzdecke, if the filter bed contained an established biological population. This suggests that Giardia cyst removal occurs not only in the schmutzdecke, but also within the depth of the sand bed.

Research performed at CSU with the diatomaceous earth test filter has confirmed the earlier observation of Hunter, et al. (13) showing that fewer total coliforms passed through the DE filter as finer grades of diatomite were used. The CSU results, showing that removals of standard plate count organisms, total coliforms, and turbidity all improve as finer grades of DE are used, are presented in Table 6. Because of the size differences of bacteria vs. the fine clays in Horsetooth water, the removal of turbidity-causing particles was less than the removal of microorganisms for most grades of DE. The finest grade tested, Filter Cel®, was very effective for turbidity removal. This grade usually is not used in potable water treatment because of the high head loss caused by small particle size.

A comparison of turbidity removal by diatomaceous earth filtration and membrane filtration was made (7). A 5 μm filter permitted passage of about 98% of the turbidity, more than the 88% passed by Celite 503®. A 1.2 μm filter passed 64%, similar to the 72% passed by Celite 512®. The 9% passed by a 0.2 μm membrane filter was similar to the 2% passed by FilterCel®. These results indicate that the particle sizes passed by various grades of DE may be estimated by a comparative analysis of the effluent turbidity from membrane filters and DE filters.

Research with alum-coated DE showed that diatomite onto which aluminum hydroxide had been precipitated was more effective than uncoated diatomite for removal of standard plate count organisms, total coliforms, and turbidity. With Celite 545® turbidity removal improved from 13% to 98% and total coliform removal increased from 49% to 99.86%. Alum coating enabled the diatomite to remove smaller particles than uncoated DE could remove, The results suggest that filtered water turbidity might be more closely related to the microbiological quality of DE filter effluent when the alum coating process is used.

Field tests of DE filtration were also conducted using two mountain stream sources. A one square foot test filter was used at these sites at filtration rates of 2.4, 4.9, and 9.8 m/hr. Temperatures ranged from 3.5 to 10°C, and raw water turbidities varied from 0.55 NTU to 32 NTU. In these field tests a coarse grade of diatomite, Celite 545®, was used. Coliform removals were less effective at the 4.9 and 9.8 m/hr filtration rates than at the commonly used rate of 2.4 m/hr (42% and 57% at higher rates vs 51% to 97% obtained at 2.4 m/hr). Turbidity percent removal was much higher on these waters than on Horsetooth Reservoir water. The highest turbidity removal with the latter water was 24% when Celite 545® was used, and the average removal for all Celite 545® runs on that water was 13%. In contrast, with the other raw waters, turbidity removal ranged from 21 to 77%. Turbidity removal by diatomaceous earth filtration is in part a function of the nature of the water being treated.

Giardia cyst removal was studied with the DE test filter and Horsetooth Reservoir water. Of the 26 runs, the only one in which cysts were detected had a very high influent cyst concentration, 34,000 cysts per liter. When influent cyst concentrations ranged from 100 to 10,000 cysts/L, no cysts were detected in the filtered water. Results in Table 7 show that DE filtration is effective for cyst removal at rates of 2.4 to 9.8 m/hr. These experiments were performed with a raw water source having very fine turbidity-causing particles. The data in Table 7 very clearly show that cyst removal efficiency was not related to turbidity removal by DE filters.

Field tests for removal of Giardia cysts were performed at the Cache la Poudre River and at Dillon, Colorado, at filtration rates of 2.4 to 9.8 m/hr. Celite 545® was used in the six filter runs in which cyst removal was studied. In five of the six runs, no cysts were detected, and removals exceeded 99.93% to 99.99%, depending on influent cyst concentration and the volume of filtered water sampled for cysts. Filtered water turbidity in these runs ranged from 0.3 NTU to 1.5 NTU.

During one of the six filter runs, cysts passed through the filter. When the filter was then subjected to leak testing, existence of a leak was confirmed. The turbidity of the filtered water in the run in which the leak occurred was 7.3 NTU, and cyst removal was 99.4%. These results suggest that when DE filtration equipment is properly maintained, this process is capable of removing over 99.9% of Giardia cysts from raw water, and that process efficiency is related to proper operation and maintenance of equipment rather than to effluent turbidity.

SMALL SCALE SYSTEM RESULTS

CSU Rapid Rate Granular Media Filtration Results

The trailer mounted Water Boy® filtration plant was used to corroborate results found using the laboratory scale pilot plant. Preliminary testing without chemicals was done during the summer of 1983 using raw water from Horsetooth Reservoir. The results for Giardia cysts, total coliform bacteria and turbidity show percent removals of about 50%, 30%, and 15%, respectively. These results are virtually the same as those obtained with the laboratory scale pilot filters when they were operated without chemical pretreatment. Removals of Giardia cysts, total coliform bacteria, and turbidity were about 99.9%, 99.9/%, and 95% respectively when effective chemical pretreatment was used, e.g. sufficient to remove 80% of the influent turbidity or to produce a filtered water turbidity below 1.0 NTU, whichever gave the lower turbidity. Again, these results were virtually the same as those obtained with the laboratory scale rapid sand filtration plant. More testing is presently under way with Cache La Poudre River water as the source water for conditions of low turbidity, e.g. below 1 NTU.

McIndoe Falls, Vermont Slow Sand Filter and DE Filter Results

One of the two filter beds of the McIndoe Falls slow sand filter has been operated for research purposes most of the time since May, 1982. Ambient raw water quality data for the first eleven months of the study are shown in Table 8. The brook that serves as a raw water source drains a marsh that would probably have numerous animals, thus the coliform counts were higher than expected.

The slow sand filter's capability for removing bacteria and turbidity is shown in Table 9. Under ambient water quality conditions, bacteria removal was slightly better during times of warm water than it was when the water was cold. This is expected, because biological processes progress at faster rates as temperature increases. Turbidity removal, however, was better during cold weather. This may be indicative of a different nature of turbidity-causing particles during different seasons. The latter portion of the October-January period would be very cold, with the marsh and pond water covered by ice. Cold weather and ice cover would reduce the level of activity by marsh-dwelling animals and minimize the influence of storms on water quality.

The sewage spiking study simulated a source water contamination problem. Bacteria removals during spiking were lower than removals during operation with ambient raw water quality.

Giardia cyst removal by the 13 gpm slow sand filter appears to have been influenced considerably by the operating temperature. The 99.99% removal at 21°C (Table 10) is similar to results obtained at 15°C at Colorado State University. The 0.5°C winter temperature was colder than any studied at CSU's Engineering Research Center, and cyst removal was also lower at 93.7%. This suggests that further work should be done with very cold water.

One DE filtration test was performed at McIndoe Falls to evaluate the 10 square foot filter's ability to remove Giardia cysts. Because previous studies (3, 4) and CSU results (14) had shown that DE filtration was effective for removal of cyst models and Giardia cysts, extensive research was not needed. The filtration test at McIndoe Falls resulted in 99.97% removal of Giardia cysts. The run was short, only 0.9 hour, in part because of the grade of diatomite used, Celite 503®. Use of a coarser grade would have given a longer run. The cyst removal obtained at a filtration flow of 59 L/min was very similar to results obtained in the 1 sq ft. pilot filter studies.

CONCLUSIONS

Large membrane filters with 5 µm pores can be used to sample filtered waters for Giardia cysts.

Although the cyst recovery and counting procedures are not as precise as desired (+ 30%), because cyst removal by water filtration processes often was observed to be on the order of 2, 3, or 4 log cycles, the excellent removal efficiencies observed were really the result of filter performance and were not caused by method variability.

Adequate chemical pretreatment of raw water must be accomplished if consistently high cyst removal is an operating goal for rapid rate granular media filtration.

Efficiency of cyst removal by coagulation-filtration is related to degree of turbidity removal and to filtered water turbidity.

Diatomaceous earth filters, when properly operated, can be depended upon to remove 99.9% or more of Giardia cysts from raw water.

Slow sand filters can remove 99.9% of applied cysts at a variety of operating rates. Condition of the filter bed influences filter efficiency.

Field scale research suggests that low temperatures may cause a decline in Giardia cyst removal by slow sand filters.

REFERENCES

1. Rendtorff, R. C. The Experimental Transmission of Giardia lamblia Among Volunteer Subjects. In Proceedings of Symposium on Waterborne Transmission of Giardiasis, Edited by W. Jakubowski and J. C. Hoff. EPA-600/9-79-01. p. 65. (June 1979).

2. Craun, G. F. and McCabe, L. J. Waterborne Disease Outbreaks in the United States and the Coliform Standard. In Proceedings of Workshop on Assessment of Microbiology and Turbidity Standards in Drinking Water, Edited by P. S. Berger and Y. Argaman. EPA 570-9-83-001. p. 9. (July 1983).

3. Logsdon, G. S., Symons, J. M., Hoye, R. L., Jr. and Arozarena, M. M. Alternative Methods for Removal of Giardia cysts and Cysts Models. Journal American Water Works Association. 73:2:111. (February, 1981)

4. DeWalle, F. B., Engeset, J., and Lawrence, J. Removal of Giardia lamblia Cysts from Drinking Water Plants. EPA Project Report, In Press.

5. Davies, R. B., Fukutaki, K., and Hibler, C. P. Cross Transmission of Giardia. EPA Report No. PB 83-1177478, Cincinnati, OH (January, 1983).

6. Hewlett, E. L., Andrews, J. S., Jr., Ruffier, J., and Schaefer, F. W. III. Experimental Infection of Mongrel Dogs with Giardia lamblia Cysts and Cultured Trophozoites. The Journal of Infectious Diseases. 145:1:89 (January 1982).

7. Bellamy, W. D., Al-Ani, M. Y., Silverman, G. P., Lange, K. P., Choi, S. I., Howell, D. G., Hibler, C. P., and Hendricks, D. W. Removal of Giardia lamblia from Water Supplies, Interim Report for EPA Cooperative Agreement CR808650-01, Environmental Engineering Program, Colorado State University, November 1982.

8. Silverman, G. P., Bellamy, W. D., and Hendricks, D. W. Slow Sand Filtration of Giardia lamblia Cysts and Other Substances, Phase I Report for EPA Cooperative Agreement CR8808650-02, Environmental Engineering Program, Colorado State University, September, 1983.

9. Standard Methods for the Examination of Water and Wastewater, 15th Ed. American Public Health Association, Washington, DC 1981.

10. Al-Ani, M. and Hendricks, D. W. Rapid Sand Filtration of Giardia Cysts. Presented at American Society of Civil Engineers National Conference on Environmental Engineering, Boulder, Colorado (July 6-8, 1983).

11. Bellamy, W. D., Silverman, G. P. and Hendricks, D. W. Giardia lamblia Removal by Slow Sand Filtration, in Proceedings of Sunday Seminar on Innovative Filtration Techniques, American Water Works Association Annual Conference, Las Vegas, Nevada (June 5, 1983).

12. Edzwald, J. K. Removal of Trihalomethane Precursors by Direct Filtration and Conventional Treatment. EPA Project Report and Project Summary, Cincinnati, Ohio. In Press (1983).

13. Hunter, J. V., Bell, G. R., and Henderson, C. N. Coliform Organism Removals by Diatomite Filtration. JAWWA, 58:9:1160 (Sept. 1966).

14. Lange, K. P., Bellamy, W. D., and Hendricks, D. W. Diatomaceous Earth Filtration -- Effects of Operating Conditions on the Removal of Giardia lamblia Cysts and Other Substances, Final Report for EPA Cooperative Agreement No. CR808650-02, Environmental Engineering Program, Colorado State University, October, 1983.

Reprinted from Proc. AWWA WQTC, Norfolk, Va. (Dec. 1983).

TABLE 1

INFLUENCE OF FILTERED WATER TURBIDITY ON COLIFORM REMOVAL BY
COAGULATION AND FILTRATION, CSU

Filtered Water Turbidity Range NTU	Number of Samples	Percent of Samples with	
		Coliform ≤ 1/100 mL	Coliform Removal ≥ 99.0 %
≤ 0.10	42	69%	71%
0.11 to 1.4	46	37%	50%
> 1.4	19	47%	42%

TABLE 2

INFLUENCE OF FILTERED WATER TURBIDITY ON GIARDIA CYST REMOVAL
BY COAGULATION AND FILTRATION, CSU

Filtered Water Turbidity Range NTU	Number of Samples	Percent of Samples with	
		≤ 5 cysts/L	Cyst Removal ≥ 99.0%
≤ 0.10	23	78%	83%
0.11 to 1.4	28	36%	43%
> 1.4	3	33%	33%

TABLE 3

RAPID RATE FILTRATION RESULTS -- NO CHEMICAL PRETREATMENT

Filtration Rate m/hr	Turbidity NTU Infl.	Effl.	% Removal	Total Coliform Number per 100 mL Infl.	Effl.	% Removal	Giardia Cysts Cysts/L Infl.	Effl.	% Removal
5.1	0.22	0.16	27%	4000	3450	14%	260	66	74%
13.6	0.22	0.38	- 72%	4000	3000	25%	260	9	96%
5.0	0.22	0.26	- 18%	1300	800	38%	458	0.6	> 99.8%
13.3	0.29	0.33	- 14%	1300	520	60%	458	0.01	> 99.99%
4.9	1.7	1.77	- 1%	20,000	2000	90%	754	58	92%
4.9	1.7	1.50	12%	20,000	2000	90%	754	184	76%
12.9	1.7	1.18	31%	20,000	2000	90%	754	86	87%
12.9	1.7	1.69	1%	20,000	2000	90%	754	83	89%
4.9	2.4	1.59	34%	--	--	--	1026	0.9	99.9%
5.0	3.2	2.40	20%	59,000	50	99.9%	1012	1.1	99.9%
4.9	3.2	0.73	77%	--	--	--	975	0.5	99.9%
24.8	1.6	1.3	19%	4650	4400	5%	269	156	42%
19.2	1.6	1.3	19%	4650	5000	- 7%	269	172	36%
12.4	1.6	1.31	19%	4650	4600	1%	269	172	36%
5.8	1.6	1.35	16%	4650	< 1	99.9%	269	85	68%
24.2	1.6	1.19	26%	200	200	0%	245	125	49%
19.6	1.6	1.23	23%	200	200	0%	245	68	72%

TABLE 4

AVERAGE BACTERIA REMOVAL BY SLOW SAND FILTRATION, CSU

Filtration Rate m/hr	Total Coliform Removal	Standard Plate Count Organism Removal	Fecal Coliform Removal
0.04	>99.99%	95.5%	>99.82%
0.12	>99.95%	95.8%	>98.43%
0.40	>99.29%	92.7%	>98.63%

TABLE 5. EFFECT OF SLOW SAND FILTER CONDITIONS ON REMOVAL OF GIARDIA CYSTS, CSU

Filter Media Condition	Filtration Rate, m/hr	Samples NO Cysts Found/ No. of Samples	Effluent Concentration Cysts/liter	Cyst Removal Percentage Range
New Media, no established biopopulation	0.40	0/1	17	99.1%
New sand, old support media	0.12	1/1	<0.04	99.9%
No schmutzdecke, but	0.04	4/6	<0.04 to 5.4	98.7 to 99.9%
sand has established	0.12	4/5	<0.01 to 1.5	99.6 to 99.9%
biological population	0.40	3/4	<0.04	99.9%
Both schumtzdecke and	0.04	5/8	<0.06 to <1.0	99.7 to 99.9%
biological population	0.12	4/8	0.04 to <0.4	99.9%
established in sand	0.40	3/8	0.01 to 1.4	98.8 to 99.9%

TABLE 6

AVERAGE REMOVAL PERCENTAGES OF TOTAL COLIFORM BACTERIA, STANDARD PLATE COUNT BACTERIA, TURBIDITY AND GIARDIA CYSTS WITH VARIOUS GRADES OF DIATOMACEOUS EARTH, CSU

Grade of Diatomaceous Earth	Celite 545®	Celite 535®	Celite 503®	Hyflo Super-Cel®	Celite 512®	Standard Super-Cel®	Filter Cel®
Number of tests	15	2	10	4	2	3	2
Median Particle Size (μm)	26.0	25.0	23.0	18.0	15.0	14.0	7.5
Total Coliform Removal (%)	49.46	85.45	69.41	92.60	97.95	99.90	99.86
Standard Plate Count Removal (%)	58.42	80.51	69.10	75.29	78.63	99.11	99.85
Turbidity Removal (%)	13.36	12.50	11.97	23.56	31.14	49.77	97.64
Giardia Cyst Removal (%)	>99.7	>99.5	>99.7	>99.4	--	--	--

TABLE 7.   GIARDIA CYST AND TURBIDITY REMOVAL BY DE FILTRATION, CSU RESULTS

Celite® Grade	Number of Runs	Filt. Rate m/hr	Giardia Cyst Concentration Added to Raw Water - cyst/L	In Filtered Water - cyst/L	Number of Cysts Found in Filtered Water Sample	Cyst Removal Percentage
545	4	2.4	100 to 770	<0.3 to <0.7	None	>99.3 to >99.95
545	4	2.4	2470 to 10,000	<0.0004 to <0.11	None	>99.998 to >99.999
545	1	2.4	3.36×10^4	25	1700	99.92
545	3	4.9	100 to 500	<0.3 to <0.4	None	>99.5 to >99.93
545	1	4.9	8850	<0.4	None	>99.995
545	2	9.8	100 to 500	<0.4 to <3.3	None	>99.3 to >99.5
535	1	2.4	100	<0.4	None	>99.5
535	1	4.9	100	<0.4	None	>99.5
503	2	2.4	100 to 500	<0.3 to <0.7	None	>99.7 to >99.8
503	1	2.4	5500	<0.06	None	>99.998
503	3	4.9	100 to 500	<0.4 to <0.9	None	>99.5 to >99.8
503	1	9.8	100	<0.5	None	>99.4
Hyflo	1	2.4	100	<0.5	None	>99.5
Hyflo	1	4.9	100	<0.7	None	>99.3

TABLE 8

RAW WATER QUALITY AT McINDOE FALLS FROM 5/10/82 TO 4/8/83

	Number of Samples	Maximum	Minimum	Average
Standard Plate Count, per 1 mL	91	5100	0.5	35
Total Coliform, per 100 mL	91	3000	7	350
Turbidity, NTU	340	59	0.3	1.0
Temperature, °C *	340	25	0	10

*5°C or below 45% of time
 Above 5°C 55% of time

TABLE 9

McINDOE FALLS SLOW SAND FILTER PERFORMANCE
BACTERIA AND TURBIDITY REMOVAL

	Average Influent	Average Effluent	Average Removal
Ambient Raw Water Quality, June - September, 1982			
Std. Plate Count/1 mL	380	4.4	98.8%
Total Coliform/100 mL	1000	6	99.4%
Turbidity, NTU	1.2	0.3	75%
Ambient Raw Water Quality, October 1982 - January 1983			
Std. Plate Count/1 mL	130	2.4	98.1%
Total Coliform/100 mL	93	2	97.9%
Turbidity, NTU	1.1	0.1	91%
Raw Water Spiked with Sewage			
Std. Plate Count/1 mL (begin 9 day spike)	300	2.5	99.2%
Std. Plate Count/1 mL (end 9 day spike)	2200	150	93%
Total Coliform/100 mL (begin 9 day spike)	4500	60	98.7%
Total Coliform/100 mL (end 9 day spike)	16,000	2900	82%

TABLE 10

GIARDIA CYST REMOVAL AT McINDOE FALLS

Filtration Rate, m/hr	Temperature °C	Days Since Most Recent Filter Scraping	Number of 24 Hour Samples Collected	Cysts Recovered	Number of Samples With NO cysts	Cyst Removal
Slow Sand Filter Results						
0.08	0.5°	34	5	4032	2	93.7%
0.08	21°	35	5	32	2	99.99%

Results of Diatomaceous Earth Filter Test

Filtration Rate, m/hr	Temperature °C	Cysts Applied	Actual Cysts Recovered	Fraction of Effluent Sampled	Cyst Removal
3.8	23°	8×10^6	48	5.3%	99.97%

Removing Giardia Cysts With Slow Sand Filtration

William D. Bellamy, Gary P. Silverman, David W. Hendricks, and Gary S. Logsdon

Pilot-plant studies were undertaken to determine the efficiency of slow-rate sand filters in removing *Giardia* cysts and other substances. The filters removed virtually 100 percent of the *Giardia* cysts, 96 percent of standard plate count bacteria, and 98 percent of particles. Because of the efficiency and the passive nature of slow-rate sand filtration, this technology is especially appropriate for small water systems.

This research was undertaken to evaluate the effectiveness of slow-rate sand filtration for the removal of *Giardia lamblia* cysts. Other variables studied were turbidity, particles, total coliform bacteria, and standard plate count bacteria. These dependent variables were evaluated with respect to the influence of design and operating conditions (i.e., the independent variables) such as the hydraulic loading rate, the concentration of cysts, the concentration of bacteria, the biological maturity of sand in the filter bed, the age of the schmutzdecke, and temperature.

Design and operation of slow-rate sand filters

Slow-rate sand filtration is a passive filtration process, subject to very little control by an operator. There is no addition of chemicals or backwashing. The effective sand sizes range from 0.15 to 0.35 mm, with a uniformity coefficient less than 2. The depth of the sand bed ranges from 60 to 120 cm and is supported by 30 to 50 cm of graded gravel. Drain tiles are placed at the bottom of the gravel support to collect the filtered water. Hydraulic loading rates range from 0.04 to 0.40 m/h.

Three slow sand filtration pilot-plant units were operated in parallel for this study.

During operation, biological growth occurs within the sand bed and within the gravel support. Also, a layer of inert deposits and biological material, called the schmutzdecke, forms on the surface of the bed. Both the schmutzdecke and the biological growth within the bed have important roles in the effectiveness of slow-rate sand filtration. They may require weeks or months to develop.

Operation of a slow-rate sand filter requires two periodic tasks: (1) removal of the schmutzdecke and (2) replacement of the sand. The schmutzdecke is removed when the head loss exceeds the designed value, which may range from 1 to 2 m. After the filter is drained, the schmutzdecke is removed by scraping about 2 cm from the surface of the sand bed. The interval between scrapings depends on the contaminants present in the raw water and the hydraulic loading rate. Since operating expenses are affected by the frequency of cleaning, pilot testing is advisable to ascertain this important operating parameter.

Sand replacement is necessary after repetitive scrapings have reduced the sand bed to its lowest acceptable depth. The recommended method of replacing sand[1] is to remove all of the remaining sand down to the gravel support layer, put in new sand to one half of the designed depth, then place the sand previously removed on top of the new sand. This procedure places clean sand in the bottom half of the filter bed and the biologically active sand in the top half. This method of resanding provides for a complete exchange of sand over time and thus prevents any excessive accumulation of silt that could clog the filter bed.

Use of slow-rate sand filters

The first application of slow-rate sand filtration for large-scale community use was in England by the Chelsea Water

The small clay particles in Horsetooth Reservoir could pass through the pilot filters, causing most of the turbidity in the effluent.

PHOTO BY ROGER KREMPEL

TABLE 1

Average percent of removal of dependent variables in slow-rate sand filter columns

Dependent Variable	Number of Analyses	Range of Variable in Raw Water	Percent of Removal		
			Filter 1 at 0.04 m/h	Filter 2 at 0.12 m/h	Filter 3 at 0.40 m/h
Giardia cysts	222	50–5075 cysts/L	99.991	99.994	99.981
Total coliforms	243	0–290 000 coliforms/100 mL	99.96	99.67	98.98
Fecal coliforms	81	0–35 000 coliforms/100 mL	99.84	98.45	98.65
Standard plate count	351	$10–1.01 \times 10^9$ organisms/mL	91.40	89.47	87.99
Turbidity	891	2.7–11 ntu	39.18	32.14	27.24
Particles (6.35–12.7 μm)	39	62–40 506 parts/10 mL	96.81	98.50	98.02

Company in 1829. By the turn of the century, the effectiveness of slow-rate sand filtration in protecting public water supplies was established throughout Europe. In 1892, its effectiveness was demonstrated when 7500 cholera deaths occurred in Hamburg, Germany, which had no water treatment. Its neighbor, Altona, which used slow-rate sand filtration to treat the same supply,[2] had only a few deaths. In 1890, the Massachusetts State Board of Health performed a two-year study of the removal of *Salmonella typhi* by slow-rate sand filtration that indicated complete removal of the bacteria. Based on this study, slow-rate sand filters were installed at Lawrence, Mass., resulting in a reduction in the occurrence of typhoid in that city.[3] Reports by the filtration commission of the city of Pittsburgh, Pa., in 1899[4] and by Hazen[5] in 1913 confirmed the effectiveness of the process for removing bacteria. Recent research in England,[6-10] in the Netherlands,[11,12] in Germany,[13,14] and by the World Health Organization[1] has further demonstrated its efficiency in removing bacteria and viruses as well as numerous organic and inorganic pollutants.

Although a number of slow-rate sand filtration treatment plants have been built in the United States, most were completed in the first decades of the twentieth century. Today the technology is well established in Europe, although it has never gained as firm a foothold in the United States. It seems appropriate, however, for small communities, where less complex technology than rapid-rate sand filtration may be needed. Also, in view of the empirical evidence of its efficiency in removing bacteria, viruses, and other substances, it seemed reasonable that slow-rate sand filtration could also remove *Giardia* cysts.

Giardiasis. Public water supplies have been implicated in outbreaks of giardiasis in several areas of the United States, including Colorado,[15,16] New Hampshire,[17] Washington,[18] and New York.[19] The problem is especially acute in areas similar to the Rocky Mountains where treatment of cold, clear waters by rapid-rate sand filtration is difficult and where the water's appearance indicates that little or no treatment should be necessary. In the cases cited, *Giardia* cysts were recovered from the raw water supply, and in three of the cases, cysts were found in the finished water. In every case, the investigators of these outbreaks found deficiencies in water treatment that could have permitted cysts in the delivered water. Thus the effectiveness of water treatment has been a subject of concern in areas where giardiasis has been reported. Further, many small communities use surface water supplies and provide only chlorination without filtration. Slow-rate sand filtration could be an appropriate technology in such a context.

Experimental apparatus and methods

Apparatus. The experiments were conducted by using three slow-rate sand filtration pilot-plant units, which were operated in parallel. Figure 1 is a schematic drawing that shows equipment, appurtenances, and lines of flow. The major items of equipment were (1) a feed tank with a capacity of 1400 L; (2) three filter feed pumps; (3) three 30-cm-diameter slow-rate sand filters; and (4) one effluent holding tank.

Figure 2 is a cross section of the filters. The filters were packed with 96 cm of sand, $d_{10}* = 0.28$ mm, $d_{60}* = 0.41$ mm, supported by 46 cm of coarse sand and gravel. The effluent could be routed through a 142-mm membrane filter for *Giardia* sampling, or it could flow directly to the constant-head discharge device. For temperature control, the filters were equipped with cooling coils in the heads. Also, the filter feed tank had a built-in temperature control so that temperatures throughout the system could be maintained within the range of 3 to 20° C.

Operation. The three filters were operated continuously from August 1981 to December 1982 at hydraulic loading rates of 0.04, 0.12, and 0.40 m/h for the filters designated 1, 2, and 3, respectively. The feed tank delivered the same influent to each of the filters, thus allowing for the evaluation of responses to different hydraulic loading rates. The characteristics of the raw water changed very little during the testing and can be characterized as follows: (1) 3–9 ntu, (2) 45–65 mg total dissolved solids/L; (3) 30 mg harness/L; (4) a count of less than 200 algae/100 mL; (5) an average of less than 1 coliform colony/100 mL, with a range from 1 to 10; and (6) 1–5 standard plate count colonies/1 mL.

The other variables were (1) temperature, (2) influent bacteria concentration and cyst concentration, (3) age of the schmutzdecke, and (4) biological maturity of sand in the filter bed. These were changed systematically to determine their effects on the removal of *Giardia* cysts, bacteria, turbidity, and particles.

Temperature effects were examined by operating the system at 5 and 15° C. The highest temperature used during *Giardia* testing was 15° C. This upper limit was based on observations by C.P. Hibler of the pathology department at Colorado State University that the cysts deteriorate at higher temperatures.

Concentrations of *Giardia* cysts varied between 50 cysts/L and 5075 cysts/L. Because removal through filtration was highly effective, high influent concentrations of cysts were necessary to assure passage of a few cysts through the filter. The high concentration of cysts also permitted the discernment of possible functional relationships and represented the highest expected ambient concentra-

Figure 1. Flow schematic of model slow-rate sand filter system

---

*d$_{10}$ and d$_{60}$ indicate sieve size through which 10 percent or 60 percent of the material will pass

tion, which had been estimated as 500 cysts/L.[20]

Total coliform bacteria concentrations added to the filters ranged from almost 0 to about 300 000/100 mL. These levels resulted from adding primary settled sewage to the filter feed tank and from fecal residue accompanying the addition of *Giardia* cysts. There was no attempt to change these levels systematically. Bacteria were added to challenge the filter, because normally the raw water bacteria counts were quite low.

The effect of the schmutzdecke was tested after it had been allowed to develop and immediately after scraping. (A developed schmutzdecke is defined as one that has had at least two weeks in which to establish itself.)

The biological maturity of the sand bed indicates the degree of microbiological development throughout its depth. This condition is not measurable, but is a function of the number of weeks of undisturbed filter operation. To determine the influence of microbiological maturity, testing was done for three filter conditions: (1) a new sand bed and a new gravel support, which simulated startup of a new filter; (2) a new sand bed with microbiologically mature gravel support, which simulated a filter that had just had its sand totally replaced; and (3) a sand bed and gravel support that were both microbiologically mature, which simulated steady-state operation. Testing under the third condition was done at various filter ages, ranging from 26 to 80 weeks. The age of the filter can be used as an index of microbiological maturity for given raw water conditions. The most pertinent conditions that affect the length of time required to bring the bed to maturity are nutrient availability and temperature.

The testing program was designed to examine the effects of each of these operating conditions on the treatment efficiency of slow-rate sand filtration. This was accomplished by changing the magnitude of one variable while holding the others constant. Each such test constituted a "test run." The total length of a slow-rate sand filter run from scraping and restart to attainment of terminal head loss for water from Horsetooth Reservoir (located near Ft. Collins, Colo.) varied from several months to more than a year. Because of this long duration and the need to conduct test runs, no priority was given to determining terminal head loss.

Testing protocol. A test run consisted of filling the batch feed tank with water from Horsetooth Reservoir and then spiking the water in the tank with a known concentration of *Giardia* cysts. When additional coliforms were desired, the tank was also spiked with primary settled sewage. The feed tank was then sampled for *Giardia* cysts, total coliform

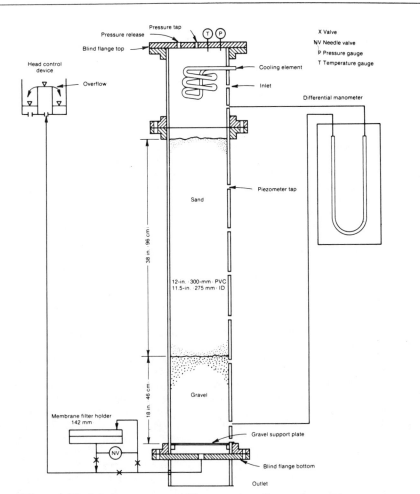

Figure 2. Model slow-rate sand filter cross section and appurtenances

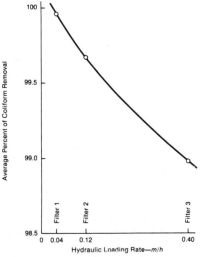

Figure 3. Coliform removal as affected by hydraulic loading in slow-rate sand filters *(plotted points are the average percent of total coliform removal, calculated for all operating and filter conditions)*

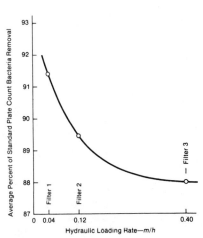

Figure 4. Standard plate count bacteria removal as affected by hydraulic loading in slow-rate sand filters *(plotted points are the average percent of removal of standard plate count bacteria for the three hydraulic loading rates, calculated for all operating and filter conditions)*

bacteria, standard plate count bacteria, particles, and turbidity. The same sampling and analyses were performed on the three filter effluents the following day to allow for the needed volume displacement within the filter column. These procedures, i.e., spiking, sampling of the feed tank, and sampling of the filter effluents, were continued daily for 3 to 11 days, depending on the particular test run. In addition, flow, temperature, and differential pressure across the sand column were measured daily.

Procurement of *Giardia* cysts. Procurement and preparation of the *Giardia* cysts was done in the following manner: (1) a fecal sample was collected from a dog suspected of having giardiasis; (2) a portion of the sample was examined by a zinc flotation procedure to ascertain whether *Giardia* cysts were present; (3) if cysts were present, the sample was weighed and added to an equal weight of cool distilled water; (4) the sample was mixed to disperse the suspension and to break apart aggregates; (5) if the sample contained an excessive quantity of organic matter, it was filtered through cheesecloth; and (6) the sample was refrigerated until it was used. The cyst concentration was determined prior to use by the Stoll dilution technique. The samples were used within two weeks of collection, even though counts have been observed to remain constant in stored concentrates for two months and longer. The *Giardia* cysts obtained from dogs are believed to be *Giardia lamblia* species, as discussed by Davies et al[21] and Hewlett et al.[22]

Measurement of *Giardia* cyst concentration. The two steps in cyst enumeration were sampling and counting. The sampling technique used for these experiments was patterned after that described by Luchtel et al.[23] It consisted of concentrating a given volume of water by filtering it through a 5-μm-pore-size, polycarbonate, membrane filter. The filter was then washed with distilled water, and the washings were processed to permit microscopic counting of cysts.

Two methods of processing for microscopic counting were employed. The first one consisted of floating the cysts in a zinc sulfate solution onto a cover slip and counting all of the cysts. The second method, the micropipette technique, consisted of reducing the sample volume to 1 mL by centrifugation, taking a 0.05-mL aliquot, and then counting the cysts in the aliquot. C.P. Hibler and coworkers experimented with various counting processing techniques and ultimately decided to use the micropipette technique.[21]

Results

The results of the testing program to determine the effects of operating conditions on removal of *Giardia* cysts and

other substances by slow-rate sand filtration are reviewed in this section. Operating conditions included the hydraulic loading rate, temperature, the concentration of bacteria, the concentration of cysts, the age of the schmutzdecke, the biological maturity of the sand bed, and the biological maturity of the gravel support. Overall removal results obtained for all parameters are outlined first; then the effects of operating variables are described.

Removals. Table 1 is a summary of the percent of removal for all dependent variables, averaged for all data, during the period of August 1981 to December 1982 for the three filter columns operated at rates of 0.04, 0.12, and 0.40 m/h. The number of samples that were obtained for each variable and the range of each are also presented.

The data show uniformly high removals of all dependent variables except for turbidity, which ranged from 27 to 39 percent removal for raw water turbidities ranging from 2.7 to 11 ntu. These turbidity removals are not as high as reported by others (e.g., the Kassler plant in Denver, Colo.[24]) because of the small clay particles that comprise the suspended matter in the raw water source used in this study, Horsetooth Reservoir. (About 30 percent of the turbidity in this water will pass through a 0.45-μm membrane filter.)

Removal of particles in the 6.35- to 12.7-μm size range varied from 96.8 to 98 percent. Thus the larger particles causing turbidity are retained by the filter whereas the smaller ones, which cause most of the turbidity, will pass through the filter. Particle measurements were taken only for a three-month period. This was because particles, such as rotifers and bacteria, grow within the filter and are continuously emitted from the filter during normal operation, making it impossible to differentiate between those particles that pass through the filter and those that are produced and sloughed from the filter.

Removal of *Giardia* cysts, total coliforms, and fecal coliforms was high. At optimum operating conditions, effluent concentrations of each parameter approached its respective detection limit.

Hydraulic loading rate. Figures 3 through 6 show well-defined relationships in which the percent of removal of coliform bacteria, standard plate count bacteria, *Giardia* cysts, and turbidity declines with increasing hydraulic loading. The plotted points were taken from the average percent removal data for all tests conducted (Table 1). For a two-way test of variance, the differences in removals shown for the three hydraulic loading rates in Figures 3, 4, and 6 are different at the 0.1 percent level, i.e., $p = 0.001$.

The plot of the percent of removal of *Giardia* cysts (Figure 5) follows the

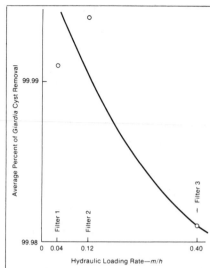

Figure 5. Removal of *Giardia* cysts as affected by hydraulic loading in slow-rate sand filters *(plotted points are the average percent of removal of* Giardia *cysts for the three hydraulic loading rates, calculated for all operating and filter conditions)*

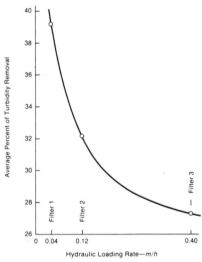

Figure 6. Turbidity removal as affected by hydraulic loading in slow-rate sand filters *(plotted points are the average percent of turbidity removal for the three hydraulic loading rates, calculated for all operating and filter conditions)*

same relationship as the others, i.e., declining removals with increasing hydraulic loading rates. The curve was drawn to follow the trend expected, even though the fit is imperfect, and there was no statistical difference shown among the three averages. The trend notwithstanding, removals, ranging between 99.98 and 99.99 percent, were uniformly high. These removals represent finding two cysts and one cyst in the effluent, respectively, per 10 000 cysts dosed in the influent. The removal percentages are not an indicator of precision of cyst sampling and counting. The process may have permitted about

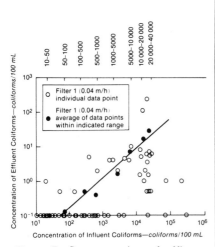

Figure 7. Concentration of effluent coliforms as a function of influent concentrations at a hydraulic loading of 0.04 m/h

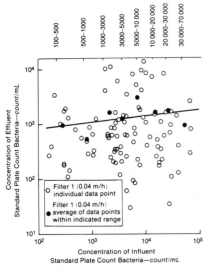

Figure 8. Concentration of effluent standard plate count bacteria as a function of influent concentrations at a hydraulic loading of 0.04 m/h

five cysts to pass through the slow-rate sand filter, whereas only one may have been detected in the sampling and analysis. The difference is between an actual removal of 99.95 percent and a measured removal of 99.99 percent. Thus, although the accuracy of sampling and analysis is not high, the measurement provides a true picture of the removal efficiency of the process.

Figures 3–6 show the unmistakable influence of the hydraulic loading rate on the percent of removal of total coliform bacteria, standard plate count bacteria, *Giardia* cysts, and turbidity. The hydraulic loading rate, therefore, should be

considered carefully during design. The advantage of reduced construction costs for a design with a high hydraulic loading rate must be weighed against increased operating costs, e.g., more frequent scraping to remove the schmutzdecke, and only slightly poorer performance.

Temperature. Table 2 compares filtration of *Giardia* cysts at 5 and 15° C, with all other conditions remaining approximately the same. The sets of comparisons are made for three hydraulic loading rates, for several ages of schmutzdecke, and for two influent concentrations of cysts.

The data indicate that the temperature at the time of filtration has no discernible influence on the efficiency of treatment for the conditions tested. This is evident when the paired results are compared. Similar results were observed for filtration of total coliform bacteria, standard plate count bacteria, and turbidity. The three filters were run continuously at the ambient temperature of 15° C; the temperature was reduced to 5° C only for the duration of a test run. This is not the same as maintaining a continuous 5°C temperature with its probable effect on microbiological populations. Further work should be done with filters operated continuously at 5° C to simulate operating at conditions under long-term cold temperatures.

Concentration of bacteria. Figures 7 and 8 show the effluent concentrations of total coliform bacteria and standard plate count bacteria as a function of influent concentrations at a hydraulic loading rate of 0.04 m/h. Although each figure shows two log cycles of scatter for all the plotted data, the plots of the averages within each range show that higher effluent concentrations may be expected with higher influent concentrations. The percent of removal, however, improves as the influent concentration increases. The scatter may be partially caused by the fact that all data were plotted without any attempt to achieve resolution for varying test conditions, e.g., microbiological maturity of the sand bed. Similar data were obtained for 0.12- and 0.40-m/h rates of operation.

Although a similar relationship could exist for influent *Giardia* cyst concentrations, none was found because of insufficient data. It was deemed more important to investigate higher priority relationships when it became apparent that effects caused by changes in influent concentrations would be extremely hard to define or would be completely masked by other variables, such as the filter's biological maturity.

Microbiological conditions. The biological conditions governing the effectiveness of the filter are (1) the degree of schmutzdecke formation and (2) the microbiological maturity of the sand bed. Figure 9 shows how these conditions

affect effluent concentrations of coliforms, i.e., the percent remaining at hydraulic loading rates of 0.04, 0.12, and 0.40 m/h. Also, shown in each of the bars are effluent coliform concentrations calculated from a hypothetical influent density of 1 million coliforms/100 mL. These figures, derived from the percent-remaining data, permit a more tangible means with which to compare results in terms of whole numbers.

To evaluate the respective roles of the schmutzdecke and the maturity of the sand bed, it is useful to examine first a filter column having a new sand bed and a new graded gravel support. This simulates a newly constructed filter during startup when there is no biological development in the sand bed and no schmutzdecke. For this condition, as indicated in Figure 9 for run 118, 15.4 percent of the coliforms remained (or 154 000 coliforms/100 mL, of a hypothetical 1×10^9 coliforms/100 mL in the influent. That is, filtration through the new sand will cause an order of magnitude reduction in coliforms.

In contrast to a filter at initial startup, a filter that has been in operation for a period of time and has a mature biological population and an established schmutzdecke is represented by runs 104, 105, and 106. These runs show that a mature filter will reduce the coliforms by 2.5–4 logs, i.e., from 1×10^9 coliforms/100 mL to 40, 1000, and 4000 coliforms/100 mL, respectively.

Schmutzdecke removal, called cleaning in practice, will result in approximately a 1-log decrease in treatment efficiency compared with operation under established conditions. This is shown by comparing runs 107, 108, and 109 at 300, 10 000 and 28 000 coliforms/100 mL, respectively, to runs 104, 105, and 106 at 40, 1000, and 4000 coliforms/100 mL, respectively.

Replacing the sand will result in almost a 1-log decrease in treatment efficiency. Run 116 shows 70 000 coliforms/100 mL remaining, compared with only 1000 coliforms/100 mL for the established condition, represented by run 105.

One additional condition tested was the effect of removing the schmutzdecke and then disturbing the sand bed, illustrated by runs 110, 111, and 112. This experiment was intended to simulate the effects of a full-scale filter operation in which the filter is drained and the sand bed is disturbed by the movement of workers and equipment over the filter surface during cleaning. The experimental disturbance was accomplished for each filter by draining it for a two-day period, removing the schmutzdecke, mixing the top 10 cm of sand, and pounding on the sand surface. This experiment caused an additional 0.5–1-log decrease in treatment efficiency com-

TABLE 2
Filter effectiveness at operating temperatures of 5° C and 15° C

Filter Number	Hydraulic Loading Rate m/h	Date 1982	Run Number	Temperature °C	Age of Schmutzdecke weeks	Influent Giardia Cyst Concentration cysts/L	Number of Effluent Samples	Effluent Volume Sampled L	Number of Cysts Detected in Effluent	Effluent Cyst Concentration cysts/L
1	0.04	3/18–3/23	54	15	3	500	5	84	13	0.305
		4/1–4/6	60	5	5	500	5	81	0	0.000
		5/17–5/24	66	15	11	50	7	175	5	0.050
		5/25–5/29	69	5	12	50	5	140	8	0.114
2	0.12	3/18–3/23	53	15	3	500	5	225	16	0.140
		4/1–4/6	59	5	5	500	5	220	5	0.035
		5/17–5/24	65	15	11	50	7	429	4	0.016
		5/25–5/29	68	5	12	50	5	345	7	0.041
3	0.40	3/18–3/23	55	15	3	500	5	346	68	0.387
		4/1–4/6	61	5	5	500	5	366	26	0.111
		5/17–5/24	67	15	2	50	7	1098	7	0.011
		5/25–5/29	70	5	3	50	5	962	8	0.017

TABLE 3
Effect of operating conditions on Giardia cyst removal by slow-rate sand filtration

Test Objective	Condition of Sand Bed and Gravel Support	Age of Schmutzdecke weeks	Length of Time of Operation weeks	Run Number	Filtration Rate m/h	Influent Cyst Concentration /L	Effluent Cyst Concentration /L	Percent Removal	Detection Limit cysts/L*	Effluent Volume Sampled L†
Effect of new sand bed and new gravel support	New sand bed and new gravel support	0	0/0	118	0.40	2000	17.05	99.15	0.046	610
	Control filter: mature sand bed and mature gravel support	10	80	119	0.40	2000	0.0	100	0.049	770
Effect of new sand bed	New sand bed and mature gravel support	0	0/67	116	0.12	3692	0.0	100	0.039	497
	Control filter: mature sand bed and mature gravel support	4	67	117	0.12	3692	0.0	100	0.040	566
Effect of schmutzdecke removal	Mature sand bed and mature gravel support	0	26	48	0.04	420	2.014	99.520	0.085	65
		0	26	49	0.40	420	5.431	98.707	0.020	270
		0	33	47	0.12	420	1.541	99.633	0.030	180
		0	41	75	0.04	50	0.0	100	0.036	314
		0	41	76	0.40	50	0.002	99.996	0.005	2239
		0	45	81	0.04	50	0.0	100	0.104	344
		0	45	82	0.40	50	0.0	100	0.042	2671
		0	48	74	0.12	50	0.0	100	0.014	803
		0	52	80	0.12	50	0.0	100	0.013	853
		0	62	107	0.04	1500	0.0	100	0.302	142
		0	62	109	0.40	1500	0.0	100	0.036	1199
		0	63	110‡	0.04	1953	0.0	100	0.151	176
		0	63	112‡	0.40	1953	0.0	100	0.026	1020
		0	69	108	0.12	1500	0.0	100	0.121	354
		0	70	111‡	0.12	1953	0.0	100	0.059	454
Effect of established schmutzdecke	Mature sand bed and mature gravel support	3	29	54	0.04	500	0.243	99.949	0.061	84
		3	29	55	0.40	500	0.321	99.936	0.015	346
		5	31	60	0.04	500	0.0	100	0.062	81
		5	31	61	0.40	500	0.111	99.978	0.014	366
		3	36	53	0.12	500	0.116	99.977	0.023	223
		5	38	59	0.12	500	0.035	99.993	0.023	220
		11	38	66	0.04	50	0.050	99.900	0.040	175
		2	38	67	0.40	50	0.011	99.978	0.006	1098
		12	39	69	0.04	50	0.114	99.772	0.037	140
		3	39	70	0.40	50	0.017	99.966	0.005	762
		11	45	65	0.12	50	0.016	99.968	0.016	429
		12	46	68	0.12	50	0.041	99.918	0.015	345
		4	49	87	0.04	1000	0.0	100	0.993	111
		4	49	88	0.40	1000	1.373	99.863	0.127	871
		5	50	90	0.04	1000	0.0	100	0.586	157
		5	50	91	0.40	1000	0.0	100	0.109	843
		4	56	86	0.12	1000	0.0	100	0.398	277
		5	57	89	0.12	1000	0.0	100	0.246	374
		16	60	101	0.04	1087	0.0	·100	0.200	138
		16	60	103	0.40	1087	0.0	100	0.024	1134
		17	61	104	0.04	5075	0.0	100	0.231	171
		17	61	106	0.40	5075	0.0	100	0.027	1440
		16	67	102	0.12	1087	0.0	100	0.081	342
		17	68	105	0.12	5075	0.0	100	0.091	435

*The detection limit calculations are described elsewhere.[20]
†This is the effluent volume that has been concentrated by a 5-μm polycarbonate membrane filter.
‡The entire filter bed was disrupted during the schmutzdecke removal process in an attempt to simulate full-scale procedures.

pared with the cleaning procedure when no disruption occurred.

The test results, summarized in Figure 9, confirm the importance of microbiological activity in the effectiveness of slow-rate sand filtration. The best treatment can be expected from a filter that has been in operation for an extended period of time. Such a filter will have a mature biopopulation within the filter bed and will have an established schmutzdecke. Figure 9 also shows that the treatment efficiency will deteriorate markedly as greater portions of the biological community are disrupted.

Table 3 shows the effects of various operating conditions on the percent of removal of *Giardia* cysts. The first two columns compare the removal of *Giardia* cysts by a control filter with the removal of *Giardia* cysts by a filter with new sand and new gravel support. No cysts were detected in the effluent of the control filter, and only 17 cysts/L were forced in the effluent of the new media filter. This demonstrates that a filter with a mature biological population is capable of removing cysts to the detectable limit, but even a new filter is capable of removing 99 percent of the influent cysts. Both filters were subjected to an influent cyst concentration of 2000 cysts/L.

Results for a similar experiment, run 116, with a new sand bed and a mature gravel support showed 0 cysts/L in the effluent compared with an influent concentration of 3692 cysts/L. This indicates that even a modest amount of microbiological growth in the sand bed or in the gravel support can provide the marginal effect needed to remove influent cysts to levels below the detection limit.

The third column of Table 3 (the effect of schmutzdecke removal) shows the results of 15 *Giardia* removal test runs on filters with freshly scraped sand surfaces, i.e., with no schmutzdecke. These test runs were arranged chronologically according to continuous filter operation, ranging from 26 to 70 weeks. Table 3 shows that removal of *Giardia* cysts to below the limit of detection was achieved in all but four of these test runs. The key difference between those tests that achieved nearly complete removal and those that did not was the degree of microbiological maturity within the sand bed. All four of the tests in which cysts passed through the filters occurred during the first 41 weeks of filter operation, indicating that when the microbiological population has developed to maturity, complete removal of *Giardia* cysts can be expected. As demonstrated by Table 3, this occurs independently of the hydraulic loading rate, the concentration of influent *Giardia* cysts, and the presence of a schmutzdecke.

The same improvement in removal of

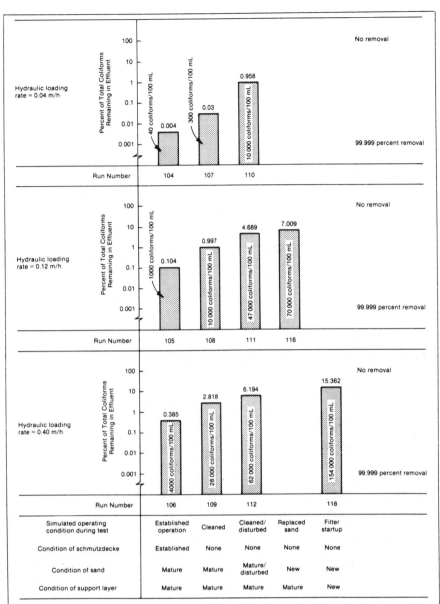

Figure 9. Effect of the condition of the schmutzdecke and sand bed on the percent of remaining total coliforms for three hydraulic loading rates *(quantities shown in each bar are concentrations of effluent coliforms calculated from a hypothetical influent density of 1 × 10⁹ coliforms/100 mL)*

Giardia cysts over time is demonstrated by results shown in the fourth column of Table 3 (the effect of established schmutzdecke), which summarizes test data in chronological order for 24 *Giardia* tests of filters with established schmutzdeckes. These results show that the removal of cysts improved steadily with time and was independent of the age of the schmutzdecke. Cysts were passed through filters with 12-week-old schmutzdeckes, whereas they were removed to below the detectable limit with four- to five-week-old schmutzdeckes when the microbiological population within the filter was given a longer time to mature. In fact, after 49 weeks of operation, removal of cysts to below the detection limit was achieved in all cases, even with influent concentrations as high as 5075 cysts/L. These results

demonstrate that the age of the schmutzdecke is not as important for *Giardia* cyst removal as the maturity of the microbiological population throughout the sand bed and the gravel support.

Summary and conclusions

The findings from the experimental program are summarized in terms of the overall removal effectiveness of slow-rate sand filtration for the parameters tested and in terms of the effect of operating conditions. The effectiveness of slow-rate sand filtration can be summarized as follows:

• *Giardia* cyst removal exceeded 98 percent for all operating conditions tested. Once a microbiological population is established within the sand bed, removal will be virtually 100 percent.

• Coliform removal exceeded 99 percent, averaged over all operating conditions. Even with new sand, coliform removal of 85 percent was attained.

• Removal of standard plate count bacteria and particles ranged from 88 to 91 percent and 96 to 98 percent, respectively.

• Turbidity removal averaged from 27 to 39 percent. This low removal was caused by the particles of fine clay characteristic of the lake water used in the testing program. Slow-rate sand filtration of other waters has resulted in higher removal of turbidity-causing particles.

Operating conditions affected the removal of *Giardia* cysts, total coliforms, and standard plate count bacteria:

• The removal of *Giardia* cysts, coliform bacteria, standard plate count bacteria, and turbidity declines as the hydraulic loading rate increases. However, even at 0.40 m/h, the removal of *Giardia* cysts and coliform bacteria was high, e.g., 99.98 percent and 99.01 percent, respectively.

• There was no discernible change in the effectiveness of the filter during short-term reductions in temperature from 15 to 5° C. The operation of a slow-rate sand filter during prolonged cold temperatures was not tested.

• The effluent concentrations of coliform bacteria and of standard plate count bacteria increase as influent concentrations increase. At the same time, the percent of removal increases. A similar relationship would be expected for the percent of removal of *Giardia* cysts, but data were not sufficient to establish it. Although this information may be of academic interest, in practical terms the percent of removal of *Giardia* cysts is influenced more strongly by the microbiological maturity of the sand bed than by the concentrations of cysts in the influent.

• A new sand bed will remove 85 percent of the coliform bacteria in the influent and 98 percent of the *Giardia* cysts in the influent. As the sand bed matures biologically, the percent of removal improves to more than 99 percent for coliform bacteria and to virtually 100 percent for *Giardia* cysts. Disturbance of the sand bed causes a reduction in the removal of coliforms but had no effect on the removal of *Giardia* cysts. Development of the schmutzdecke further improves the removal of coliform bacteria by an order of magnitude. The presence or absence of a schmutzdecke has essentially no influence on the filter's efficiency in removing *Giardia* cysts. The microbiological maturity of the sand bed is the most important variable in the removal of *Giardia* cysts and coliform bacteria. This mature condition develops in the sand bed over a matter of weeks or months, depending on raw water conditions.

Slow-rate sand filtration is an effective water treatment technology. It is passive in nature and requires little action on the part of the operator. Because of its effectiveness and its passive nature, slow-rate sand filtration should be an especially appropriate treatment for small water supplies.

Acknowledgments

This research was conducted under the auspices of USEPA Cooperative Agreement No. CR808650-02 between the US Environmental Protection Agency and Colorado State University. This article does not necessarily reflect the views of the USEPA. The authors thank the staff of the Drinking Water Research Division, Municipal Environmental Research Laboratory, Cincinnati, Ohio, for advice and support for the project; Charles P. Hibler, Colorado State University, for conducting the analyses of samples for *Giardia* cysts; Donna Howell, Colorado State University, for microbiological and cyst analyses; and the late Sumner M. Morrison for advise on microbiological testing.

References

1. HUISMAN, L. & WOOD, W.E. Slow Sand Filtration. World Health Organization, Geneva, Switzerland (1974).
2. FOX G.T. & LEKKAS, T.D. Slow Sand Filters. *Wtr. Services*, 82:984:113 (1978).
3. MCCARTHY, W.J. Construction and Development of Slow Sand Filters at Lawrence, Mass. *Jour. New England Water Works Assn.*, 89:1:36 (Jan. 1975).
4. Report of the Filtration Commission of the City of Pittsburgh. R. Pitcairn, chairman. Pittsburgh, Pa. (Jan. 1899).
5. HAZEN, A. *The Filtration of Public Water Supplies.* John Wiley & Sons, London (1913).
6. BURMAN, N.P. Slow Sand Filtration. H_2O, 11:16:348 (1978).
7. LLOYD, B. The Construction of a Sand Profile Sampler: Its Use in the Study of the *Vorticella* Populations and the General Interstitial Microfauna of Slow Sand Filters. *Water Res.*, 7:7:963 (July 1973).
8. POYNTER, S.F.B. & SLADE, J.S. The Removal of Viruses by Slow Sand Filtration. *Progress in Water Technol.*, 9:1:75 (Jan. 1977).
9. SLADE, J.S. Enteroviruses in Slow Sand Filtered Water. *Jour. Inst. of Water Engrs. & Scientists*, 32:6:530 (June 1978).
10. TAYLOR, E.W. Forty-fifth Rept. on the Results of the Bacteriological, Chemical, and Biological Examination of the London Waters for the Years 1971–1973. Metropolitan Water Board, London (1974).
11. NOTERMANS, S.; HAVELAAR, A.H.; & SCHELLART, J. The Occurrence of *Clostridium botulinum* in Raw Water Storage Areas and Their Elimination in Water Treatment Plants. *Water Res.*, 14:11:1631 (Nov. 1980).
12. SSF Research and Demonstration Project on Slow Sand Filtration, Nagpur, India. Natl. Envir. Engrg. Res. Inst., Project Rept., Phase I. The Hague, the Netherlands (Mar. 1982).
13. SCHOTTLER, V. Removal of Trace Elements by Slow Sand Filter. *Wasser Abwasser Forsch.*, 9:3:88 (Mar. 1976).
14. SCHMIDT, K. Behavior of Special Pollutants in Slow Sand Filters Used in Artificial Recharge of Groundwater. 17th Congress Intl. Assoc. for Hydraulic Res., 6:731, Baden-Baden, Federal Republic of Germany (Aug. 15, 1977).
15. BLAIR, J.R. *Giardia lamblia* Contamination at Estes Park, Colo. Unpubl. Rept. Colo. Dept. of Health, Denver (June 1979).
16. BLAIR, J.R. *Giardia lamblia* Sampling in the Resort Areas of Vail and Aspen, Winter 1980. Unpubl. Rept., Colo. Dept. of Health, Denver (June 1980).
17. LIPPY, E.C. Tracing a Giardiasis Outbreak at Berlin, N.H. *Jour. AWWA*, 70:9:512 (Sept. 1978).
18. KIRNER, J.C.; LITTLER, J.D.; & ANGELO, L.A. A Waterborne Outbreak of Giardiasis in Camas, Wash. *Jour. AWWA*, 70:1:34 (Jan. 1978).
19. SHAW, P.K. ET AL. A Community-wide Outbreak of Giardiasis With Evidence of Transmission by a Municipal Water Supply. *Annals Intl. Medicine*, 87:4:426 (Apr. 1977).
20. BELLAMY, W.D. ET AL. Removal of *Giardia lamblia* From Water Supplies. Interim Rept. 5836-82-4, USEPA Cooperative Agreement CR808650-01, Colorado State Univ., Ft. Collins (Nov. 1982).
21. DAVIES, R.B.; FUKUTAKI, K; & HIBLER, C.P. Cross Transmission of *Giardia*. USEPA Rept. PB83-117 747, Cincinnati, Ohio (Jan. 1983).
22. HEWLETT, E.L. ET AL. Experimental Infection of Mongrel Dogs With *Giardia lamblia* Cysts and Cultured Trophozoites. *Jour. Infectious Diseases*, 145:1:89 (Jan. 1982).
23. LUCHTEL, D.L.; LAWRENCE, W.P.; & DEWALLE, F.B. Electron Microscopy of *Giardia lamblia* Cysts. *Appl. & Envir. Microbiol.*, 40:4:821 (Apr. 1980).
24. Slow Sand Filtration at Denver. Rept. 38, *Research News*, AWWARF, Denver (Nov. 1982).

About the authors: *William D. Bellamy is a sanitary engineer with CH2M Hill, 1301 Dove St., Newport Beach, CA 92660. He conducted the research on which this article is based* while working on his PhD dissertation at Colorado State University, Ft. Collins. Prior to returning to CSU to obtain his PhD, he was employed by Aramco in Saudi Arabia and by Texaco in the United States as a project engineer designing water and wastewater treatment systems. Gary Silverman is an environmental engineer with the USEPA, 215 Fremont, San Francisco, CA 94105. David W. Hendricks is professor of civil engineering at Colorado State University, Ft. Collins, CO 80523. Gary S. Logsdon is research sanitary engineer with the USEPA, 26 W. Saint Clair St., Cincinnati, OH 45268.

Reprinted from *Jour. AWWA*, 77:2:52 (Feb. 1985).

EPA Project Summary

Slow Sand Filter and Package Treatment Plant Evaluation: Operating Costs and Removal of Bacteria, *Giardia*, and Trihalomethanes

Gordon R. Pyper

Gordon R. Pyper is with
Dufresne-Henry, Inc., North
Springfield, VT.

Gary S. Logsdon is the EPA
Project Officer (see below).

The complete report, entitled
"Slow Sand Filter and Package
Treatment Plant Evaluation:
Operating Costs and Removal of
Bacteria, Giardia, and
Trihalomethanes," will be
available only from:

National Technical Information
 Service
5285 Port Royal Road
Springfield, VA 22161
Telephone: 703-487-4650

The EPA Project Officer can be
contacted at:

Water Engineering Research
 Laboratory
U.S. Environmental Protection
 Agency
Cincinnati, OH 45268

A study was conducted to evaluate two simple methods of water filtration for small water systems: Slow sand filtration and pressure diatomaceous earth (DE) filtration. The study addresses the concerns of small water systems with regard to Giardia cysts, bacteria, trihalomethanes (THMs) and operating costs. Objectives are (1) to determine effectiveness of the two filtration systems for removing bacteria, turbidity, and Giardia cysts under various loading conditions, (2) to spike bacteria and Giardia cysts into the raw water under various loadings to determine the breakthrough point, (3) to determine the level of technical expertise needed to operate the systems, (4) to obtain operating and maintenance data and costs, and (5) to evaluate the potential for the formation of THMs with the two filtration systems.

The study was conducted at McIndoe Falls, Vermont. Raw water was of high quality with respect to most parameters, but the source was a highly organic impoundment site. Raw water values were generally low for the principal parameters studied (total coliforms, standard plate count, Giardia cysts, particles, and water temperature).

Slow sand filtration provided dependable water treatment with little attention required, but capital cost was high. Bacteria and Giardia cysts were removed very dependably at warm temperatures (above 5° to $10^{\circ}C$) but less efficiently at lower temperatures. Turbidity was below 1 NTU 99.19 percent of the time.

Pressure DE filtration also reduced Giardia cysts and bacteria dependably, but the system required full-time, skilled operation when running and careful attention to every detail.

Both systems failed to reduce THM precursors significantly, and both systems incurred comparable costs for producing small quantities of water.

This Project Summary was developed by EPA's Water Engineering Research Laboratory, Cincinnati, OH to announce key findings of the research project that is fully documented in a separate report of the same title.

Introduction

This study compared two simple methods of water filtration that could be used for small water systems. Many small water utilities in traditionally cold, clear water areas of the country have continuously used such water without filtration or other treatment, and some do not even chlorinate. Chlorination might be considered a minimum requirement for bacterial control, but with the increasing occurrence of Giardia cysts in surface water supplies, chlorination adequate to inactivate coliform bacteria may not provide adequate protection. Filtration is needed also.

This study addresses the concerns of the small water system with particular emphasis on turbidity, bacteria, and Giardia cysts. The principal objectives of this study were:

1. To determine the effectiveness and efficiency of slow sand filtration and pressure diatomaceous earth (DE) filtration on removal of bacteria, turbidity, and Giardia cysts under various loading conditions.

2. To spike bacteria and Giardia artificially into the raw water under various loading conditions to determine when the loadings might produce a breakthrough in the systems and contribute bacteria and Giardia to the effluent.

3. To observe, record, and evaluate the level of technical expertise required to operate the systems (observations were to be based primarily on ambient operating conditions).

4. To obtain operating data in terms of hours and costs associated with operating and maintaining the system (costs related to chemical additions, cleaning and restoring the systems to operation, and similar well-defined operational requirements).

5. To evaluate the potential for and formation of trihalomethanes (THMs) in the untreated water and compare the results with those from the effluent of a slow sand filter and a DE package treatment plant treating the same source of supply.

This study was conducted at McIndoe Falls, Vermont, at the site of the slow sand filtration plant. The water source was a small brook fed from a marsh. The raw water was of high quality with respect to most parameters, but the source was a highly organic impoundment site. The water could be described as low in color, turbidity, chloride, manganese, calcium, hardness, alkalinity, nitrate, and suspended solids; pH was neutral. The water had moderate to heavy levels of iron (0.1 to 1.5 mg/L), total Kjeldahl nitrogen (17 to 56 mg/L), ammonia nitrogen (7 to 25 mg/L), and sodium (1 to 155 mg/L). No primary standard chemicals were found to be above MCL limits.

The principal parameters considered during the study were turbidity, total coliforms, standard plate count, Giardia cysts, particles (7- to 12-um range), and water temperature. Raw water values for these parameters were generally low. Raw water turbidity averaged 1.4 NTU during the study (83 percent of the samples were 2.0 NTU or less), but some high spikes occurred during storms or during road work in the impoundment area. Concentrations of total coliform and standard plate count bacteria were influenced by rain and snow storms. Levels averaged 296/100 mL for total coliforms and 185/mL for standard plate count bacteria. Fifty percent of the total coliform samples had values of 100/100 mL, and the standard plate count had values of 80/mL or fewer. The average particle concentration (7- to 12-um range) was 12,780 per mL, and 50 percent of the samples contained 5300 per mL or fewer. Water temperatures tended to be cold most of the time except in the middle of summer. The average temperature was 9.7°C, but 38 percent of the readings were 2°C or below from about April 4 to December 15.

Slow Sand Filtration

The slow sand filter was operated under normal ambient conditions and under special biological stress conditions. The rate of filtration was maintained at a constant value of 0.08 m/hr (2 million gallons per acre per day [mgad]) throughout the study because summer flows from the source were not dependable above this rate.

Total Coliforms and Standard Plate Counts

The total coliform and standard plate count results for ambient conditions are summarized in Figures 1 and 2 and Table 1. Under ambient loading conditions, reductions averaged 80 percent for total coliforms and 90 percent for standard plate count bacteria. However, 90 percent of the total coliform and standard plate count samples showed reductions of 80 percent or more. Also, 60 percent of the effluent samples contained total coliform concentrations of 1/100 mL or fewer, and standard plate count values were 2/mL or fewer. Eighty percent of the effluent samples showed total coliform values of 7/100 mL or fewer and standard plate count values of 4/mL or fewer. These results showed relatively dependable bacterial quality in the effluent with rather variable raw water quality. The slow sand filter did not show an immediate response to sudden improvements in raw water quality. When the bacterial concentration rapidly declined in the raw water, the concentration of bacteria in the filtered water may have been close to or greater than the concentration in the raw water for a day or so. This circumstance would cause very low removal percentages, or negative removal percentages (increases).

Table 1 - Percent Reduction of Bacteria by Slow Sand Filtration*

Parameter	Number of Samples	Mean	Maxiumum	Minimum
Total coliforms/100 mL	67	79%	99.99%	-60%
Standard plate count bacteria	67	89%	99.99%	-200%

*Ambient operation for all periods from 6/18/82 to 5/4/84.

Recovery of Filter After Scraping

Normal recovery after filter scraping was an important consideration for filter operation. Bacterial quality of the water did suffer immediately after filter cleaning, particularly during cold water conditions. About 2 days after cleaning, reductions decreased from approximately 95 percent to 20 percent for total coliforms, and from about 90 percent to -300 percent for standard plate count bacteria. In warmer water situations, the reduction in total coliforms dropped to about 55 percent in about 7 days and to approximately 93 percent for the standard plate count in the same time period. This disruption of the treatment capability was much less severe. These results demonstrated that temperature must be considered when evaluating any biological impact on slow sand filtration.

Spiking the Filter After Cleaning --

During the summer months when the water was warm (22°C), the filter was spiked with bacteria immediately after it was cleaned. The filter showed very little disruption of bacterial treatment capability after being cleaned, even with heavy spikes of total coliform and standard plate count organisms. The filter effluent averaged 7/100 mL for total coliforms 8 to 10 days after cleaning and 2 to 3/100 mL 15 to 20 days after cleaning. The standard plate count results were similar. The effluent averaged 2/mL for standard plate count bacteria 8 to 10 days after cleaning, and it decreased to 1/mL 20 days after cleaning.

Turbidity After Filter Cleaning --

Turbidity reductions after filter cleaning were similar to bacterial results. Under warmer ambient water conditions, turbidity reductions tended to remain at 92 percent to 95 percent removal throughout the recovery period. In cold water conditions, removals dropped from 92 percent to 95 percent to about 50 percent in 1 to 5 days. Recovery during cold weather tended to take 10 to 20 days and was much more erratic than warm weather results. However, under the worst of the cleaning conditions, the filtered water never exceeded 0.9 NTU except for the start-up condition when the filter had just been cleaned. The erratic response would be expected because the filter had not been used for several years and thus represented a biologically immature sand bed. The results for all values, including turbidity, were very erratic and irregular for about 100 days after this initial cleaning, a result completely different from subsequent cleaning and normal operating results.

Particle Count After Filter Cleaning --

Information on reductions of particles (7- to 12-um range) proved to be very erratic for this water. Reductions were at times in the 90 percent to 95 percent range; but at other times they were -100 percent to -200 percent, with no particular pattern to their changes. For normal ambient operation, the average particle reduction was about 45 percent, and during recovery from cleaning, averages were about the same. Recovery from cleaning with bacterial spiking demonstrated severe impacts, with an average reduction of -8 percent. Little correlation was apparent with bacteria, turbidity, or Giardia removal. Slime growths in the water were believed to contribute to this erratic behavior, as such organisms occasionally clogged the particle counter and did continuously clog the automatic turbidimeters, often within a day after cleaning.

Bacterial Spiking

Bacterial spiking was also evaluated under normal operating conditions. The temperature influence was again demonstrated during cold water conditions (1°C). Effluent bacterial reductions deteriorated steadily during spiking and showed signs of breakdown in treatment about 10 days after the spike started. During warmer water conditions (9°C), the effluent hardly showed any disruption for either total coliforms or standard plate count, with bacterial loads of 1,000 to 10,000/100 mL for total coliforms and 100 to 1000/mL for standard plate count organisms.

Cyst Reduction

The removal of Giardia lamblia cysts was an important consideration in this research. Eight spikes of fluorescent-tagged cysts were applied to the slow sand filter during this study, including a series of 5 spikes applied at 1-month intervals. After each spiking, 8 percent of the filter effluent was sampled each day on a continuous basis. The sampling periods ranged in length from 6 days to 5 months; the 5-month sampling period was done during the series of 5 cyst spikes. Cyst removal was excellent, even though some cysts did appear in the effluent. Cyst reduction through the filter is summarized in Figure 3 and Table 2. Removals tended to be best (99.9 percent) during warm water conditions and less effective (99.5 percent) during cold water conditions, except for one result in cold water that yielded only a 93.7 percent reduction. This somewhat lower reduction occurred during seweage spiking (water temperature $0.5^{\circ}C$) and it appeared that the biological capabilities of the filter were stressed almost to the limit since removals of both bacteria and Giardia cysts showed degradation in their removal patterns. This hypothesis should be investigated further because it has important ramifications when considering cold water situations that might involve sewer breaks or other contamination with Giardia cysts. During the study, water temperatures were $2^{\circ}C$ or below for 3.5 months or more per year.

Turbidity

Turbidity reductions under normal ambient conditions were uniformly good. The results are summarized in Table 3. The average slow sand filter effluent turbidity value was 0.22 NTU and 90 percent of the effluent samples showed 0.5 NTU or less.

Trihalomethane Precursors

THM precursor reduction was studied for both warm and cold water conditions. Precursor reduction was evaluated by generating THMs under conditions of excess free chlorine residuals. No appreciable reduction in precursors occurred when passing through the slow sand filter.

Diatomaceous Earth Filtration

Diatomaceous earth filtration was evaluated using pressure (0.74 and 0.93 m^2, or 8 and 10 ft^2 septum area filters. Filtration rates averaged 2.4 and 4.3 m/hr using Celite 503.®* At low ambient bacterial loads (5/100 mL of total coliforms and 30/mL or fewer standard plate count bacteria), the reduction in bacteria appeared to be low.

---

*Mention of trade names or commercial products does not constitute endorsement or recommendation for use.

Table 2. Giardia Cyst Removal with the Slow Sand Filter

Days Since Most Recent Filter Scraping	Number of 24-Hour Samples Collected	Cysts Recovered	Number of Samples With No Cysts	Cyst Removal (%)
34	6	4032	3	93.7
88	26	3214	9	99.62
117	26	3503	10	99.46
50	38	4090	19	99.36
144	28	485	2	99.91
174	32	51	19	99.99
82	7	8	6	99.98
35	5	42	2	99.98

Spike Date	Sample Dates	Temperature (^{o} C)	Cysts Applied
2/28/83	3/1/83 – 3/6/83	0.5^o	2.1×10^6
1/16/84	1/17/84 – 2/14/84	0.5^o	2.55×10^7
2/14/84	2/15/84 – 3/12/84	0.5^o	2.31×10^7
12/8/83	12/9/83 – 1/16/84	0.75^o	2.3×10^7
3/12/84	3/13/84 – 4/9/84	0.75^o	2.55×10^7
4/9/84	4/10/84 – 5/11/84	7.5^o	2.31×10^7
5/16/83	5/17/83 – 5/23/83	11°	2.1×10^6
8/8/83	8/9/83 – 8/12/83	21°	8×10^6

*Filtration rate is 0.08 m/hr.

Table 3. Turbidity Removal with the Slow Sand Filter

Sample Point	Number of Observations	Turbidity (NTU)		
		Average	Maximum	Minimum
Influent	674	1.4	59	0.2
Effluent	701	0.22	8.0	0.05

Total Coliforms and Standard Plate Counts

The average reduction was 87 percent for total coliform and 89 percent for standard plate count bacteria. Of these runs, 50 percent showed 92 percent reduction in total coliforms and 90 percent or more reduction in standard plate count bacteria. Filtration at 4.3 m/hr demonstrated slightly lower reductions in bacteria compared with the 2.4 m/hr rate. The average reduction was 77 percent for total coliforms and 87 percent for standard plate count bacteria at the high rate of filtration compared with the 92 percent and 88 percent, respectively, at the low rate; but this difference did not constitute a significant variation.

Bacterial Spiking

Under bacterial spiking conditions, the average reduction was 98 percent for total coliforms and 93 percent for standard plate count bacteria. Of these runs, 50 percent reduced total coliforms 98 percent or more and standard plate count 94 percent or more.

Turbidity

Turbidity reduction was fairly consistent during the DE runs. The average reduction was 71 percent with an average effluent of 0.5 NTU.

Giardia Cyst Spiking

One Giardia spike of 8×10^6 cysts produced 99.97 percent removal--a result consistent with the DE results previously reported by other researchers. No other Giardia spike applications were conducted because cyst supplies needed to be conserved and because other work had shown similar results. The results of this study are shown in Table 4. The reduction of 99.97 percent was excellent at the 4.3 m/hr filtration rate.

THM Precursors

The DE filter did not affect the THM precursor reduction, and the results were similar to the values for the slow sand filter.

For this water body, feed rates of 3 to 4 mg/L appeared to produce the best results when related to pressure buildup. The higher filtration rate (4.4 m/hr) appeared to have a slight advantage over the lower rate (2.4 m/hr) filtration.

Table 4.
Giardia Cyst Removal by the Diatomaceous Earth Filter with Celite 503*®

Filtration Rate m/hr	Temperature °C	Cysts Applied	Actual Cysts Recovered	Portion of Effluent Sampled	Cyst Removal
3.8	23o	8×10^6	48	5.3%	99.97%

*Mention of trade names or commercial products does not constitute endorsement or recommendation for use.

Grade of Diatomaceous Earth

DE grade effects were not studied extensively. From very limited studies, it did appear that a finer grade might have been advantageous for this water in that body feed rates on the order of 28 mg/L produced much longer operation time than the 3 mg/L at the same rate of filtration. The longest operation times were provided by the coarse grade at body feed values of about 3 to 4 mg/L and by the fine grade at about 28 mg/L or possibly more.

Water Plant Operation

Operation and cost information was accumulated during the course of this study and analyzed to evaluate slow sand filter needs and compare them with DE operation.

Cleaning requirements for the slow sand filter can be expressed by the relationship:

$$Y = 1.6 + 3.5x \pm 1.0$$

where y = person hours to clean, and x = removed sand volume (m^3). A considerable amount of variation existed in a single determination. The type of installation and operating conditions affect the cleaning results.

The length of the slow sand filter runs could not be extrapolated from head loss information. For this full-scale filter and the particular water source, filter runs could be expected to range from 100 to 250 days, but plots of head loss versus time tended to be very flat for many months and then suddenly increase exponentially to limiting head loss values. Many more studies over 5 to 10 years would be required to provide sufficient data for predicting a pattern, if a pattern is possible.

Operation time data were recorded during the study to determine the time required to obtain and record turbidity, temperature, and chlorine residuals and also to sample for bacteria and make chlorine solutions. The mean time requirements are as follows: 1.46 hr for reading, testing, and recording turbidity and temperature; 1.54 hr for bacteria, turbidity, and temperature; 0.38 hr for chlorine residual; and 0.20 hr for chlorine preparation. Results will vary considerably depending on facilities and personnel.

Production costs for water were evaluated. If the slow sand filter used in the study were constructed new in 1984 and operated at full capacity (0.08 m/hr), water would cost $4.60/1,000 gal. A similar-capacity DE pressure filter might produce water at a comparable cost. The DE studies did not provide sufficient operating data to permit extrapolating DE costs in a meaningful manner. The research included too many operational variables to permit development of information that would be comparable to that from a functioning treatment plant. The DE filtration research operating data were not comparable to the slow sand filter operating data.

The stated cost values are produced costs, not delivered ones. These costs are high. For small systems, however, they are comparable to costs that could be incurred when individuals drill wells and install private water systems that can produce high quality equalling or exceeding the Safe Drinking Water Act's quality requirements.

Conclusions

Slow Sand Filtration

1. Slow sand filtration provided dependable water treatment with a minimum of attention, but capital cost was high.

2. Turbidity was below 1 NTU 99.19 percent of the time. After the first 100 days of operation, the effluent turbidity values were below 1 NTU 99.68 percent of the time. Turbidity values were 0.2 NTU, or less, 72 percent of the time.

3. Slow sand filtration reduced total coliforms to 10/100 mL, or fewer, 86 percent of the time under ambient load conditions.

4. The standard plate count bacteria were reduced to 10/mL, or fewer, 94 percent of the time under ambient load conditions.

5. Massive spikes of total coliform and standard plate count bacteria were removed from raw water at temperature conditions above 5^0 to 10^oC.

6. Slow sand filtration was not as efficient in removing bacteria at temperatures below 5^oC, particularly around 0° to 1^oC.

7. Giardia cysts were removed very dependably; 99.98 percent removals or better were achieved under warm temperature conditions.

8. Giardia cysts were not as completely removed at low temperatures; at temperatures below 7^oC, removals were 99.36 to 99.91 percent.

9. Heavy applications of bacteria and Giardia cysts to the filter at the same time under cold conditions produced signs of competition for the biological treatment capability. Giardia cyst removal was reduced to 93.7 percent, and reduction of total coliforms and standard plate count bacteria dropped to 43 percent and to 79 percent to 82 percent, respectively.

10. Slow sand filtration did not produce any significant reduction of THM precursors.

11. Erratic particle reduction in the 7- to 12-um range did not compare with the Giardia cyst removal results.

12. Particle reduction did not provide a dependable method of predicting Giardia cyst removal in this full-scale operating filter experiment with this particular water.

13. The mature filter recovered from cleaning within 2 weeks to provide dependable bacteria and turbidity removal. Limited data showed that at times under warm weather conditions, the effluent water contained satisfactory bacteria and turbidity concentrations before 2 weeks had elapsed.

14. A minimum of 1.5 hr of operation were required each day to run the system properly and meet monitoring requirements.

Pressure Diatomaceous Earth Filtration

1. Pressure DE filtration removed Giardia cysts dependably using Celite 503 with 99.97 percent reduction.

2. Total coliforms were reduced 86 percent or more in 70 percent of the samples, and standard plate count bacteria were reduced 80 percent or more in 70 percent of the samples.

3. Eighty-six percent of the average run values for total coliforms did not exceed 8/100 mL, and 82 percent of the average run values for standard plate counts did not exceed 12/mL.

4. The average bacterial content in the effluent under ambient conditions was 38/100 mL for total coliforms and 6/mL for standard plate count bacteria.

5. Under spiking conditions, the average reduction was 97.6 percent for total coliforms and 92.7 percent for standard plate count bacteria. Eighty percent of the average run values showed total coliform reductions of 95.8 percent or more and standard plate count reductions of 87.5 percent or more.

6. Under spiking conditions, effluent total coliforms averaged 122/100 mL (107/100 mL or fewer for 77 percent of the runs), and standard plate counts averaged 47/mL (7/mL or fewer for 77 percent of the runs).

7. Pressure DE filtration provided rapid cycle time and flexible filter water production capability.

8. The system required full-time operation when running and careful attention to every detail of operation.

9. Highly skilled operators are needed for dependable production of the most satisfactory water the treatment process can produce.

10. The costs for producing small quantities of water are comparable with those for the slow sand filter.

11. The process is labor-, energy- and materials-intensive, as compared with slow sand filtration.

12. Particle reductions in the 7- to 12-um range were erratic for this water. Slime organisms may have contributed to the erratic results.

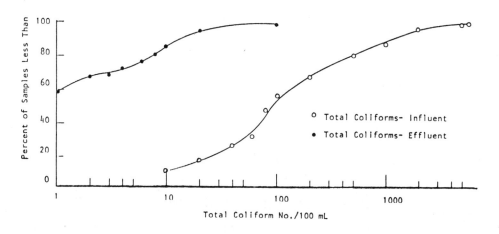

Figure 1.   Total coliforms in influent and effluent of slow sand filter during
normal operation, filtration rate of 0.08 m/hr.

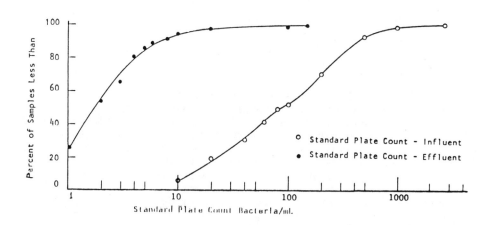

Figure 2.   Standard plate count bacteria in influent and effluent of slow sand filter
during normal operation, filtration rate of 0.08 m/hr.

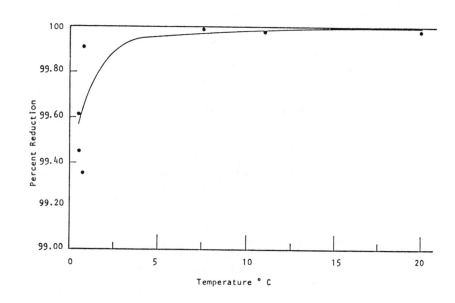

Figure 3.   Slow sand filter reduction of Giardia cysts at various temperatures
(filtration rate of 0.08 m/hr).

EPA Project Summary

Removal of *Giardia lamblia* Cysts by Drinking Water Treatment Plants

Foppe B. DeWalle, Jogeir Engeset, and William Lawrence

Foppe B. DeWalle, Jogier Engeset, and William Lawrence are with the University of Washington, Seattle, WA 98195.

Gary S. Logsdon is the EPA Project Officer (see below).

The complete report, entitled "Removal of Giardia lamblia *Cysts by Drinking Water Treatment Plants," (Order No. PB 84-162 874) will be available only from:*

*National Technical Information Service
5285 Port Royal Road
Springfield, VA 22161
Telephone: 703-487-4650*

The EPA Project Officer can be contacted at:

*Municipal Environmental Research Laboratory
U.S. Environmental Protection Agency
Cincinnati, OH 45268*

A pilot study was conducted at the University of Washington to evaluate the removal of *Giardia lamblia* cysts and cyst-sized particles from Cascade Mountain waters. Methods included coagulation/sedimentation and filtration, or direct filtration using three 2.3-L/min (0.6-gpm) pilot treatment units and diatomaceous earth (DE) filtration with a 3.8-L/min (1-gpm) DE pilot filter. Results were verified by use of a 75-L/min (20-gpm) pilot unit (Waterboy, Neptune Microfloc*) in field trials at Hoquiam and Leavenworth, Washington.

Granular media filtration tests yielded greater than 99.9% removal of spiked cysts under optimum conditions, although removal percentages decreased greatly at lower spiking levels. Both the pilot unit and the field unit established the importance of a minimum alum dosage (10 mg/L), an optimum pH range, and intermediate flow rates of 4.9 to 9.8 m/hr (2 to 4 gpm/ft²). Effluent turbidity and cyst-sized particles passing the filter increased rapidly when the above conditions were not attained or when sudden changes occurred in plant operation. When no coagulants were used during filtration, the process removed only 48% of the spiked cysts and 47% of the turbidity. A cyst spike in the pilot unit in Hoquiam using alum as coagulant resulted in an 81% cyst removal, and the spike at Leavenworth using a polymeric flocculant gave a 72% removal. Producing a low-turbidity filter effluent with alum or polymeric flocculant was difficult when the water temperature was less than 3°C. Further research on low-temperature direct filtration is necessary to improve the removal efficiency under these conditions.

DE filtration proved effective both for turbidity, particle, and cyst removal. The addition of 0.0075 mg/L nonionic polymer showed some improvement in efficiency. Cyst removals generally ranged from 99% to 99.99%.

This Project Summary was developed by EPA's Municipal Environmental Research Laboratory, Cincinnati, OH, to announce key findings of the research project that is fully documented in a separate report of the same title (see Project Report ordering information at back).

Introduction

A study was undertaken to evaluate the removal of *Giardia lamblia* cysts and cyst-sized (8- to 12-μm) particles by drinking water plants. The first phase of the study was devoted to a laboratory-scale evaluation of *Giardia* removal efficiency by coagulation, flocculation, and filtration. In addition, a diatomaceous earth filter was tested. The second phase consisted of a pilot- and full-scale evaluation of *Giardia* removal from drinking water plants where cysts might be present in the raw water.

All laboratory water treatment plant experiments were conducted with unfiltered Seattle tap water to which cysts were added. The cysts that were used to spike the water were isolated from the feces of human giardiasis patients. The cysts were recovered from the spiked water using membrane filtration techniques. *Giardia* cysts present in the membrane retentate were enumerated with a hemacytometer and a Coulter Counter.

This project was of considerable significance in the State of Washington because waterborne giardiasis outbreaks had occurred in Camas in 1976 and in Leavenworth in 1980. The latter community was selected as one of the sites for field testing with the 75-L/min mobile pilot plant.

Experimental Procedures

The *Giardia* cysts were collected from the feces of human giardiasis patients. The cysts were separated from the feces using a sucrose gradient technique. During testing, the cyst suspension was added to the influent of the unit process to be studied. The cysts were recovered from the process effluent by filtering 10 to 100 L through a 5-μm-pore-size, 293 mm Nuclepore membrane filter at 10 psi. The cysts were removed from the filter by agitation, and the washwater was concentrated to 5 mL by centrifugation.

The average recovery of cysts at initial concentrations ranging from 10^3 to 10^5 cysts/mL measured with the hemacytometer was 20% using the Millipore* membrane and 85% with the Nuclepore membrane, indicating that the membrane structure may have had an effect when using the 293-mm membrane. At concentrations below 1 cysts/mL, the recoveries became highly variable because of the low number of cysts that could be enumerated. For example, the recovery of duplicate runs at 0.1 cyst/mL was 75% and 23%, respectively.

Two counting techniques were used for the enumeration of the *G. lamblia* cysts. The first technique was microscopic counting using different counting chambers, and the second was an electric current displacement technique using a ZBI Coulter Counter and Channelyzer (Coulter Electronics, Hialeah, Florida) calibrated to measure particle densities in the *Giardia* size range (8 to 12 μm). Both methods exhibited some nonlinearity for the more diluted suspensions. Although the Coulter Counter was more precise and had a lower detection limit, only the hemacytometer method was specific for cysts. The Coulter Counter could not differentiate between cysts and other particles of a similar size, so counts from that system are referred to as cyst-sized particles.

Tests for electrophoretic mobility (EM) and zeta potential (ZP) were carried out to determine how they varied for formalin-fixed *Giardia* cysts at different pH values using a Zeta Meter (Zeta Meter, Inc., New York, New York). The ZP values for the fixed *G. lamblia* cysts clearly showed a decreasing potential at decreasing pH values. But, even at low pH values, the cysts retained their negative charge. The ZP was always more negative

---

* Mention of trade names or commercial products does not constitute endorsement or recommendation for use.

---

than −20 mv in the range of pH 5 to 10. Thus alum and cationic polymers would be appropriate coagulants.

The pH was measured with a Model 5 Corning pH meter (Corning Glass, Corning, New York) standardized daily with pH buffer. Turbidity was measured with a continuously recording low-range Hach 1720A turbidity meter (Hach Chemical, Loveland, Colorado) and standardized daily as suggested in the manual. To verify the readings of the flow-through turbidimeters, grab samples of the influent and effluent were analyzed daily on a DRT-100 (H.F. Instruments, Ft. Meyers, Florida) bench-top turbidimeter.

The pilot plant built at the University of Washington (UW) for this study had rapid mixing, flocculation, sedimentation, and filtration in three parallel treatment trains with flow rates of 2.3 L/min each. The 10.8-cm diameter Plexiglas filter columns contained 50.8 cm of 0.92-mm effective size (e.s.) anthracite with a uniformity coefficient (u.c.) of 1.28 and 25.4 cm of 0.40-mm e.s. sand (u.c. = 1.30). The columns had headloss taps at 10.2-cm (4-in.) intervals.

The DE test filter was a 0.1-m² (1-ft²) pressure filter operated at 3.8 L/min (1.0 gpm). The operation of the filter consisted of three steps: Precoating, filtration, and filter cleaning. Precoating was accomplished by adding a measured amount of diatomite to tap water and recirculating the slurry through the filter. During the runs, body feed diatomite was added to the raw water. The thickness of the filter cake increased during the run. Filter runs were terminated when headloss exceeded 30 psi. The filter cake was removed from the septum, and the spent diatomite was discharged to waste. Septum and filtration chamber were carefully sluiced to make the filter ready for a new precoating.

Several different grades of diatomite (obtained from the Manville Products Corp., Denver, Colorado) were used. The amounts of precoat material applied to the septum ranged from 0.5 to 1.2 kg/m² (0.1 to 0.24 lb/ft²). The results indicated that 1.0 kg/m² (0.2 lb/ft²) would be an adequate precoat thickness for all grades of diatomite, including Celite 560, the coarsest in the group.

The last part of the study was used to compare the laboratory results with field data by using a mobile pilot plant. In addition, the mobile pilot plant results were compared with results of the treatment plants at Hoquiam and Leavenworth. The trailer-mounted unit was a U.S. Environmental Protection Agency (EPA) drinking water pilot plant, a Waterboy-27 (Neptune Microfloc, Corvallis, Oregon) that was modified by extending the length of the sand filter compartment by 83.8 cm (33 in.) to provide for more

---

headloss buildup. The treatment train consisted of three static in-line mixers, Model 2-50-541-5 (Kenecs, Danver, Massachusetts), a flocculator with 8 min of detention time, and a mixed media filter. The filter had 45.7 cm (18 in.) anthracite, (e.s. 1.0 to 1.1 mm, u.c. < 1.7), 22.9 cm (9 in.) of sand (e.s. 0.42 to 0.55 mm, u.c. < 1.8), 7.6 cm (3 in.) of fine garnet (e.s. 0.18 to 0.28 mm, u.c. < 2.3), 7.6 cm (3 in.) of coarse garnet, (1 to 2 mm size), 10.2 cm (4 in.) of 0.95 cm (3/8-in.) gravel, and 12.7 cm (5 in.) of 1.9-cm (3/4-in.) gravel (Neptune Microfloc, Corvallis, Oregon). Capacity was approximately 75 L/min, giving a filtration rate of 12 m/hr (5 gpm/sf).

Granular Media Filtration Results

Granular media filtration was studied first in experiments with the UW pilot plant. The raw water source was unfiltered Seattle tap water from the Tolt Reservoir. The first seven runs for *Giardia* cyst removal were made using the coagulation/sedimentation unit followed by filtration. Runs made thereafter were direct filtration runs bypassing the sedimentation unit. The cyst spiking results of the first seven runs generally showed high (96% to 99.9%) cyst removals at cyst levels of 23 to 1100 cysts/L. The removal efficiency seemed to decrease at low spiking levels. Removal was only 30% at a cyst concentration of 2.3 cysts/L in raw water. Results are listed in Table 1.

A large number of runs were made without addition of cysts. The filter runs were conducted at different alum dosages, pH values, and flow rates. The main parameters measured during the testings were removal of particles in the *Giardia* size range, turbidity removal, length of filter run, headloss buildup at different depths in the filter, and particle distribution at different filter depths. The importance of using an adequate alum dose was established. Best removal of cyst-sized particles and turbidity occurred in the 7- to 15-mg/L range of alum doses. Highest particle removals were obtained at pH 6.5, although acceptable results could be attained in a pH range of 5.6 to 7.0. Sudden pH changes could cause filtrate quality to deteriorate. When the filtration rate was increased above 17 m/hr, filtered water quality deteriorated and headloss buildup increased sharply.

After the direct filtration runs had established operating conditions for effective removal of cyst-sized particles, additional runs were made with *G. lamblia* cysts. During most of these filter runs, about 20 million cysts per run were added to the raw water. These tests established the importance of good coagulation practice. Results appear in Table 2.

Table 1. *Performance of Each Filter Run with Cysts Added Directly to Filter*

Run No	Filter Loading Rate m/hr	pH	Cyst Dosage	% Cyst-sized Particle Removal	% Cyst Removal
3	5	6.7	984/L Filter B, 622/L Filter C	93.1	99.96
4	6	6.3 (no lime)	1093/L Filter C	91.5	99.95
5	6	7.2 (lime used)	23/L Filter C	31	96.74
6	6	6.4 (no lime)	30/L Filter C	95.6	96.67
7	6	7.2 (lime used, low spiking level)	2.3/L Filter C	95.2	30

At optimum conditions, cyst removal was consistently high. An alum dosage of 12 mg/L, pH 6.2, and a filter loading rate of 4.9 m/hr (2 gpm/ft²), would give a 99.73% removal of cysts at the end of the 1-hr filter ripening period. Later in the run, cyst reduction was 99.94% and the effluent turbidity was constant at 0.02 ntu. The influent turbidity was 1.2 ntu. A filter run with a higher loading rate to 9.8 m/hr (4 gpm/ft²) did not show any adverse effect on the filter's ability to remove cysts.

A reduction in the coagulant dosage led, as expected, to an increase in the number of cysts passing through the filter. When the alum dose was 4 mg/L, the filter-ripening period increased to 2 hr, and only 64% of the cysts added to the plant after 2.5 hr of operation were removed in the filter. The effluent turbidity was 0.5 ntu, but was slowly decreasing. The effluent turbidity at 72.5 hr remained relatively high at 0.4 ntu, whereas the cyst removal had increased to 91.8%. As expected, when no coagulant was added to the water, the filter performed poorly with regard to both cyst removal and turbidity reduction. More than half the cysts (52%) passed through the filter, and the effluent turbidity remained relatively high.

For the raw water being studied, cyst removal exceeded 99% at pH 5.6 and pH 6.2, but it dropped to just over 95% when pH was raised to 6.8.

An examination of Table 2 shows that filtered water turbidity could be used as a guide to cyst removal efficiency. Cyst removal exceeded 99.0% ten times. In seven of the ten instances, the filtered water turbidity was less than 0.10 ntu. Cyst removal was less than 99% on five occasions, and

each time the filtered water turbidity exceeded 0.10 ntu (actual range was 0.19 to 0.52 ntu).

To confirm information developed in the UW pilot plant, field tests were performed with the EPA's 75-L/min mobile pilot plant at Hoquiam and at Leavenworth. In each case, the data were also compared with actual plant operating data to provide guidance to local plant operators. Thirty pilot plant runs were made at Hoquiam between May and September 1980. Another 19 runs were made at Leavenworth from September through November 1980.

At Hoquiam, the most effective particle and turbidity removal occurred with an alum dose of at least 8 mg/L. The optimum pH was 6.7, with acceptable performance in the range of 6.4 to 7.0, according to pilot plant results. Actual treatment plant practice before the study resulted in turbidity and cyst-sized particle removals that were somewhat erratic. The high removal variability was primarily due to fluctuations in pH during coagulation and flocculation. The pH ranged from 6.6 to 7.4, and the lower removals were observed at the higher pH values. The apparent inability of alum to affect the overall performance of the plant at dosages above those found effective in the pilot plant was related to the treatment plant's operation at high pH values. Some operational changes at the plant were considered as a result of the pilot plant work. One change, a closer monitoring of the raw water pH as it entered the flocculator, was well under way toward the end of the study, and it included reducing the amount of soda ash added to the raw water. Instead, additional soda ash was

Table 2. *Cyst Removal During Direct Filtration at UW Pilot Plant*

Run No.	Alum Coagul. and Dosage (mg/L)	pH	Filter Loading Rate (m/hr)	Total Dosage	Filter Influent Dosage	Cyst Removal (%)	Elapsed Time (Hr:Min)	Influent Turbidity (ntu)	Effluent Turbidity (ntu)	Turbidity Removal (%)
72	None	6.5	6.1	2.0·10⁶	6.6·10⁶	48	4:30	0.73	0.39	47
73	12.0	6.2	6.1	12.7·10⁶	3.8·10⁶	99.73	1:15	1.24	0.03	98
			4.3	20.0·10⁶	4.2·10⁶	99.943	26:00	1.19	0.19	84
74	12.0	6.2	9.7	15.5·10⁶	7.3·10⁶	99.936	1:00	1.37	0.04	97
76	12.0	6.2	9.4	19.0·10⁶	8.7·10⁶	99.979	7:00	1.14	0.02	98
77	7.0	6.2	9.7	19.2·10⁶	9.8·10⁶	99.75	1:00	1.94	0.24	88
			8.6	20.4·10⁶	9.4·10⁶	99.87	16:00	0.81	0.03	96
78	4.0	6.2	10.1	21.4·10⁶	10.7·10⁶	64	2:30	1.31	0.52	60
			8.6	20.2·10⁶	8.8·10⁶	91.8	72:30	1.35	0.37	73
79	12.0	6.8	9.7	21.5·10⁶	10.3·10⁶	95.4	1:00	0.95	0.28	71
			8.3	21.5·10⁶	8.4·10⁶	99.41	10:00	1.02	0.04	96
80	12.0	5.6	9.7	20.4·10⁶	10.0·10⁶	99.83	1:00	1.73	0.03	98
81	12.0	5.6	9.7	20.4·10⁶	9.8·10⁶	99.84	7:00	1.78	0.02	99
	CatFloc		9.7	20.1·10⁶	9.8·10⁶	95.9	1:00	0.92	0.23	75
82	T-1 5.0	6.4	9.7	20.1·10⁶	9.8·10⁶	99.911	21:00	0.80	0.27	66

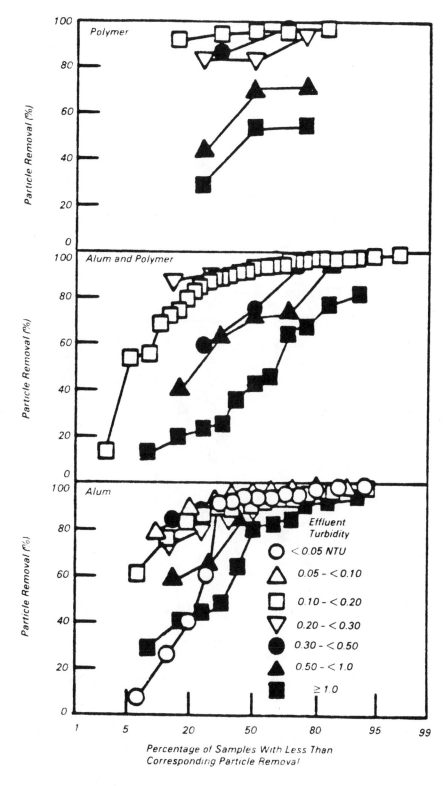

100
80
60
40
20
0

Particle Removal (%)

Polymer

100
80
60
40
20
0

Particle Removal (%)

Alum and Polymer

100
80
60
40
20
0

Particle Removal (%)

Alum

Effluent
Turbidity

○ < 0.05 NTU

△ 0.05 – < 0.10

□ 0.10 – < 0.20

▽ 0.20 – < 0.30

● 0.30 – < 0.50

▲ 0.50 – < 1.0

■ ≥ 1.0

1   5   20   50   80   99

Percentage of Samples With Less Than
Corresponding Particle Removal

Figure 1. *Relationship between effluent turbidity and particle removal at Hoquiam.*

added to the clearwell as a corrosion control measure to increase the pH before distribution.

A relationship was established between effluent turbidity and particle removal (Figure 1). An effluent turbidity below 0.05 ntu corresponded with a median (50% of the values) particle removal of 95.1%, and an effluent turbidity between 0.05 and 0.1 ntu was associated with a 94.3% particle removal. A surprisingly large number of samples (33%) with an effluent turbidity below 0.05 ntu had particle removals below 90%. This result was observed especially in the beginning of the run, directly after the filter ripening or during the running phase when the influent particle concentration declined temporarily. The polymer plus alum and polymer runs did not produce effluent turbidities below 0.1 ntu, but high median particle removals of 95.3% and 92.6%, respectively, were noted for effluent turbidities between 0.10 and 0.20 ntu. These results also demonstrate the need to produce filtered water with very low turbidity.

The study also evaluated the removal of actual *G. lamblia* cysts by the pilot unit. Cysts recovered from human stool specimens were added to the raw water ahead of any chemical addition during an 8-min spike. The cysts were recovered from the influent and effluent using a membrane filtration technique. Of the 1.67×10^6 cysts added to the raw water, 1.06×10^3 remained in the water just before entering the filter according to the membrane filtration technique. The filter effluent contained a total of 2.6×10^2 cysts, representing at least an 81% cyst removal. This corresponded with a 99% removal of particles in the 8- to 12-μm range as determined by the particle counter and a 95% turbidity removal.

At Leavenworth, most of the pilot plant testing was conducted at a filtration rate of 10.7 m/hr (4.4 gpm/ft²), which corresponded to the maximum filter loading rate at the full-scale plant. The optimum alum dosage during September and October was 15 mg/L, resulting in a 90% reduction in turbidity. The corresponding cyst-sized particle removal was 96%, with the maximum 98% occurring at an alum dosage of 13 mg/L. The pH optimum was 6.7. At pH 6.4 and 7.1, both turbidity and particle removals were less than optimum. During November, the average water temperature dropped from 8.5°C to 3°C, and alkalinity declined from 24 to 12 mg/L. This water was more difficult to treat. At times, as much as 70% of the alum added as coagulant was passing through the filter. The reason may have been that the 8 min of flocculation time available in the pilot plant was not adequate for the very cold water. Because alum was not very effective at 3°C, a cationic polymer was tried. Removals were comparable with those experienced with alum at this low temperature. Maximum turbidity removal of 59% was realized at a polymer dose of 0.2 to 0.4 mg/L, compared with 43% removal with no coagulant. Particle removal decreased at low temperatures from 45% to 12% when 0.2 to 0.4 mg/L of polymer was added. To achieve optimum particle removal under these conditions, a polymer dosage of 3.5 mg/L was required. But this dose decreased turbidity removal to 35%, apparently a result of the polymer's restabilization of very fine colloids that were smaller than the size being measured by the particle counter but large enough to cause light scatter (turbidity).

To determine the ability of the pilot plant to remove cysts, 1.25 x 10⁶ cysts were added to the raw water over a 20-min period, and the filter influent and effluent were sampled and analyzed for cysts. Before the cyst addition, a salt solution had been added and traced through the plant to determine suitable sampling times. The plant was operated at a 10.7-m/hr (4.4 gpm/ft²) filtration rate with 1.2 mg/L Cat Floc T as the coagulant. The raw water turbidity was 0.33 ntu, and the temperature was 1°C. During the cyst addition, the effluent turbidity was 0.19 ntu, a 42% reduction. The three filter influent samples recovered a calculated 867 cysts, and 242 cysts were recovered from the effluent corresponding with a 72% removal. Cyst-sized particle removal was 53%.

Removal of turbidity and particles at Leavenworth was greater when alum was used as the coagulant instead of a cationic polymer. Lower filtered water turbidities were attained with alum (Figure 2), and the lowest turbidity ranges more frequently were associated with very high removal percentages.

The Leavenworth treatment plant used cationic polymer as the primary coagulant. Both the raw water and filtered water turbidities were below 1.0 ntu, but some equipment problems associated with filter control values prevented the plant from operating at maximum efficiency. These problems were later corrected, and new equipment was installed to improve the rapid mixing process.

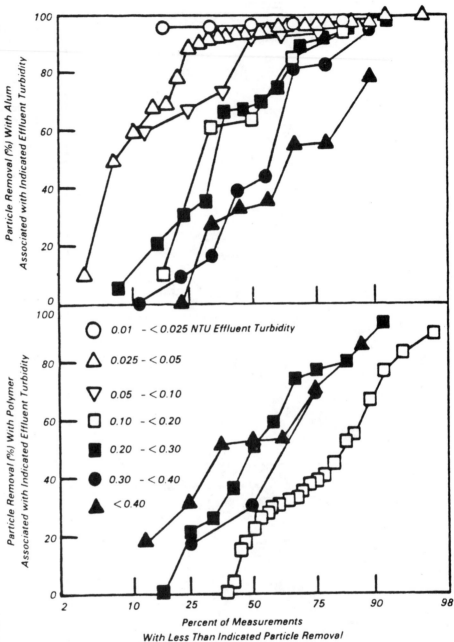

Figure 2. Frequency distribution of particle removal at different effluent turbidities during alum and polymer treatment at Leavenworth.

Diatomaceous Earth Filtration Results

Numerous DE filter runs were performed without addition of cysts to develop general information on filter performance. The results from the initial runs showed that the cyst-sized particle removal by the DE filter was generally better than the reduction in turbidity. This result was expected, because particles 6 µm or smaller are generally more abundant than particles in the 7- to 12-µm range. Of the several types of filter aids tested, the best performers were the finer grades, especially in the very beginning of the run. Later in the run, when the cake was thick, no grade of DE specifically outperformed the others for removal of turbidity and cyst-sized particles.

The most noticeable difference between these runs was the rate of headloss buildup, which was slowest for the coarsest grades. This result was also manifested by longer filter runs. The length of the run depended not only on the type of diatomite used, but to a significant degree on the amount of body feed added to the filter. The body feed rate ranged from 10 to 40 mg/L. Though the raw water used for these runs was of high quality, with turbidity normally ranging from 0.6 to 0.9 ntu during this time of the year, body feed rates of 10 or 40 mg/L resulted in shorter filter runs than did application of 20 mg/L of body feed. This result indicated that the 10 mg/L was an inadequate body feed dose, but 40 mg/L was excessive.

During one of the runs when Hyflo Super-Cel was used as a filter aid, a nonionic polymer (Magnifloc 985N, American Cyanamid Co., Wayne, New Jersey) was added to the raw water. The most noticeable effect of the 0.0075-mg/L polymer addition was a significant improvement in the effluent quality in the very beginning of the run. As the run progressed, the efficiency of the DE filter seemed to be similar to earlier runs where no polymer had been added. When the run was terminated, it was approximately 25% shorter than similar runs where no polymer was added. The single most important factor for the decrease in run duration was assumed to be the polymer addition.

In the DE filtration research on *Giardia* cyst removal, the cysts were added to the raw water at the same location as the body feed, either as a slug or as a constant continuous dosage. The slug contained a total of 3.0×10^6 cysts, added in 10 sec. For continuous addition, the cysts were metered into the raw water line with a peristaltic pump. Raw water cyst concentrations used during these runs ranged from 1.5×10^5 to 9.0×10^6 cysts/L. Because data on DE filter run length and removal of turbidity and cyst-sized particles had already been obtained, runs with *Giardia* cysts were conducted only long enough to dose the cysts and obtain information on cyst removal efficiency.

The filter effluent sampling schedule was determined from a series of tests in which a salt solution was added to the raw water in place of cysts. The conductivity of the filter effluent was monitored continuously to determine (1) how fast a 10-sec slug would pass through the filter, and (2) the time required to reach a constant effluent concentration when a continuous dosage was added to the filter influent. Results showed that the entire slug would have reached the filter effluent in 10 min. Thus to trap all the cysts escaping the filter, a 38-L sample would need to be collected. This volume was not an unreasonably large one to process by the technique developed for this study. When a constant dosage was added to the raw water, the effluent concentration had attained its maximum and constant level after 10 min. By adding cysts for 15 min and sampling the filter effluent during the last 5 min, a 19-L sample containing an average effluent concentration of cysts was collected.

All cyst runs used Hyflo Super-Cel as a filter aid. Based on results from preceding runs without cyst addition, a 1.0-kg/m² (0.2-lb/ft²) precoat and 20-mg/L body feed

was judged most suitable for the 3.8-L/min filtration rate and raw water quality. During two of the four runs, a 0.0075-mg/L dosage of the 985N polymer was added to the raw water for the duration of the run. Results appear in Table 3. Diatomaceous earth filtration proved to be very effective for *Giardia* cyst removal, with more than 99% of the cysts removed in every one of the 12 filtered water samples analyzed in this series of tests.

Conclusions

High cyst removals (above 99%) can be attained by properly operated granular media filters, but close attention must be given to maintaining the unit processes at optimum conditions.

When a plant operates with alum as the coagulant, an adequate alum dose is essential for effective removal of turbidity and cysts. Close monitoring of pH during the coagulation and flocculation processes is also necessary. For the waters studied in this investigation, the pH optimum under most conditions was 6.7. Changes in raw water quality could cause a shift in the pH optimum, as was demonstrated during the field operation.

The polymers tested as primary coagulants did not perform as well as alum. On the other hand, the filter runs were longer, and the necessity for close pH monitoring experienced during alum treatment was not required.

UW pilot plant data on cyst removal and mobile pilot plant data on removal of cyst-sized particles suggested that the highest removal percentages for cysts and for 8- to 12-μm particles usually tended to be associated with production of low turbidity water. Best particle removal results were seen for filtered water with turbidities of 0.05 ntu or lower.

An increase in filter-loading rate beyond the 10 m/hr (4.1 gpm/ft²) used for many of the filter runs could be detrimental. The effluent water quality would normally suffer, and because of the more rapid headloss buildup and hence shorter filter run, the amount of water produced per run was often less than at the more moderate filtration rates.

Low temperatures can greatly affect the coagulation and flocculation process when alum is used. A similar phenomenon can occur with polymers. Increasing the flocculation time beyond the 8 min that were available in the mobile pilot plant might have improved the plant's ability to treat the cold water at Leavenworth. The optimum operating conditions attained during the period of low water temperature would most likely have changed with an increase in the flocculation time.

Diatomite filtration was very effective in removing *G. lamblia* cysts, even in the very beginning of the filter run, when the precoat acted as the only barrier. As the run progressed and the body feed caused the thickness of the filter cake to increase, the removal of cysts improved. The only decrease in the filter's ability to trap the cyst particles was recorded when the dosage at the end of the run was increased six times. This decrease in performance was less evident when a polymer was added to the raw water. The polymer addition generally improved the removal efficiency, but it tended to shorten the filter run because of a more rapid rate of headloss buildup, especially toward the end of the run.

The full report was submitted in fulfillment of Grant No. R806127 by the University of Washington under the sponsorship of the U.S. Environmental Protection Agency.

Table 3. *Filter Runs with Cysts Using DE Filter*

Run No.	Polymer Added	Cyst Addition			Cyst-Sized Particle Removal (%)
		Number Added	Elapsed Time (Hr:Min)	Removal (R) (%)	
		Slug	Continuous (Cysts/L)		
63	No	$3.0 \cdot 10^6$		0:05	99.35 < R < 99.78	98.4
		$3.0 \cdot 10^6$		0:20	R > 99.65	95.1
		$3.0 \cdot 10^6$		2:00	R > 99.65	97.9
			$1.5 \cdot 10^6$	2:30	99.61 < R < 99.96	98.8
			$9.0 \cdot 10^6$	3:00	99.03 < R < 99.10	92.3
64	Yes	$3.0 \cdot 10^6$		0:05	99.61 < R < 99.96	98.1
		$3.0 \cdot 10^6$		0:20	R > 99.65	98.2
		$3.0 \cdot 10^6$		2:00	R > 99.65	98.2
			$1.5 \cdot 10^6$	2:30	R > 99.65	97.9
			$9.0 \cdot 10^6$	3:00	99.48 < R < 99.56	96.3
65	No	$4.5 \cdot 10^6$		3:00	99.83 < R < 99.94	87.1
66	Yes	$4.5 \cdot 10^6$		3:00	99.87 < R < 99.99	95.0

EPA Project Summary

Filtration of *Giardia* Cysts and Other Substances
Volume 1: Diatomaceous Earth Filtration

Kelly P. Lange, William D. Bellamy, and David W. Hendricks

Kelly P. Lange, William D. Bellamy, and Davis W. Hendricks are with Colorado State University, Fort Collins, CO 80523.

Gary S. Logsdon is the EPA Project Officer (see below).

The complete report, entitled "Filtration of Giardia Cysts and Other Substances: Volume 1: Diatomaceous Earth Filtration," (Order No. PB 84-212 703) will be available only from:

National Technical Information Service
5285 Port Royal Road
Springfield, VA 22161
Telephone: 703-487-4650

The EPA Project Officer can be contacted at:

Municipal Environmental Research Laboratory
U.S. Environmental Protection Agency
Cincinnati, OH 45268

How effective is filtering drinking water through diatomaceous earth to remove *Giardia lamblia* cysts, total coliform bacteria, standard plate count bacteria, turbidity, and particles? We evaluated the process for a range of operating conditions and simulated ambient conditions. Hydraulic loading rates imposed were 2.44, 4.88, and 9.76 m/hr (1,2, and 4 gpm/ft²). Seven grades of diatomaceous earth were used. Temperatures were from 5° to 19°C; concentrations of *Giardia* cysts ranged from 50 to 5000 cysts/L; and bacteria densities were varied from 100 to 10,000/100 mL.

The results of this study showed that diatomaceous earth filtration is an effective process for water treatment. *Giardia* cyst removals were greater than 99.9 percent for all grades of diatomaceous earth tested, for hydraulic loading rates of 2.44 to 9.76 m/hr, and for all temperatures tested. Percent reduction in total coliform bacteria, standard plate count bacteria, and turbidity are influenced strongly by the grade of diatomaceous earth used. The coarsest grades of diatomaceous earth recommended for water treatment (e.g., C-545®)* will remove greater than 99.9 percent of *Giardia* cysts, 95 percent of cyst-sized particles, 20 to 35 percent of coliform bacteria, 40 to 70 percent of heterotrophic bacteria, and 12 to 16 percent of the turbidity from Horsetooth Reservoir water. The use of

the finest grade of diatomaceous earth (i.e., Filter-Cel®), or alum coating on the coarse grades, will increase the effectiveness of the process, resulting in 99.9 percent removals of bacteria and 98 percent removals of turbidity.

This Project Summary was developed by EPA's Municipal Environmental Research Laboratory, Cincinnati, OH, to announce key findings of the research project that is fully documented in a separate report of the same title (see Project Report ordering information at back).

Introduction

This study was conducted at Colorado State University under a cooperative agreement with the U.S. Environmental Protection Agency (EPA) to determine the effectiveness of diatomaceous earth filtration for removal of *Giardia* cysts. At the same time, removals of turbidity, total coliform bacteria, standard plate count bacteria, and particles were determined. Operating conditions examined included the grade of diatomaceous earth, hydraulic loading rates, influent concentrations of bacteria and *Giardia* cysts, head loss, run time, temperature, and the use of alum-coated diatomaceous earth.

Giardia lamblia is a protozoan prevalent in the clear, cool waters characteristic of the Rocky Mountain region. This organism causes giardiasis, a harmful but nonfatal intestinal disease. Many communities use water from these Rocky Mountain streams, which are

---

*Mention of trade names or commercial products does not constitute endorsement or recommendation for use.

considered pristine pure because they look aesthetically pleasing and will meet the 1-NTU turbidity water quality standard. How to treat these waters has become an important concern over the last few years as outbreaks of giardiasis have occurred. Economical and effective filtration systems are needed to remove *Giardia* cysts. Designs appropriate for small water systems are particularly needed.

Diatomaceous earth filtration was introduced in 1942 as a technology for water treatment. The process was adopted by the U.S. Army for field use in 1944 after being shown effective for removal of *Endamoeba histolytica* cysts. The basic principles of the process were outlined in the 1940's and 1950's, and further studies were made in the 1960's.

The diatomaceous earth filtration process consists of three basic operations: (1) precoating, (2) filtering, and (3) cleaning. In precoating, an initial filter cake consisting of a 3- to 5-mm layer of powder-sized diatomaceous earth filter medium is applied to a support membrane called a septum. The cake is applied by circulating flow from the precoat tank through the filter, causing the slurried diatomaceous earth to be deposited on the filter septum. In filtering, the second step of the process, raw water combined with bodyfeed passes through the filter cake. The bodyfeed consists of a filter medium slurry metered into the raw water stream during filtration. The continuous addition of bodyfeed maintains the filter cake permeability. In the third operation, cleaning, the filter cake is removed from the septum and discarded.

Operating parameters for diatomaceous earth filtration include the grade of diatomaceous earth (a commercial designation of particle size), hydraulic loading rate, precoat thickness, bodyfeed concentration, terminal headloss, and run time. Chemical coating of the diatomaceous earth may be used under some circumstances to improve removal effectiveness.

The protozoan *Giardia lamblia* is of interest because it causes giardiasis. This organism has been identified as a pathogen only recently. Giardiasis is considered a serious problem in mountainous and forested regions of the United States where the organism is endemic; the *Giardia* cysts shed by dogs, humans, and animals, such as the beaver, are believed to be of the *Giardia lamblia* species.

The *Giardia lamblia* trophozoite (Figure 1a) can reside in the intestine of a variety of warm-blooded animals. The cyst (Figure 1b) is the form shed and transmitted. An infected person may shed up to 900 million cysts per day. The cyst form of the organism is hardy and may remain viable for a long period (2 months, for example), particularly in cold water. Infection is caused by ingestion; an infective dose may be from 1 to 10 cysts and the incubation period is 1 to 2 weeks. A surface water supply source is a vehicle for cyst transmission.

Materials and Methods
Design of Tests

The objective of the experimental program was to evaluate the removal effectiveness of diatomaceous earth filtration for the dependent variables (*Giardia* cysts, total coliform bacteria, standard plate count bacteria, turbidity, particle counts, and headloss) as a function of the independent variables (grade of diatomaceous earth, run time, headloss, hydraulic loading rate, temperature, influent coliform concentration, and alum coating). This was achieved by changing a given independent variable over a range of magnitudes and observing the effect on the filtration performance (i.e., the dependent variables).

Tests were terminated either at the predetermined head loss across the septum of 40 psi, or after a given period of operating time. The precoat application was established at 1 kg/m² (0.2 lb/ft²) for all tests. The bodyfeed concentration was established by using successive concentrations until a linear headloss versus time relationship was found.

The pilot plant was used mostly in the laboratory, but final confirming tests were conducted in the field. The filter, the main element of the system (Figure 2), consisted of a 1-ft² septum enclosed in a pressure housing. The septum used in this work was stainless steel wire mesh of 110 x 24 wires per in.². The operations were controlled by the ancillary pumps, valves, and gauges.

Experimental Procedures

Giardia testing began with the processing of *Giardia* cysts from dog fecal samples. The processing consisted of adding the infected feces to distilled water, straining the feces and then making a count of cysts in the fecal concentrate.

A known concentration of the *Giardia* concentrate was then added to a 1400-L filter feed tank. This tank was a modified milk cooler that could be maintained at 2 to 15 ±1°C. The filter feed tank was filled with Horsetooth Reservoir water and cooled before the addition of the cysts. Primary settled sewage was added also to increase the concentration of total coliform bacteria.

Preparation for a test run began with the precoat step. After precoating, 10 mg/L chlorine was added for disinfection, and the recycle of precoat water was continued for 10 min. The chlorine was purged by operating in the filtering mode for 30 min. Sampling was started after the 30-min washout period.

Samples were obtained from the filter feed tank and from the effluent side of the filter for measurements of turbidity, particle counts, total coliform bacteria, standard plate count bacteria, and

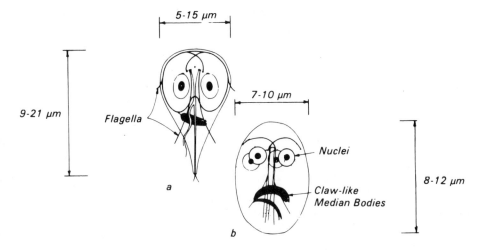

Figure 1. Sketches of a) trophozoite and b) cyst stages of Giardia lamblia *(Jakubowski and Hoff, 1979).*

Giardia cysts. Grab samples were collected for all parameters except *Giardia* cysts. Other measurements included elapsed time from beginning of run, headloss, hydraulic loading rate, and water temperature.

The *Giardia* cyst sampling technique used a 142-mm-diameter polycarbonate membrane filter with a 5- μm pore size to remove and concentrate the cysts from the sampled water. After the sample was concentrated, the membrane filter was washed and the wash water was analyzed for cysts by microscopic counting. The influent water in the filter feed tank and the diatomaceous earth effluent were both sampled for *Giardia* cysts in this manner. The influent sample volume ranged between 2 and 10 L, and the effluent sample volume ranged between 86 and 174/L.

Field tests to verify laboratory findings were conducted in April and May 1983. The April test used raw water from the Cache La Poudre River, and the May tests used water from Straight Creek and Laskey Creek at the Dillon, Colorado water treatment plant.

Leak Testing

During the initial phases of experimentation, the need to determine whether a leak was present in the filter septum or its manifold became apparent. Thus a technique was developed to test for leaks. First the filter was precoated with 2 kg/m^2 of the finest grade of diatomaceous earth, Filter Cel. This grade was determined to be capable of removing 100 percent of the applied coliform bacteria. Then the filter was operated at 1 gpm/ft^2 for 1 hr with a high influent coliform concentration. If any coliforms were detected in the effluent, the equipment was assumed to have a mechanical problem resulting in a leak, and the problem was corrected. This technique was used as a quality control measure throughout all testing.

Results

The diatomaceous earth filtration process was found to be effective for removing *Giardia* cysts under virtually all operating conditions tested No cysts were detected in the filter effluent when normal water treatment practices were simulated. Note, however, that removals of bacteria, turbidity, and particles in the 6.35-μm to 12.67-μm size range were functionally dependent on: (1) grade of diatomaceous earth, (2) use of chemicals, (3) hydraulic loading rate, and (4) influent concentrations.

Overall Process Effectiveness

Giardia cyst removals were greater than 99.9 percent, regardless of grade of diatomaceous earth, filtration rate, temperature, duration of test, or influent concentration of *Giardia* cysts (when cysts are fewer than 10,000/L). The single breakthrough occurred at a high influent concentration of 33,600 cysts/L.

Figure 3 illustrates the uniformly high removals of *Giardia* cysts. Testing was done only for the water treatment grades, since removals would have been at least as much for the finer grades.

The grade of diatomaceous earth was not a factor in removal of *Giardia* cysts, even with grades C-545 and C-535. These grades create a filter cake with reported median pore sizes of 17 and 13 μm, respectively, which are larger than *Giardia* cysts. However, many pores were apparently smaller than the cysts and blocked their passage.

Removal of coliform bacteria, standard plate count bacteria and turbidity approached 100 percent for the smallest grades of diatomaceous earth and fell below 40 percent for the coarsest grades (Figure 3). Removals of total coliforms and standard plate count bacteria would follow this trend regardless of the water source. Turbidity removal, however,

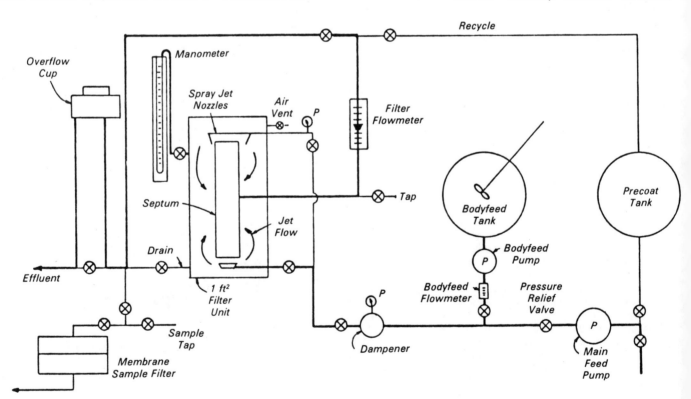

Figure 2. Layout of diatomaceous earth filtration pilot plant.

would depend on the size of the particles making up the turbidity; improved removal would result if the turbidity consisted of larger particles.

Effects of Operating Conditions on Removals

Operating conditions examined include grade of diatomaceous earth, hydraulic loading rate, *Giardia* cyst concentration, bacteria concentration, temperature, duration of filtration run, and alum coating of the diatomaceous earth. The influence of each condition on removals of turbidity, bacteria, *Giardia* cysts, and particles is reported in the following paragraphs.

Grade of Diatomaceous Earth

As mentioned earlier, removals of *Giardia* cysts are greater than 99.9 percent for all earth tested, but removals of turbidity, standard plate count bacteria, and total coliform bacteria are strongly influenced by grade. Removals of particles in the 6.35- to 12.67-μm size range were uniformly high (87 to 94 percent) for the water treatment grades of diatomaceous earth, thus indicating no correlation of particle removal to grade.

Hydraulic Loading Rate

Some scatter occurs in the data on the effects of hydraulic loading on removals of particles, standard plate count bacteria, coliform bacteria, and turbidity; but the trends are toward declining removals with increasing hydraulic loading rate. Percent removals of total coliform bacteria were affected the most, standard plate count bacteria declined nominally, and particles and turbidity were affected only moderately. The fine clays constituting most of the raw water turbidity passed readily through the C-503 and C-545 diatomaceous earth grades at all hydraulic loading rates. Hydraulic loading rate had no detectable influence on removal of *Giardia* cysts, since all but one test resulted in complete removal of the influent cysts. Because the cysts are larger than some of the pores in the filter cake (which vary statistically), they will be blocked by some pore as they are convected by the flow within the cake.

Giardia Cyst Concentrations

No discernible relationship existed between the *Giardia* cyst influent concentration and cyst removal. Influent *Giardia* cyst concentrations ranged from 500 to 10,000 cysts/L, with one excep-

tion of 33,600 cysts/L. Only for the latter case were cysts detected in the effluent stream. This result indicates that *Giardia* removal for expected ambient influent concentrations will be virtually 100 percent.

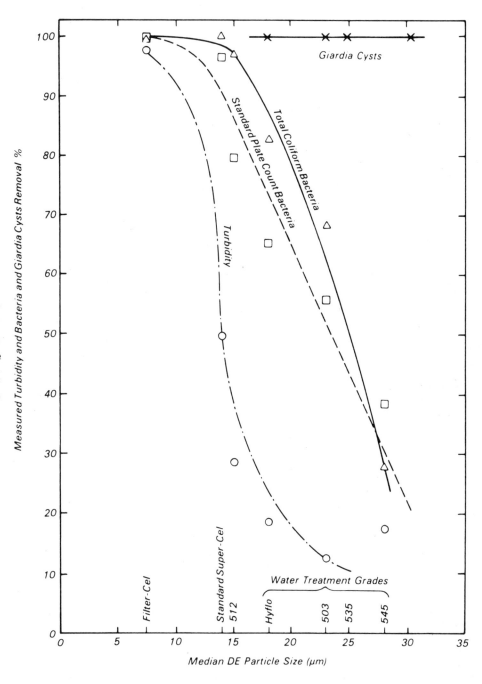

Figure 3. *Removals of turbidity, bacteria, and* Giardia *cysts for different grades of diatomaceous earth. Turbidity and bacteria data points represent avaerages for six 5-hr test runs.* Giardia *cyst data are averages for 32 test runs of various durations.*

Bacteria Concentration and Turbidity

The influent concentration of total coliforms has a stronger influence on removals for the water treatment grades of diatomaceous earth than for the finer grades. This phenomenon is reasonable, since the finer grades can completely strain coliform bacteria.

Turbidity did not vary enough in the waters tested to determine whether a relationship existed between influent levels of bacteria and their removal. For the laboratory tests with Horsetooth Reservoir water, the influent turbidity levels ranged only between 4.5 and 5.4 NTU.

Most turbidity in Horsetooth water was caused by particles smaller than 1 μm. Consequently, a 1-NTU turbidity standard could be attained only with the finest grade of diatomaceous earth or by using alum-coated diatomaceous earth. Removals of coliform bacteria were therefore higher, since they were larger than much of the turbidity. The small turbidity particles found in Horsetooth Reservoir are not necessarily characteristic of all waters, since many sources have been treated to meet a 1-NTU standard with water treatment grades of diatomaceous earth and without chemicals.

Temperature

Water temperature did not appear to affect any of the parameters measured. Because the removal process in diatomaceous earth filtration is physical, slightly poorer removals might be expected at lower temperatures because shear forces are higher. This possible effect was not noticeable, however, since it was masked by the significant changes in removal caused by variations in influent concentration and flow rate.

Alum Coating

Alum-coated diatomaceous earth removed significantly more total coliform bacteria, standard plate count bacteria, and turbidity than did the uncoated diatomaceous earth. Coarse grades of diatomaceous earth were used (e.g., C-545 and C-503), and both the precoat and bodyfeed were coated while in slurry form. Removals of total coliform bacteria ranged from 96 to 99.9 percent compared with 30 to 70 percent without alum coating. Removals of standard plate count bacteria ranged from 79 to 99.5 percent compared with 38 to 56 percent without alum coating. Turbidity removals ranged from 66 to 99 percent compared with 11 to 17 percent without alum coating. These results demonstrate a marked improvement in treatment with alum coating and illustrate how diatomaceous earth filtration can be applied in otherwise marginal situations.

Note, however, that not all water sources can be treated to a 1-NTU standard with the normal water treatment grades of diatomaceous earth. To obtain a 1-NTU effluent for these special cases, the advantages of alum coating must be weighed against using a finer grade of diatomaceous earth. Both techniques will reach the desired goal, but only pilot plant studies and economic considerations will determine which approach should be taken.

Field Testing

Field tests were conducted to confirm the laboratory results. Water was obtained from the Cache La Poudre River and from the raw water intake at the Dillon water treatment plant. Turbidity conditions were about 4 NTU for tests conducted April 17, 1983, at the Cache La Poudre River, and about 0.6 NTU for the May 1983 tests at Dillon. Despite the use of these different water sources, no *Giardia* cysts were detected in any filtered water samples. Removals of turbidity, total coliform bacteria, and total plate count bacteria were all consistent with laboratory results.

Conclusions

1. Diatomaceous earth filtration is virtually 100 percent effective in *Giardia* cyst removal for all grades of diatomaceous earth over a wide range of conditions.

2. Grade of diatomaceous earth is the most important factor in the removal of bacteria and turbidity. Removals effected by coarse and fine grades were, respectively, 17 and 98 percent for turbidity, 28 and 99.9 percent for coliform bacteria, and 38 and 99.8 percent for standard plate count bacteria.

3. Increasing the hydraulic loading rate causes a decrease in removals of bacteria and turbidity for the water treatment grades of diatomaceous earth. The effect was strongest for coliform bacteria and weakest for turbidity. Hydraulic loading rate showed no effect on the removal of *Giardia* cysts.

4. Water temperature did not influence the effectiveness of diatomaceous earth filtration, as demonstrated by testing over the range of 3.5° to 15°C. The results are not, however, conclusive.

5. Bacteria removal decreased with increased influent concentrations of bacteria, especially for the coarser grades of diatomaceous earth. A three-log increase in influent coliforms reduced removals from 77 to 39 percent for C-545; but for C-512, a two-log increase in coliforms reduced removals only from 96 to 92 percent.

6. Bacteria removals decreased slightly with increasing filtration time-- from 87 to 79 percent in 5.5 hr for C-503 and from >99.98 to 99.92 percent for Standard Super-Cel.

7. Alum-coated diatomaceous earth filtration removed significantly more turbidity and bacteria than diatomaceous earth filtration with no alum. The use of alum coating increased removals from 17 to 99 percent for turbidity, from 30 to 96 percent for total coliform bacteria, and from 56 to 99.5 percent for standard plate count bacteria.

8. Increased removals of turbidity and bacteria can be accomplished either by chemically coating the diatomaceous earth or by using a smaller grade.

9. Field testing with two different raw waters yielded the same results as laboratory tests.

10. Pilot-plant testing should be done before implementing any full-scale application of diatomaceous earth filtration. Applicability, design criteria, and operating conditions cannot be determined without pilot tests.

11. Periodically, a diatomaceous earth filtration system should be checked for leaks by applying Filter-Cel and then filtering a coliform-contaminated water. In production of potable water, this should be done only as a part of a routine performance evaluation program in which careful controls are set up to ensure that a cross connection is not possible.

The full report was submitted in fulfillment of Cooperative Agreement No. CR808650-02 by Colorado State University under the sponsorship of the U.S. Environmental Protection Agency.

EPA Project Summary

Filtration of *Giardia* Cysts and Other Substances Volume 2: Slow Sand Filtration

William D. Bellamy, Gary P. Silverman, and David W. Hendricks

William D. Bellamy is with CH2M-Hill Consulting Engineers, Newport Beach, CA 92660, and Gary P. Silverman is with the U.S. Environmental Protection Agency, San Francisco, CA 94105. David W. Hendricks is with Colorado State University, Fort Collins, CO 80523.

Gary S. Logsdon is the EPA Project Officer (see below).

The complete report, entitled "Filtration of Giardia Cysts and Other Substances, Volume 2: Slow Sand Filtration," will be available only from:

National Technical Information
 Service
5285 Port Royal Road
Springfield, VA 22161
Telephone: 703-487-4650

The EPA Project Officer can be contacted at:

Water Engineering Research
 Laboratory
U.S. Environmental Protection
 Agency
Cincinnati, OH 45268

Slow sand filtration was evaluated for a range of operating conditions and simulated ambient conditions using 1-ft-diameter laboratory filters in two phases of experimentation. The objective was to determine its effectiveness as a process in drinking water treatment for removal of Giardia lamblia cysts, total coliform bacteria, standard plate count bacteria, turbidity, and particles.

During Phase I experiments, three filters were operated for 16 months at hydraulic loading rates of 0.04, 0.12, and 0.40 m/hr using raw water from Horsetooth Reservoir, located adjacent to the Engineering Research Center at Colorado State University. In Phase II experiments, six filters were operated for 12 months, all at hydraulic loading rates of 0.12 m/hr, each under a different operating condition (e.g., depth of sand, size of sand, disinfection of raw water, nutrient addition, sand size, and temperature).

Phase I results showed removals of Giardia cysts that exceeded 99.9 percent for the three hydraulic loading rates used. The most important operating condition was the development of a biopopulation within the sand bed. Cysts removals were about 99.0 percent with new sand, but as the biopopulation matured (after about 40 weeks), removals were 100 percent, qualified by detection limits. Removals of total coliform bacteria related well to the development of the biopopulation within the sand bed, showing 90 percent removal for a new sand bed operated at 0.40 m/hr hydraulic loading rate, and 99.99 percent removal for a mature sand bed and established schmutzdecke operated at 0.04 m/hr. Removal of the schmutzdecke caused removals to decline to 99.9 percent, but recovery to 99.99 percent removal occurred within a few days.

Removals of standard plate count bacteria usually ranged from 88 to 91 percent. Because the sand bed comprising the filter develops an internal microbiological population, organisms were continuously sloughed from within the sand bed, causing significant counts of standard plate count bacteria in the effluent. Particle count removals in the size range of 6.35 to 12.7 um ranged from 96 to 98 percent. Also because of the sloughing of material from the filter bed, significant numbers of particles occurred in the effluent. Turbidity removal was usually 27 to 40 percent. The mineral particles that made up the turbidity within the Horsetooth Reservoir consisted mostly of particles 1 um or smaller, which passed readily through the filters.

Phase II testing was done using total coliform bacteria as the primary measure of effectiveness and periodic spiking with Giardia cysts. Removals of total coliform bacteria ranged from 60 percent for the filter maintained with no biological activity (e.g., chlorinated between tests) to 99.9 percent for the filter with nutrients added. Coliform removal for the control filter averaged 97 percent. Using a larger sand size (0.62 instead of 0.29 mm) caused a decline in removal rates, as did using a sand depth of 48 instead of 97 cm. Operation at 2^0 instead of 17°C caused a decline in removals to 92 percent compared with 99 percent for the control filter. Removals of Giardia cysts were 100 percent for all tests conducted (again, qualified by detection limits).

This Project Summary was developed by EPA's Water Engineering Research Laboratory, Cincinnati, OH to announce key findings of the research project that is fully documented in a separate report of the same title (see Project Report ordering information).

Introduction

This report is the second of three volumes describing the research conducted under EPA-CSU Cooperative Agreement No. CR808650-02. The first is entitled, "Filtration of Giardia Cysts and Other Substances. Volume 1: Diatomaceous Earth Filtration," EPA-600/S2-84-114, Sept. 1984, and the third is entitled, "Filtration of Giardia Cysts and Other Substances. Volume 3: Rapid Rate Filtration: EPA-600/2-85-027, May 1985.

Objective

This report describes the results of experimental research to evaluate the effectiveness of slow sand filtration for removal of Giardia lamblia cysts. Other variables studied were turbidity, particles, total coliform bacteria, and standard plate count bacteria. These dependent variables were evaluated with respect to the influence of the independent variables--design and operating conditions. The research had two phases. Phase I operating conditions included hydraulic loading rate, cyst concentration, bacteria concentration, biological maturity of sand in the filter bed, age of schmutzdecke, and temperature. Phase II operating conditions included depth of sand, size of sand, temperature, disinfection of raw water, and nutrient addition to raw water.

Design and Operation of Slow Sand Filters

Slow sand filtration is a passive filtration process--that is, it is subject to very little control by an operator. The process involves no chemical addition or backwash. For recommended design, effective sand size ranges from 0.15 to 0.35 mm, with a uniformity coefficient of less than 2. The sand bed depth ranges from 60 to 120 cm and is supported by graded gravel 30 to 50 cm deep. Drain tiles are placed at the bottom of the gravel support to collect the filtered water. Hydraulic loading rates range from 0.04 to 0.40 m/hr.

During operation of the slow sand filter, biological growth occurs within the sand bed and the gravel support. Also, a layer of inert deposits and biological material called the schmutzdecke forms on the surface of the sand bed. In the literature, the effectiveness of slow sand filtration is attributed mostly to the role of the schmutzdecke, but this research has found that both the schmutzdecke and the biological growth within the sand bed have important roles in the effectiveness of slow sand filtration. The latter may require weeks or months to develop, depending on the nutrients within the raw water.

Operation of a slow sand filter requires two periodic tasks: (1) cleaning by removal of the schmutzdecke, and (2) replacing of the sand. Schmutzdecke is removed when the headloss exceeds the designed value, which may range from 1 to 2 m. After draining the filter, the schmutzdecke is removed by scraping about 2 cm from the surface of the sand bed. The removal interval depends on the rate of accumulation of material, which in turn is related to the contaminants present in the raw water and the hydraulic loading rate. Since operating expenses are affected by the frequency of schmutzdecke removal, pilot testing is advisable to determine this important operating parameter. Replacing sand is necessary after repetitive scrapings have reduced the sand bed in the filter to its lowest acceptable depth.

A number of slow sand filtration treatment plants have been built in the United States, but most were completed in the first decades of the century. Today, the technology is well established in Europe, though it never gained a firm foothold in the United States. The process seems well suited to small communities that need a technology less complex than rapid sand filtration.

Experimental Apparatus and Methods

Apparatus

The Phase I experimentation was conducted using three 1-ft-diameter pilot plants units operated in parallel. The filters were packed with 96 cm of sand $(d_{10} = 0.28$ mm, $d_{60} = 0.41$ mm) supported by 46 cm of coarse sand and gravel. The effluent was routed through a 142-mm, 5-um-pore-size membrane filter for Giardia sampling, or it could flow directly to the constant head discharge device. For temperature control, the filters were equipped with cooling coils in the heads, and the filter feed tank had a built-in temperature control. Temperatures were maintained constant throughout the system within the range 3° to $17^{\circ}C$.

Six slow sand filter columns were used in Phase II arranged in a circular configuration about an operating platform. The constant head tank in the center distributed equal flow to each filter by means of orifices. The six columns were operated continuously for a 12-month period.

The effluent flow could be directed through the constant head outflow device or through a 142-mm membrane filter used for Giardia sampling. A cooling element was used for two filters to maintain temperatures at 5° and $2^{\circ}C$. The feed water to the six filters was maintained at $17^{\circ}C$ by means of cooling and heating elements located in a 1200-L feed tank.

Operation

The three slow sand filters in Phase I were operated continuously from August 1981 to December 1982 at hydraulic loading rates of 0.04, 0.12, and 0.40 m/hr for filters designated 1, 2, and 3, respectively. The common feed tank delivered the same influent to each of the slow sand filters, thus allowing for the evaluation of process response to different hydraulic loading rates.

The other operating variables studied were (1) temperature, (2) influent bacteria concentration and cyst concentration, (3) age of the schmutzdecke, and (4) biological maturity of the sand in the filter bed. These variables were changed systematically to determine their effect on removals of Giardia cysts, bacteria, turbidity, and particles.

Temperature effects were examined by operating the system at 5° and 15°C. The highest temperature permitted during Giardia testing was 15°C. This upper limit was based on observation that the cysts deteriorate at higher temperatures.

Giardia cyst concentrations were varied between 50 and 5,075 cysts/L. Because the filtration removal processes were highly effective, high influent cyst concentrations were necessary to ensure passage of a few cysts through the filter. The high cyst concentration also permitted discernment of possible functional relationships and encompassed the highest expected ambient concentration, which was estimated as 500 cysts/L.

Total coliform bacteria ranged from almost 0 to about 300,000/100 mL. These latter levels were the result of adding primary settled sewage to the filter feed tank and from fecal residue accompanying Giardia cyst addition. No attempt was made to change these concentrations systematically. The bacteria were added to challenge the filter, since the raw water bacteria counts were normally quite low.

The effect of the schmutzdecke was determined by testing after it had been allowed to develop and immediately after scraping. A developed schmutzdecke is defined as one that has had at least two weeks to establish itself.

The biological maturity of the sand bed indicates the degree of microbiological development throughout its depth. This condition is not measurable, but is a function of the number of weeks of undisturbed filter operation. To determine the influence of microbiological maturity, testing was done for three filter conditions: (1) new sand bed and new gravel support (which simulated start up of a new filter), (2) new sand bed with microbiologically mature gravel support (which simulated a filter that had just had its sand totally replaced), and (3) sand bed and gravel support that are both microbiologically mature (which simulated steady-state operation). Testing under the third condition was done at various filter ages, ranging from 26 to 80 weeks. The age of the filter can be used as an index of microbiological maturity for given raw water conditions. The most pertinent conditions that affect the length of time to bed maturity are nutrient availability and temperature.

A test run consisted of filling the batch feed tank with lake water and then spiking the water in the tank with a known concentration of Giardia cysts. When additional coliforms were desired, the tank was also spiked with primary settled sewage. The feed tank was then sampled for Giardia cysts, total coliform bacteria, standard plate count bacteria, particles, and turbidity. The same sampling and analyses were performed on the three filter effluents the following day to allow for the needed volume displacement within the filter column. This procedures (i.e., spiking, sampling of the feed tank, and sampling of the filter effluents) was continued daily for 3 to 11 days, depending on the particular test run.

The six pilot filters in Phase II were operated in parallel with a common raw water source. Filter No. 1 was operated as the control, providing a basis for comparison with the other filters. Each of the other columns was operated with one of the process variables having a different magnitude the control. The other variables were maintained the same as the control.

194

With the six filters, three levels of biological activity were studied along with three other variables--sand bed depth, sand size, and temperature. Filter 1, the control, had the amount of biological development that would typically occur with the Horsetooth Reservoir water. Filter 3 was subjected to 5 mg/L residual chlorine between runs to minimize biological growth. Sterile synthetic sewage was aded to Filter 4 to promote additional biological growth within the sand bed. Filter 2 had a 48-cm sand bed depth (instead of the 97 cm of the other filters). Filter 5 was packed with sand having d_{10} size 0.615 mm (instead of the 0.287-mm size in the other filters). Filters 5 and 6 were operated at 5°C continuously (instead of the 17°C of the other filters).

To evaluate the effects of process variables, the filters were spiked with a laboratory culture of total coliform bacteria. Filter No. 3, which was disinfected by a sodium hypochlorate solution, was purged of disinfectant with sodium thiosulfate before such tests. Effluent samples from the six filters were obtained once each day during the test period. This series of measurements, together with the spiking, constituted a test run. Such test runs were conducted at various times, usually weekly, throughout the 11-month period of continuous filter operation. In addition to the total coliform testing, removals of turbidity and standard plate count bacteria were monitored routinely. Tests with Giardia cysts, concentrated from dog feces, were conducted on a more limited basis to minimize the fouling of the sand surface caused by debris and fats present in the Giardia cyst concentrate.

Results

Phase I Removals

Table 1 summarizes the removals from Phase I averaged for all data over the period August 1981 to December 1982 for the three filter columns operated at 0.04, 0.12, and 0.40 m/hr. The number of samples obtained for each variable and the range of each are included.

The data showed uniformly high removals for all dependent variables except turbidity, which ranged from 27 to 39 percent for raw water turbidities ranging from 2.7 to 11 NTU. These turbidity removals are not as high as reported by others (e.g., the Kassler plant in Denver) because of the small clay particles that make up the suspended matter in the raw water source, Horsetooth Reservoir. About 30 percent of the turbidity in this water will pass through a 0.45-um membrane filter.

Removals of Giardia cysts, total coliforms, and fecal coliforms were all high. At optimum conditions of operation, effluent concentrations of each approached their respective detection limits.

Hydraulic Loading Rate --

Well-defined relationships can be seen from the data in Table 1, in which removals of coliform bacteria, standard plate count bacteria, Giardia cysts, and turbidity decline with increasing hydraulic loading rate. For example, average removals of total coliform bacteria declined from 99.991 percent at 0.04 m/hr to 99.981 percent at 0.12 m/hr. Though hydraulic loading rate has an influence on filtration efficiency, the effect is not great enough to warrant establishing a firm design criterion for this parameter. Rather, the concern should be with respect to economic aspects. For example, the advantage of reduced construction costs for a design with a high hydraulic loading rate must be weighed against increased operating costs caused by the need for more frequent schmutzdecke removals. Performance would be only slightly poorer at the higher hydraulic loading rate.

Microbiological Conditions --

The biological conditions governing the process effectiveness of the filter are: 1) the degree of schmutzdecke formation, and 2) the microbiological maturity of the sand bed. Figure 1 illustrates how these conditions affect coliform effluent concentrations (i.e., the percent remaining at hydraulic loading rates of 0.04, 0.12, and 0.40 m/hr). Also, each of the bars shows effluent coliform concentrations calculated from a hypothetical influent density of 1 million coliforms per 100 mL. These figures are derived from the percent remaining data and permit a more tangible means for comparing results in terms of whole numbers.

To evaluate the respective roles of the schmutzdecke and the maturity of the sand bed, it is useful to examine first a filter column with a new sand bed, including a new, graded gravel support. This simulates a newly constructed filter during startup when there is no biological development in the sand bed and no schmutzdecke. For this condition of new sand (as indicated in Figure 1 for Run 118), 15.4 percent coliforms remained, or 154,000 coliforms/100 mL remained from a hypothetical 1 million coliforms/100 mL in the influent. In other words, filtration through the new sand will cause an order of magnitude reduction.

In contrast to the initial startup of a filter is the filter that has been in operation for a period of time and has a mature biological population and an established schmutzdecke. Such a case is represented by runs 104, 105, and 106, which show that a mature filter will reduce the coliform concentration by 2.5 to 4 logs, or from 1 million coliforms/100 mL to 40, 1000, and 4000 coliforms/100 mL, respectively.

Schmutzdecke removal will result in approximately a 1-log decrease in treatment efficiency when compared with operation under established conditions. This result can be demonstrated by comparing runs 107, 108, and 109 (at 300, 10,000, and 28,000 coliforms/100 mL, respectively) with runs 104, 105, and 106 (at 40, 1000, and 4000 coliforms/100 mL, respectively).

Replacing sand will result in almost a 2-log decrease in treatment efficiency. Run 116 shows 70,000 coliforms/100 mL remaining, compared with only 1000 coliforms/100 mL for the established condition represented by run 105.

One additional condition tested was the effect of removing the schmutzdecke and then disturbing the sand bed, as illustrated by runs 110, 111, and 112. This experiment was intended to simulate the effects of a full-scale filter operation in which the filter is drained and the sand bed is disturbed by the movement of men and equipment over the filter surface during cleaning. The experimental disturbance was accomplished for each filter by draining the filter for a 2-day period, removing the schmutzdecke, mixing the top 10 cm of sand, and pounding on the sand surface. This experiment caused an additional 0.5- to 1-log decrease in treatment efficiency compared with the filter cleaning procedure when no disruption occurred.

The test results as shown in Figure 1 confirm the importance of the role of microbiological conditions in the treatment effectiveness of slow sand filtration. The best treatment can be expected from a filter that has been in operation for an extended period of time. This filter will have a mature biopopulation within the filter bed and will have an established schmutzdecke. The treatment efficiency will deteriorate markedly as greater portions of the biological community are disrupted as shown in Figure 1.

The data in Table 2 show the effects of the various filter operating conditions on removals of Giardia cysts. The first two rows compare removals of Giardia cysts between a control filter and one that has new sand and new gravel support. No cysts were detected in the effluent of the control filter, and only 17 cysts/L were found in the effluent of the new media filter. This result demonstrates that a filter with a mature biological population can remove cysts to the detectable limit, and that even a new filter can remove 99 percent of the influent cysts. Both filters were subjected to an influent cyst concentration of 2000 cysts/L.

Results for a similar experiment with a new sand bed and mature gravel support (Run 116) showed zero cysts/L in the effluent, compared with an influent cyst concentration of 3692 cysts/L. This result indicates that even a modest amount of microbiological growth in the sand bed, or indeed in the gravel support, can provide the marginal effect needed to cause removal of influent cysts to levels below the detection limit.

The third section of Table 2 presents the results of 15 Giardia removal test runs on filters with freshly scraped sand surfaces (i.e., no schmutzdecke development). These test runs were arranged chronologically according to continuous filter operation, ranging from 26 to 70 weeks. Table 2 shows that removal of Giardia cysts to below the detection limit was achieved in all but four of these test runs. The key difference between those tests that achieved nearly complete removal and those that did not was the degree of microbiological maturity within the sand bed. All four of the tests in which cysts were passed occurred during the first 41 weeks of filter operation, indicating that when the microbiological population has developed to maturity, complete removal of Giardia cysts can be expected. As demonstrated in Table 2, this result occurs independently of hydraulic loading rate, influent Giardia cyst concentration, and pressure of a schmutzdecke.

The same improvement in Giardia cyst removal with time is demonstrated by results shown in the fourth part of Table 2, where the test data are summarized in chronological order for 24 Giardia tests with filters having established schmutzdeckes. These results show that the removal of cysts improved steadily with time and was independent of schmutzdecke age. Cysts were passed through filters with 12-week-old schmutzdeckes, whereas they were removed below the detectable limit with 4- to 5-week-old schmutzdeckes when the microbiological population within the filter was given a longer time to mature. In fact, after 49 weeks of operation, cyst removal below detection limit was achieved in all cases, even with influent cyst concentrations as high as 5075 cysts/L. These results demonstrate that the age of the schmutzdecke is not as important for Giardia cyst removal as the maturity of the mcirobiological population throughout the sand bed and gravel support.

Phase II Removals

Design --

Phase II testing was designed to ascertain the effects of sand size, sand bed depth, and sustained low temperatures on removals of Giardia cysts and other parameters. No experiments to determine removals of Giardia cysts were done with the chlorinated filter and the nutrients-added filter, to avoid the influence of such testing on their performance (e.g., the influence of nutrients and debris).

The results of the Phase II Giardia removal tests demonstrated that removal was not affected by increasing the effective sand size to 0.615 mm, by continuous operation at 5°C, or by reducing the sand bed depth to 0.48 m. Each of the filters (1, 2, 5, and 6) had a mature biological population and the same influent water, and each was operated at hydraulic loading rates of 0.12 m/hr. Cysts were not detected in any of the effluent samples.

To induce cyst breakthrough, another filter column was newly packed with 0.615 mm sand, operated at 0.47 m/hr, and then challenged during the first 2 days of operation with 2770 cysts/liter. This test resulted in passage of 26 cysts/liter through the filter. Even under these extreme conditions, removal was 99 percent. Testing with new sand was also carried out during rapid sand experimentation, which was the subject of Volume 3 of this study. Testing with 0.43 mm sand at hydraulic loading rates in excess of 14 m/hr and without chemical addition resulted in four of eight test runs with removals of less than 50 percent.

Effects of Process Variables on Filter Performance in Phase II --

The effects of process variables on filter performance were evaluated by using the percent removals of total coliform bacteria as the measure of efficiency. Removals of standard plate count bacteria and turbidity were determined also, but they were not suitable for this purpose because heterotrophic bacteria were shed by the filter as a result of internal growth, and turbidity removal from Horsetooth Reservoir water had little relation to operation because of its unique an nonrepresentative behavior, as reported in the Phase I results.

Biological Community--Phase II investigations were designed to study the effects of low, natural, and accelerated biological activity on filter performance, as represented by Filters 3, 1, and 4, respectively. For low biological activity, growth was prevented in Filter 3 by maintaining a 5-mg/L chlorine residual between test runs and dechlorinating with sodium thiosulfate a test run. Augmented biological activity was created by Filter 4 by continuously adding sterile synthetic sewage to the filter. Filter 1 was a control filter that used raw water from Horsetooth Reservoir with no alteration; this filter represented the natural condition.

The results of the Phase II testing demonstrated that as the activity of the biological community increased from minimal biological community for the chlorinated filter to augmented activity for the nutrients-added filter, the removals of coliforms, standard plate count bacteria, and turbidity increased significantly. For Filters 3, 1, and 4, removals were 60, 98, and 99.9 percent for total coliform bacteria; -89, -41, and 58 percent for standard plate count bacteria; and 5, 15, and 52 percent for turbidity. These results demonstrate the unmistakable influence of biological activity on filter performance.

Temperature--Decreasing the temperature from 17° to 5° or 2°C decreased removals of coliform bacteria and standard plate count bacteria from about 99 percent nominally to 90 percent nominally for each. The filtration efficiency was not reduced as sharply as expected. The literature has reported sharp reductions in percent removals as a result of lower temperatures.

Sand Bed Depth--The removals of total coliform bacteria were 97 percent at a sand bed depth of 1 m and 95 percent at 0.5 m. This result indicates that bacterial removal is not overly sensitive to sand bed depths above 0.5 m. In practice, this result means that a series of schmutzdecke removals with the resulting attrition of the sand bed from 1 m to 0.5 m will not seriously impair the efficiency of the filtration process.

Sand Size--To discern better the role of effective sand size, three filters were packed with sand with effective sizes of 0.62, 0.28, and 0.13 mm. Eighteen test runs using pure cultures of total coliform bacteria were then conducted parallel with each filter. Each of these filters had a mature biological population. The coliform removal improved from 96.0 to 98.6 to 99.4 percent for effective sand sizes of 0.615, 0.278, and 0.128 mm, respectively.

Summary of Results

Findings from the experimental program are summarized first in terms of the overall removal effectiveness of slow sand filtration for the parameters tested, and second in terms of the effects of operating conditions. The effectiveness of slow sand filtration for removing the parameters tested is summarized as follows:

1. Giardia cyst removal exceeded 98 percent for all operating conditions tested. Once a microbiological population is established within the sand bed, removal will be virtually 100 percent.

2. Coliform removals exceeded 99 percent on the average over all operating conditions. Even with new sand, coliform removals were 85 percent.

3. Removals of standard plate count bacteria and particles ranged from 88 to 91 percent and from 96 to 98 percent, respectively.

4. Turbidity removals averaged from 27 to 39 percent. This low removal was caused by the fine clay turbidity particles characteristic of the lake water used in the testing program.

Operating conditions affected removals of Giardia cysts, total coliforms, and standard plate count bacteria in the following ways:

1. Hydraulic loading rate. Removals of Giardia cysts, coliform bacteria, standard plate count bacteria, and turbidity declined with increasing hydraulic loading rate. However, even at 0.40 m/hr, removals of Giardia cysts and coliform bacteria were high--99.98 percent and 99.01 percent, respectively.

2. Temperature - The Phase II experiments for mature filters showed that Giardia removals were uniformly 100 percent for continuous operation at both 17° and at 5°C. However, the removals of total coliform bacteria declined from 97 percent at 17°C to 87 percent at 5°C. Effluent concentrations of standard plate count bacteria were 100 times higher at 2°C than at 5°C.

3. Influent concentration of bacteria and Giardia cysts. Effluent concentrations of coliform bacteria and standard plate count bacteria increased with increasing influent concentrations. At the same time, removals increased. A similar relation would be expected for removals of Giardia cysts, but data were not sufficient to establish it. Though this information may be of academic interest, removals of the above are influenced more strongly by the microbiological maturity of the sand bed than by influent concentrations.

4. Conditions of the sand bed. A new sand bed removed 85 percent of influent coliform bacteria and 98 percent of influent Giardia cysts. As the sand beds matured biologically, removals improved to greater than 99 percent for coliform bacteria and virtually 100 percent for Giardia cysts. Disturbance of the sand bed caused reduced coliform removals, but it had no effect on Giardia cyst removals. Development of the schmutzdecke further improved removals of coliform bacteria by an order of magnitude. The presence or absence of a schmutzdecke has essentially no influence on Giardia cyst removal efficiency. The microbiological maturity of the sand bed is the most important variable in removal of Giardia cysts and coliform bacteria. This mature condition develops over a matter of weeks or months, depending on raw water conditions.

5. Sand Bed Depth - Coliform removals averaged 97 percent for the control filter with a bed depth of 1.0 m and declined only to 95 percent for the filter with a bed depth of 0.5 m. These results demonstrate that the bed depth càn be reduced to 0.5 m by repeated schmutzdecke removals without significant impairment of filtration removal efficiency.

6. Sand Size - Removals of Giardia cysts were 100 percent for all sand sizes tested. Removals of total coliform bacteria declined from 99.4 percent for 0.128-mm sand to 96.0 percent for 0.615-mm sand. Though results showed that sand size had a functional influence on bacteria removals, the removals were high even with the 0.615-mm sand. So the argument for using smaller sand is not strong from the standpoint of removal effectiveness. Instead, the argument for using an effective sand size of about 0.35 mm is economic. The schmutzdecke will penetrate to a greater depth with larger sand, necessitating removal of more sand during schmutzdecke removal and resulting in higher operating costs. Thus using the smaller sand size is preferable when the choice is economically favorable and if the trade off in higher headloss is not appreciable.

7. Biological Activity - Phase II results showed that the average coliform removals for Filter 3, which was chlorinated between test runs and had no biological community, were only 60 percent. For the control filter, the average coliform removal was 98 percent. Filter 4, which had nutrients added, showed an average coliform removal of 99.9 percent. These results augment those of Phase I and establish unequivocally the importance of the biological community and its level of activity within the sand bed.

Conclusions

Slow sand filtration is an effective water treatment technology as determined by removals of total coliform bacteria and Giardia cysts. Furthermore, the process is passive in nature, requiring little action on the part of the operator. This technology should be considered as an alternative when water treatment systems are being selected. Pilot plant testing should be done, however, to determine the technical feasibility of each alternative. The selection should be made by economic analysis and judgments of how appropriate the technologies are in terms of community attitudes toward operation.

The full report was submitted in fulfillment of EPA Cooperative Agreement No. CR808650-02 by Colorado State University under the sponsorship of the U.S. Environmental Protection Agency.

TABLE 1. Average Percent Removals for Dependent Variables
          in Slow Sand Filter Columns

| Dependent Variable | Total Number of Analyses | Range of Variable in Raw Water | Percent Removal of Parameter | | |
			Filter 1 v=0.04 m/hr	Filter 2 v=0.12 m/hr	Filter 3 v=0.40 m/hr
Giardia cysts	222	50-5,075 cysts/liter	99.991	99.994	99.981
Total coliforms	243	0-290,000 coliforms/100 ml	99.96	99.67	98.98
Fecal coliforms	81	0-35,000, coliforms/100 ml	99.84	98.45	98.65
Standard plate count	351	10-1,010,000 organisms/ml	91.40	89.47	87.99
Turbidity	891	2.7-11 NTU	39.18	32.14	27.24
Particles (6.35- 12.7 μm	39	62-40,506 particles/10 ml	96.81	98.50	98.02

TABLE 2. Effect of Operating Conditions on _Giardia_ Cyst Removal by Slow Sand Filtration

Test Objective	Condition of Sand Bed and Gravel Support	Age of Schmutz-Decke (Weeks)	Length of Time of Operation (Weeks)	Run Number	Filtra-tion Rate (m/hr)	Influent Cyst Conc. (cysts/liter)	Effluent Cyst Conc. (cyst/liter)	Percent Removal (%)	Detec-tion Limit (cyst/liter)	Effluent Volume Sampled (liters)*
Effect of New Sand Bed and New Gravel Support	New Sand Bed/ New Gravel Support	0	0/0	118	0.40	2000	17.05	99.15	0.046	610
	Control Filter: (Mature Sand Bed/Mature Gravel Support)	10	80	119	0.40	2000	0.0	100	0.049	770
Effect of New Sand Bed	New Sand Bed/ Mature Gravel Support	0	0/67	116	0.12	3692	0.0	100	0.039	497
	Control Filter: (Mature Sand Bed/Mature Gravel Support)	4	67	117	0.12	3692	0.0	100	0.040	566
Effect of Schmutz-decke Removal	Mature Sand Bed/ Mature Gravel Support	0	26	48	0.04	420	2.014	99.520	0.085	65
		0	26	49	0.40	420	5.431	98.707	0.020	270
		0	33	47	0.12	420	1.541	99.633	0.030	180
		0	41	75	0.04	50	0.0	100	0.036	314
		0	41	76	0.40	50	0.002	99.996	0.005	2239
		0	45	81	0.04	50	0.0	100	0.104	344
		0	45	82	0.40	50	0.0	100	0.042	2671
		0	48	74	0.12	50	0.0	100	0.014	803
		0	52	80	0.12	50	0.0	100	0.013	853
		0	62	107	0.04	1500	0.0	100	0.302	142
		0	62	109	0.40	1500	0.0	100	0.036	1199
		0	63	110+	0.04	1953	0.0	100	0.151	176
		0	63	112+	0.40	1953	0.0	100	0.026	1020
		0	69	108	0.12	1500	0.0	100	0.121	354
		0	70	111+	0.12	1953	0.0	100	0.059	454
Effect of Established Schmutz-decke	Mature Sand Bed/ Mature Gravel Support	3	29	54	0.04	500	0.243	99.949	0.061	84
		3	29	55	0.40	500	0.321	99.936	0.015	346
		5	31	60	0.04	500	0.0	100	0.062	81
		5	31	61	0.40	500	0.111	99.978	0.014	366
		3	36	53	0.12	500	0.116	99.977	0.021	223
		5	38	59	0.12	500	0.035	99.993	0.023	220
		11	38	66	0.04	50	0.050	99.900	0.040	175
		2	38	67	0.40	50	0.011	99.978	0.006	1098
		12	39	69	0.04	50	0.114	99.772	0.037	140
		3	39	70	0.40	50	0.017	99.966	0.005	762
		11	45	65	0.12	50	0.016	99.968	0.016	429
		12	46	68	0.12	50	0.041	99.918	0.015	345
		4	49	87	0.04	1000	0.0	100	0.993	111
		4	49	88	0.40	1000	1.373	99.863	0.127	871
		5	50	90	0.04	1000	0.0	100	0.586	157
		5	50	91	0.40	1000	0.0	100	0.109	843
		4	56	86	0.12	1000	0.0	100	0.398	277
		5	57	89	0.12	1000	0.0	100	0.246	374
		16	60	101	0.04	1087	0.0	100	0.200	138
		16	60	103	0.40	1087	0.0	100	0.024	1134
		17	61	104	0.04	5075	0.0	100	0.231	171
		17	61	106	0.40	5075	0.0	100	0.027	1440
		16	67	102	0.12	1087	0.0	100	0.081	342
		17	68	105	0.12	5075	0.0	100	0.091	435

* This is the effluent volume that has been concentrated by a 5-μm polycarbonate membrane filter.

+ The entire filter bed was disrupted during the schmutzdecke removal process in an attempt to simulate full-scale procedures.

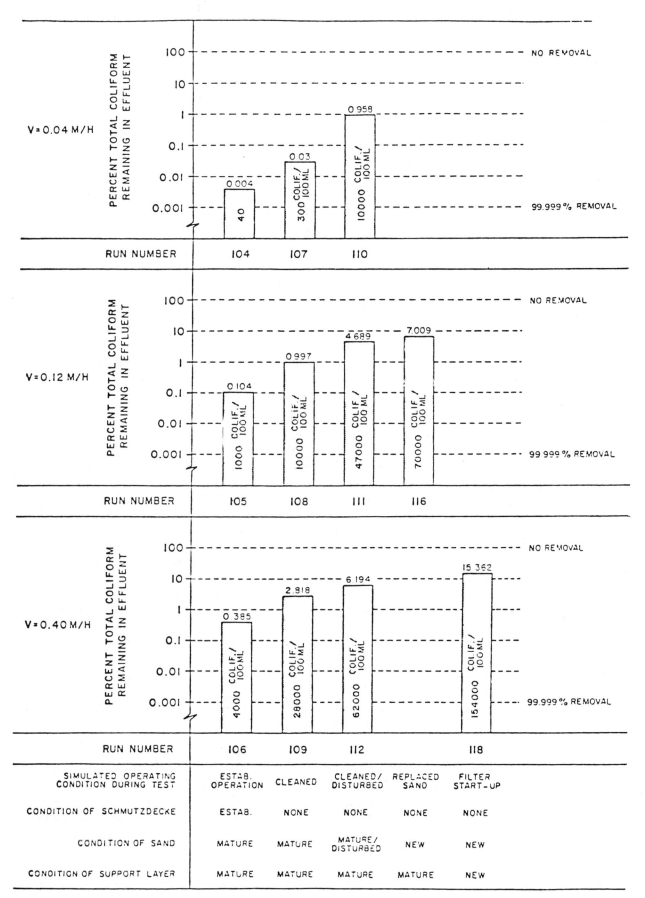

Figure 1. Effect of schmutzdecke and sand bed conditions on percent of remaining of total coliforms for three hydraulic loading rates.

EPA Project Summary

Filtration of *Giardia* Cysts and Other Substances
Volume 3: Rapid-Rate Filtration

Mohammed Al-Ani, John M. McElroy, Charles P. Hibler and David W. Hendricks

Mohammed Al-Ani is with the Water Board of the Scientific Research Council, Bagdad, Iraq. John M. McElroy, is with CH2M-Hill Consulting Engineers, Bellevue, WA 98009-2050. David W. Hendricks and Charles P. Hibler are with Colorado State University, Fort Collins, CO 80523.

Gary S. Logsdon is the EPA Project Officer (see below).

The complete report, entitled "Filtration of Giardia Cysts and Other Substances, Volume 3: Rapid-Rate Filtration," will be available only from:

National Technical Information Service
5285 Port Royal Road
Springfield, VA 22161
Telephone: 703-487-4650

The EPA Project Officer can be contacted at:

Water Engineering Research Laboratory
U.S. Environmental Protection Agency
Cincinnati, OH 45268

Rapid-rate filtration was evaluated for a range of operating conditions using waters having turbidity levels of less than 1 NTU and temperatures ranging from 0° to 17°C. The object was to determine its effectiveness as a process in drinking water treatment for removal of Giardia lamblia cysts, total coliform bacteria, standard plate count bacteria, turbidity, and particles from low-turbidity, low-temperature water.

Results showed that when the filter is operated as a strainer (i.e., when no chemical coagulation is used), removals ranged from 0 to 50 percent. Improvement was not significant when ineffective coagulants or improper dosages were used. Effective coagulation (that adequate to reduce turbidity from about 0.5 NTU to about 0.1 NTU) was capable of removing 95 to 99.9 percent of Giardia cysts and 95 to 99.9 percent of total coliform bacteria. Two coagulant aids were found in this research that provided effective coagulation.

The filtration efficiency was unaffected by mode of filtration. The in-line mode was as efficient as the direct filtration mode. Testing at 3° and at 17°C showed no discernible differences in percent removals. Increasing the hydraulic loading rate from 8 to 41 cm/min (2 to 10 gpm/ft^2) showed no discernible effect until the latter rate was reached. The work showed that efficient filtration of Giardia cysts and other substances present in low-turbidity waters requires careful selection of coagulants and the use of proper dosages. Turbidity reduction can be used as a measure of efficiency. Routine use of pilot plants, operated side-by-side with full-scale plants, is advocated for this purpose. The pilot plant can be spiked with bacteria as another means to evaluate the effectiveness of coagulation.

This report is the third (last) describing the research conducted under EPA-CSU Cooperative Agreement No. CR808650-02. The first was entitled "Filtration of Giardia Cysts and Other Substances. Volume 1: Diatomaceous Earth Filtration," EPA-600/S2-84-114, September 1984, and the second was entitled "Filtration of Giardia Cysts and Other Substances. Volume 2: Slow Sand Filtration," EPA-600-S2-85-026, April 1985.

This Project Summary was developed by EPA's Water Engineering Research Laboratory, Cincinnati, OH, to announce key findings of the research project that is fully documented in a separate report of the same title.

Introduction

Background

Giardiasis is an intestinal disease caused by ingestion of cysts of the protozoan Giardia lamblia. In recent years, reports of waterborne outbreaks of giardiasis in the United States have become increasingly frequent. Some 53 waterborne outbreaks and 20,039 cases were reported during the period 1965 through 1981. Most outbreaks have been reported in small mountain communities in the western and northeastern United States, but during the period 1983-84, cases were reported in Pennsylvania as well.

In the western United States, the cyst is known to occur in ambient waters with turbidity levels of less than 1 NTU. Though there is no reason to believe that cysts are not found pervasively in all kinds of waters, it is important to point out that they do occur in low-turbidity waters and that many outbreaks have been associated with these clear waters. The conventional wisdom is that such waters are likely to be benign, since even without treatment, they may nearly conform to well-established standards--the 1-NTU turbidity standard and the coliform standard, for example. Yet public systems using low-turbidity raw waters may meet all of these standards and still be the source of a giardiasis outbreak.

A major process problem for treatment plants using low-turbidity waters is that it is common practice not to use chemical pretreatment, though polymers are sometimes used as filter aids. These waters may already meet the 1 NTU turbidity standard and coliform counts are often very low, e.g. 10 organisms/100 mL. Thus, they are not as easily used measures of process efficiency as they are in other surface waters, such as those found at lower elevations. Also, the low turbidity waters are more difficult to treat. Without chemical treatment, however, the rapid-rate filtration process is simply a method of straining and its use deviates from one of the basic tenets of rapid-rate filtration--that chemical pretreatment be integral to the process. Without such chemical treatment, substances in the raw water such as Giardia cysts and bacteria can and do pass through the filter. Though such waters have occasionally been treated successfully, chemical pretreatment knowledge is not adequate to effect rapid-rate filtration of low-turbidity, low-temperature waters containing Giardia cysts.

Objectives

The purpose of the research was to determine how to remove Giardia lamblia cysts from water supplies by rapid-rate filtration when raw waters have turbidity levels of less than 1 NTU. The main objectives were (1) to determine a chemical pretreatment for low-turbidity water that results in efficient rapid-rate filtration and (2) to determine the respective roles of process variables on removal efficiencies of Giardia cysts, turbidity, and bacteria. Process variables included chemical pretreatment (coagulant selection and dosages), filtration mode (conventional, direct, or in-line), media, hydraulic loading rate, and temperature. A third objective was to determine whether a surrogate indicator

was feasible to assess the treatment efficiencies for removal of Giardia cysts.
Methods

Pilot Plants

The research was based on two physical models--laboratory-scale and field-scale, rapid-rate filtration pilot plants. The laboratory-scale pilot plant was a dual-train, conventional, rapid-rate filtration plant built to be operated under pressure. The raw water to be processed was from the Cache La Poudre River when water with less than 1 NTU turbidity could be obtained. When this was not possible, an artificial low-turbidity water was prepared by treating water from the Horsetooth Reservoir with diatomaceous earth filtration to remove turbidity-causing particles without changing chemical quality. The raw water was stored in a 1400-L, temperature-controlled milk cooler and then pumped by a positive displacement pump (with dampener to control pressure surges) to three rapid-mix basins in series. Each basin was a 12.7-cm cube (inside dimensions). The stirring paddles had four rectangular blades 1.25 cm wide, 1.25 cm high, and 2.54 cm from the center of stirring shaft to outer edge of blade. The maximum rotational speed of the shaft was 600 rpm, yielding a calculated G value of 400 sec^{-1} at 20°C. The rotational speed could be varied by means of a rheostat controlling the motor speed. Chemicals were metered by positive displacement pumps capable of metering flows as low as 0.2 mL/min. Flows were measured volumetrically using 50-mL graduated burettes, which served also for chemical storage.

Four filter columns 183 cm long were installed from a manifold that permitted one to four filters to be operated in any combination. Two of the filters were 5 cm in diameter, and two were 10 cm. Copper coils were installed in the top of the filters for temperature control. Tailwater elevation was controlled by overflow cups, which were maintained above the media. Headloss across the media was measured by a mercury manometer. Air scrubbing and backwash were provided. The experiments used both a single sand medium (76 cm deep) and dual media of anthracite and sand (45 cm and 30 cm deep, respectively).

The field-scale pilot plant was a 1.3-L/sec (20-gpm) trailer-mounted package water treatment plant. During periods of low-turbidity water, this unit was located adjacent to the Cache La Poudre River at Fort Collins Water Treatment Plant No. 1.

Both the laboratory-scale and the field-scale pilot plants could be operated in three modes of filtration: conventional (rapid mix, flocculation, sedimentation, filtration), direct (rapid mix, flocculation, filtration), and in-line (rapid mix, filtration). The in-line mode was used for the research, except for the beginning exploratory work to ascertain the effect of the filtration mode.

Experimental Design

The purpose of the laboratory-scale pilot plant was to ascertain the effect of selected process variables on a group of dependent variables. The purpose of the field-scale pilot plant was to conduct confirming tests for several of these variables. The process variables examined included the effects of temperature, coagulant types and dosages, hydraulic loading rate, and media on removals of Giardia cysts, total coliform bacteria, and turbidity, with some measurements of removals of standard plate count bacteria and particles. Field-scale tests were conducted under low-turbidity low-temperature river conditions to verify the effects of no coagulation, nonoptimum coagulation, and optimum coagulation on removals of turbidity, total coliform bacteria, and Giardia cysts.

Giardia Cyst Sampling and Analysis

Giardia cysts in the filter effluents were concentrated by polycarbonate membrane filters with 5-um pore sizes. The entire flow from the laboratory-scale, rapid-rate filter column was passed through a 142-mm-diameter membrane filter. To sample the field-scale pilot plant, a portion of the effluent was passed through a 293-mm-diameter membrane filter. The total volume of water passed through the membrane filter was limited by the pressure increase across the filter. Sampling was terminated when the pressure became about 5 psi. Usually about 20 L was passed through the 142-mm membrane filter, but 140 L was passed in one test. When the sampling was ended, the membrane filter holder was removed from the filter effluent line, and the membrane was removed and washed to remove cysts and anything else that had collected on the membrane. The sample was then sent to a laboratory, where the cysts were counted using a micropipette technique.

To determine removals of Giardia cysts under a given set of filtration conditions, the milk cooler (used as a reservoir for the laboratory-scale pilot plant) was spiked with several million cysts (determined to be Giardia lamblia) obtained from the feces of infected dogs. To spike the field-scale pilot plant, a cyst concentrate was metered into the influent pipe where it was mixed as the result of the turbulent flow through four elbows.

To determine the concentration of cysts in the milk cooler, a sample stream was pumped through the 5-um, 142-mm polycarbonate filter. For the field-scale pilot plant, the influent cyst concentration was sampled by pumping a portion of the flow (after the four mixing elbows) through the 293-mm membrane filter.

Sampling and Analysis of Turbidity, Bacteria, and Particles

Sampling for turbidity, bacteria, and particles was done with grab samples. For the laboratory-scale pilot plant, these samples were taken from water in the milk cooler, after spiking, and from the filter effluent stream after 30 min, 60 min, and (for some tests) 2 hr of operation. Cuvettes were used for the turbidity samples, and sterile 250-mL plastic bottles were used for bacteria sampling. For particle sampling, 500-mL bottles were used. They were cleaned and rinsed with distilled water passed through 0.2-um filters. For the field-scale pilot plant, samples were taken after the mixing elbows for the influent stream and from the discharge line for the filtered water.

Results

Removals

Table 1 summarizes the removals of turbidity, standard plate count bacteria, total coliform bacteria, particles, and Giardia cysts for low-turbidity water at low temperatures (2° to 4°C). Results of 21 test runs are shown for three categories of chemical pretreatment: (1) none, in which no chemicals were used, (2) nonoptimum, in which a nonoptimum chemical dose was used as determined by turbidity removal, and (3) optimum, in which an optimum chemical dose was used as measured by turbidity removal. Raw water turbidity levels were 0.4 to 0.7 NTU for water hauled from the Cache La Poudre River for the pilot plant tests, and 0.2 to 0.6 NTU for the water obtained by diatomaceous earth filtration of Horsetooth Reservoir water (referred to in this report as artificial low turbidity water).

The effects of chemical pretreatment can be seen in Table 1. For the eight tests with no coagulant dosage, removals of all parameters except particles were uniformly low. No explanation exists for the higher percent removals of particles. For the nonoptimum chemical dose, the percent removals were generally higher but not uniformly high. For the optimum chemical dosage, percent removals of all parameters were uniformly high, ranging from about 80 to greater than 99.9. The exception, Run 106, was for a polymer used commonly as a filter aid for low-turbidity waters. The result is added to the table to illustrate that high removals cannot be expected for some polymers. For the optimum chemical dosage, filtered water turbidity was generally about 0.05 NTU; the percent removals ranged from 82 to 93.

The field-scale pilot plant runs (117 and 129) in Table 2 show results obtained for tests with no coagulant dosage in which Giardia cysts and coliform bacteria were injected into low-turbidity, low-temperature raw water. Without a coagulant, coliform bacteria removals were 20 and 15 percent, respectively. The Giardia cyst removal of Run 117 was only 30 percent. No Giardia removal data are reported for Run 129 because the cysts had questionable identities for analysis. The effluent turbidity was greater than the influent turbidity for each of these runs without chemicals.

Runs 123 to 128 were classified as nonoptimum. Results for removals of turbidity, total coliform bacteria, and Giardia cysts were not significantly different than those for tests without coagulant dosage. For example, in Runs 123 and 124, removals of Giardia cysts were 45 and 40 percent, respectively; coliform removals were 20 and 50 percent, and turbidity removals were less than 1 percent. The coagulant aid commonly used in the Rocky Mountain Region was simply not appropriate for the low-turbidity waters.

Run 138 was classified as optimum. Removals in all three categories were high (i.e., 95 percent for Giardia cysts, 98 percent for coliform bacteria, and 42 percent for turbidity). Coagulant chemicals used for Run 138 were 7.0 mg/L of alum as $Al_2(SO4)314H2O$ and 2 mg/L of Magnifloc 572-C.* These coagulant dosages and chemicals were found to be effective in bench-scale and laboratory-scale testing. For this test, the raw water turbidity was 0.7 NTU, the effluent turbidity was 0.4 NTU, and the water temperature was $<1^oC$. Most important was the selection of the polymer used as a coagulant aid with alum.

The data in Tables 1 and 2 show that with proper chemical pretreatment, removals of turbidity, standard plate count bacteria, total coliform bacteria, particles, and Giardia cysts were uniformly high--generally greater than 80 percent for turbidity and 98 percent for other parameters. Run 138 in Table 2 for field-scale testing showed a nominal deviation from this general finding, having only 42 percent turbidity removal but 95 percent removal of Giardia cysts and 98 percent removal of total coliforms. For nonoptimum chemical dosages, results were more variable, with both high and low removals. With no chemical pretreatment, removals were markedly lower for all parameters except particles, which ranged from 81 to 99 percent. Using an effective polymer is important also, as demonstrated by the results of Runs 127 and 128, which used some of the polymers found to be inefficient in filtration. These results show that with proper chemical pretreatment, rapid-rate filtration will generally remove 95 to 99.9 percent of Giardia cysts.

The data in Tables 1 and 2 show that rapid-rate filtration will work only as a simple strainer when no chemicals are used and will pass appreciable percentages of turbidity, bacteria, and Giardia cysts. The critical importance of proper chemical coagulation is demonstrated. From these results, little doubt exists that the rapid-rate filtration process can be effective if the proper chemicals are selected and if they are used at proper dosages.

Effects of Process Variables

The process variables investigated were chemical pretreatment (coagulant selection, coagulant dosages, and mode of filtration), comparison of single and dual media, hydraulic loading rate, and temperature. The effects of these variables on turbidity removal was the main focus because it indicated removals of both bacteria and Giardia cysts.

Coagulant Selection --

Alum alone was not effective as a chemical coagulant for low-turbidity waters unless a high dosage was used, (e.g., 15 to 50 mg/L as $Al_2(SO_4)_314H_2O$). Furthermore, the polymers tested were not effective when used alone. Thus attention was focused on selection of a polymer that could be an effective coagulant aid when used with alum. To screen polymers and determine dosages, turbidity reduction was used as the measure of effectiveness. Turbidity should be reduced from about 0.5 NTU in raw water to 0.1 or 0.05 NTU in filtered water. The search for an effective coagulant aid was wholly trial and error. The idea was to test different polymers as coagulant aids and then to stop when an effective one was found. Nine polymers were tested using the laboratory-scale pilot plant. Two of these, Magnifloc 572C® and Magnifloc 573C®, were determined to be effective as measured by turbidity removal. Some recommended polymers found effective elsewhere by others were not effective for the low-turbidity raw waters. All chemicals were added to mixing basin except 8102®, which was injected into the pipeline ahead of the filter. This procedure simulated the practice of using the polymer as a filter aid. The effect of no chemical addition (i.e., using the filter as a strainer) should be noted. Without the use of coagulant chemicals, or with the use of unsuitable coagulant chemicals, removals of turbidity are erratic.

The data showed that a combination of alum and 572C® or 573C® will remove 85 percent or more of the Giardia cysts. Many of the data show removals greater than 99.9 percent. Without chemicals, removals are likely to be in the 0 to 50 percent range, which corroborates the turbidity removal results for runs with no chemicals.

Dosages of Coagulants --

Data obtained showing filtered water turbidity as a function of alum and 572C® dosages showed that when alum is used with Magnifloc 572C® as a coagulant aid, filtered water turbidity levels can be reduced to 0.05 NTU nominally (as compared with nominal raw water turbidity of about 0.5 NTU). The response surface does not seem to be strongly sensitive to either alum or polymer dosage, but it does show that either alum alone or polymer alone is not effective. The polymer dosage used most often in the research was 1 to 2 mg/L, with alum dosages of 3 to 7 mg/L as $Al_2(SO_4)_314H_2O$. When the dose of an effective type of polymer exceeded 1 mg/L and the alum dose ranged between 3 and 15 mg/L, Giardia removals were generally high.

Mode of Filtration

Coagulation and flocculation of the low-turbidity waters did not produce any visible floc unless high alum dosages were used. Therefore conventional filtration using sedimentation was not used. The experimental work began by using direct filtration (rapid mix, flocculation, and filtration). Shortly thereafter, the in-line filtration mode, which included rapid mix followed by filtration, was also tried. Two comparisons of in-line and direct filtration, with all conditions the same for each, resulted in filtered water turbidity levels of about 0.1 NTU for each. Based on these data, all further test runs were performed using the in-line mode of filtration.

Media --

Three test runs were conducted to compare filtered water turbidity levels for the same conditions with single medium (76 cm sand) and dual media (30 cm sand and 45 cm anthracite). The first comparison was for runs with no chemical pretreatment. With raw water turbidities of 0.5 NTU, filtered water turbidities were 0.4 NTU for both the single and dual media. Headloss was 92 cm of water for the single media and 54 cm of water for the dual media after 50 min of operation. Water temperatures were 2° to $4^{o}C$. The second comparison was conducted using alum and 573C® chemical pretreatment at optimum dosages with respect to turbidity removal. Effluent turbidities were 0.04 NTU for both, and again, headloss was higher for the single media. Based on these results, the dual media was preferred because of lower headloss.

Hydraulic Loading Rate --

Tests conducted to determine the effect of hydraulic loading rate on removals of turbidity, standard plate count bacteria, total coliform bacteria, and Giardia cysts at optimum chemical dosages showed little influence on removals of total coliform bacteria. Even at 25 m/hr (10 gpm/ft^2), the removal is 99 percent. Similar influences are seen for removals of standard plate count bacteria and Giardia cysts. The influence is stronger, however, for removals of turbidity. Another series of tests was conducted under nearly the same conditions, except the sand in the dual media was obtained from the Loveland, Colorado, Water Treatment Plant. The trends were virtually the same. These results indicate that the hydraulic loading rate has little influence on percent removals of Giardia cysts and bacteria in the range of 4.9 to 19 m/hr (2 to 8 gpm/ft^2). An influence begins to become discernible, however, at 25 m/hr (10 gpm/ft^2). Pilot testing should be conducted to ascertain the influence of hydraulic loading rate for the conditions at hand in a particular situation.

Temperature --

The removals of turbidity, standard plate count bacteria, and total coliform bacteria measured at operating temperatures of 5° and $18^{o}C$, for four conditions of chemical pretreatment showed either no difference for the two temperatures or conflicting trends. If there is an influence, it does not seem to be great. The data showed no conclusive trends.

Surrogate Indicators for Giardia Cyst Removal

Sampling of raw or filtered water to recover Giardia cysts requires passing a large volume of water through a membrane filter or through a fiber-wound filter with a pore size small enough to strain the cysts. Whatever cysts were in the sample will be retained by the filter if its pores are smaller than the cysts. Once the sample is obtained, measurement of Giardia cyst concentration requires skilled technique to process, identify, and count the cysts. Routine measurement of Giardia cysts is not likely to be incorporated into water treatment practice. Thus a surrogate indicator for removals of Giardia cysts is desirable. Several were investigated, including turbidity, particles, standard plate count bacteria, and total coliform bacteria. Histogram plots were developed to relate removals of Giardia cysts with removals of turbidity and also removals of Giardia cysts with removals of total coliform bacteria. Similar histograms could have been developed using particles or standard plate count bacteria, but for the sake of brevity, this was not done. Also the use of particle counting was not continued throughout the project since considerable effort was required. Considering the effort and quality of the relationships obtained, turbidity and total coliform bacteria were the most useful surrogate parameters.

For example, one of the histograms showed that if turbidity removal is high, removal of Giardia cysts will be high also. Specifically, the plot showed 44 observations when turbidity removal was greater than 70 percent for low turbidity raw water. Of these 44 observations, 37 show removals of Giardia cysts exceeding 99 percent. In other words, if turbidity removal exceeds 70 percent and if filtered water turbidity is lower than 0.10 NTU, the probability is 0.85 (37/44) that removals of Giardia cysts would equal or exceed 99 percent.

A similar histogram was constructed using coliform bacteria as the surrogate. The histogram showed that if high removals of total coliform bacteria occur by the filtration process, then high removals of Giardia cysts can also be expected. Though the histogram indicated that removals of total coliform bacteria would be a good indicator for removals of Giardia cysts, this use may not be practical for water treatment plants using mountain streams as a source of supply. In such streams, concentrations of total coliform bacteria are usually less than 100 organisms/100 mL. To evaluate filtration performance, a pilot filter should be operated alongside the full-scale filter and spiked with raw sewage. Turbidity, on the other hand, is easy to measure.

To summarize, removals of turbidity could be used to monitor plant performance routinely. Periodic evaluations could be done by spiking a pilot column with coliform bacteria. Both are recommended when filtering low-turbidity waters.

CONCLUSIONS

This research shows that proper chemical pretreatment is imperative if the rapid-rate filtration process is to be effective when using low-turbidity waters. Most important is selection of proper coagulant polymers to use with the primary coagulant, such as alum. The range of dosages must also be proper to achieve high reductions of turbidity. With proper chemical pretreatment, removal of all parameters can be expected to exceed 70 percent for turbidity, 99 percent for bacteria, and 95 percent for Giardia cysts. With no chemical pretreatment, removal of Giardia cysts, bacteria, and turbidity can be expected to range from 0 to 50 percent. The turbidity rule of thumb of 70 percent removal pertains to low turbidity waters nominally about 0.5 NTU for the raw water. If raw water turbidity levels are greater than 10 NTU, the rule does not apply since the turbidity reductions should be sufficient to meet standards. If the raw water turbidity is about 0.1 NTU, it would be very difficult to use this rule of thumb as the turbidity reduction may not be easily detectable.

The roles of other process variables were not as important as chemical pretreatment. In-line filtration was as effective as direct filtration. Single-medium (sand) and dual media (anthracite and sand) both have the same efficiencies in reducing turbidity and bacteria. Hydraulic loading rate has very little effect on removals when it ranges between 5 and 19.2 m/hr (2 and 8 gpm/ft^2). At 25.2 m/hr (10 gpm/ft^2), a moderate effect is indicated. Investigation of temperature influence showed no trend in removals of turbidity, bacteria, and Giardia cysts at 5°C compared with removals at 18°C. Further work is recommended in this area.

Analysis of data by means of histograms showed that both removals of turbidity and total coliform bacteria could serve as surrogate indicators of removals of Giardia cysts. If percent removal of turbidity is 70 or greater, for example, reducing turbidity from say 0.5 NTU to 0.15 NTU, the probability is 0.85 that removal of Giardia cysts will exceed 99 percent. Pilot filter columns spiked with coliform bacteria are recommended for use alongside the full-scale filters to evaluate proper filtration of low-turbidity waters.

TABLE 1. Effect of Chemical Pretreatment on Removal of Turbidity, Standard Plate Count Bacteria, Total Coliform Bacteria, Particles, and Giardia Cysts from Water with Artificial Low-Turbidity, and Low-Temperature Cache La Poudre River Water, Subject to In-Line, Laboratory-Scale Rapid-Rate Filtration*,+

Run No.	Filter V (cm/min)‡/	Filter Media §/	Raw Water Source **/	Raw Water Temp. (°C)	Pretreatment (Chemicals Used) Dosage Category ++/	Pretreatment Species	Pretreatment Dosage (mg/l)	Filter Effectiveness (Percent Removal) Turbidity	Standard Plate Count	Total Coliform	Particle Count	Giardia Cysts
46	8.46	Sand(L)	IHE	3.0	None	None	0.0	-27.3	-51.3	13.8	86	7.6
47	22.59	Dual(L)	IHE	3.0	None	None	0.0	-72.4	-66.6	25.0	99	96.3
49	22.20	Sand(L)	IHE	2.0	None	None	0.0	-13.8	20.6	60.0	99	>99.9
48	8.26	Dual(L)	IHE	3.0	None	None	0.0	-18.2	-108.2	38.4	94	99.9
119	41.40	Dual(F)	CLP	3.0	None	None	0.0	18.8	9.7	5.3	90	41.9
120	32.00	Dual(F)	CLP	3.0	None	None	0.0	18.8	16.1	-7.5	86	36.4
121	20.70	Dual(F)	CLP	3.0	None	None	0.0	18.1	16.1	1.1	82	36.3
122	9.60	Dual(F)	CLP	3.0	None	None	0.0	15.6	99.6	99.9	82	68.3
69	22.69	Dual(L)	IHE	3.0	Nonoptimum	alum/573c	15.0/1.1	73.6	78.8	99.9	-142.4	99.2
82	22.45	Dual(L)	IHE	3.0	Nonoptimum	alum/572c	8.8/0.6	61.0	38.0	++/	99.6	++/
114	7.8	Dual(F)	CLP	3.0	Nonoptimum	alum/572c	23.7/1.2	69.0	96.8	99.0	58.9	95.3
50	8.20	Sand(L)	IHE	4.0	Optimum	alum/572c	2.1/0.9	88.9	82.3	>99.9	98.6	97.8
51	23.48	Dual(L)	IHE	3.0	Optimum	alum/572c	4.1/1.7	86.1	85.4	83.0	98.9	99.1
52	8.45	Dual(L)	IHE	4.0	Optimum	alum/572c	2.1/1.2	91.7	95.6	>99.9	99.2	99.7
53	23.19	Sand(L)	IHE	4.0	Optimum	alum/572c	3.4/2.1	82.6	++/	>99.9	93.8	99.5
70	22.20	Dual(L)	IHE	3.0	Optimum	alum/573c	7.6/1.3	88.7	98.4	99.5	81.9	99.4
81	8.35	Dual(L)	IHE	3.0	Optimum	alum/572c	6.8/0.9	85.4	97.8	99.9	98.3	++/
104b	8.26	Dual(F)	CLP	3.5	Optimum	alum/572c	13.4/0.6	82.4	99.5	79.8	98.6	98.7
106	8.47	Dual(F)	CLP	3.5	Optimum	8102N	0.5	-43.1	99.9	>99.9	87.0	39.5
107b	8.38	Dual(F)	CLP	3.0	Optimum	alum/572c	11.3/0.5	92.7	96.7	90.0	95.4	>99.9
118	9.37	Dual(F)	CLP	3.0	Optimum	alum/572c	23.7/1.4	85.5	98.4	99.4	++/	97.6

*/ Artificial water was obtained by diatomaceous earth filtration of Horsetooth Reservoir water; filtered water turbidity ranged from 0.2 to 0.6 NTU.

+/ The term "in-line" filtration is the designation for treatment train comprising rapid mix and filtration (no flocculation or sedimentation).

‡/ The term "V" designates a hydraulic loading rate, which equals flow divided by area of filter.

§/ Sand (L) means the media was all sand and was obtained from Loveland Treatment Plant at Big Thompson Canyon. Dual (L) means the bed comprised 30 cm sand from Loveland and 45 cm anthracite having the trade name Philterkal Special No.1 (produced by Reading Anthracite Coal Company, Pottsville, PA. 17901). Bed depth was 76 cm. Dual (F) means that the bed contained 30 cm of sand obtained from Fort Collins Treatment Plant No.2 and 45 cm of Philterkal Special No.1(R) anthracite.

**/ IHE is water obtained from Horsetooth Reservoir, filtered by diatomaceous earth to give low turbidity, (0.2 to 0.6 NTU). CLP is low turbidity raw water obtained from the Cache La Poudre River during the period September to April when raw water turbidity was generally 0.4 to 0.7 NTU.

++/ "Optimum" and "none optimum" are designations of coagulant dosages producing turbidities of filterd water that are minimum and greater than minimum, respectively.

‡‡/ No sample taken.

TABLE 2. Turbidity, Giardia, and Coliform Results of Using Low-Turbidity Raw Water*/ for Field-Scale, Rapid-Rate Filtration Pilot Plant

Run No.	Coagulant Dosage Category	Coagulants Used Chemical Species†/	Chemical Dose (mg/l)‡/	Water Temperature (°C)	Turbidity Influent**/ (NTU)	Effluent††/ (NTU)	Percent Removal	Giardia Cyst††/,***/ Influent§§/ cysts/liter	Effluent***/ cysts/liter	Percent Removal	Coliforms†††/ Influent§§/ No./100 mL	Effluent***/ No./100 mL	Percent Removal
117	None	None	0	2	0.4	0.6	<1	260	180	30	15000	12000	20
129	None	None	0	1	0.6	0.7	<1	** ‡‡‡/	**	**	3500	3000	15
123	Nonoptimum	8102	0.1	1	0.6	1.1	<1	325	180	45	6900	5700	20
124	Nonoptimum	8102	0.4	1	0.6	0.85	<1	325	100	40	6900	3500	50
125	Nonoptimum	Alum	0.4	1	0.55	1.0	<1	1300	850	35	9000	8500	10
126	Nonoptimum	Alum	5.0	1	0.55	1.0	<1	1300	950	30	9000	6500	30
127	Nonoptimum	Alum/8102	3.0/0.2	1	0.9	1.1	<1	175	100	45	*	*	*
128	Nonoptimum	Alum/8102	3.0/0.4	1	0.9	0.9	<1	175	125	30	*	*	*
138	Optimum	Alum/572-C	7.0/2.0	<1	0.7	0.4	42	1300	70	95	10000	150	>98

*/ Cache La Poudre River water having raw water turbidities less than 1 NTU.

†/ Nalco 8102, Magnifloc 572-C.

‡/ Alum doses are mg/L as $Al_2(SO_4)_3 \cdot 14H_2O$.

**/ Influent turbidity before to contaminant injection.

††/ Effluent turbidity after 1 hr of filtration.

§§/ Detected cyst concentrations, sampling influent stream after mixing by four elbows and before injection of coagulants. Samples were analyzed by micropipette technique. Membrane filters used were Nucleopore polycarbonate 5-micrometer pore size, 293-mm diameter.

***/ Procedures were the same as used for influent sampling and analysis.

†††/ Double asterisk indicates cysts were of questionable viability; single asterisk indicates cysts no data, missed dilution range.

PILOT TESTING AND PREDESIGN OF TWO
WATER TREATMENT PROCESSES FOR REMOVAL OF
GIARDIA LAMBLIA IN PALISADE, COLORADO

James L. Ris, P.E.
Project Engineer
Henningson, Durham and Richardson, Inc.
Denver, Colorado

Ivan A. Cooper, P.E.
Project Manager
Henningson, Durham and Richardson, Inc.
Denver, Colorado

W. Russell Goddard
Town Administrator
Town of Palisade
Palisade, Colorado

BACKGROUND

The Town of Palisade, Colorado has a projected municipal water supply demand of five million gallons per day by the year 2000. Palisade's water supply comes from a combination of springs, creeks, and a manmade reservoir filled with water piped from the springs and creeks. The entire water supply drainage basins are on the Grand Mesa near Grand Junction, Colorado.

Presently, treatment of the Town's raw water supplies consist of microstraining and chlorination. Turbidity of the raw water supply ranges from 3 to 5 NTU during spring runoff periods of the year to 1 NTU or less for the remainder of the year. The Colorado Department of Health requires that all surface water supplies used for potable water supplies be treated by conventional water treatment processes including flocculation, sedimentation, filtration, and disinfection prior to distribution to a municipal water supply system. Complete conventional water treatment is required due to concern for the potential and actual occurrences of Giardiasis. Non-conventional water treatment processes may be used, but significant evidence that those processes will remove Giardia lamblia cyst must be submitted to the Colorado Department of Health to obtain its concurrence.

Two non-conventional treatment process schemes were piloted in Palisade to develop a cost-effective water treatment system which meets the drinking water criteria of the Colorado Department of Health. The pilot testing determined the ability of each process to remove Giardia lamblia cyst-sized particles and turbidity, and the findings were used to develop design criteria. The design criteria were used to predesign and estimate construction and operating costs for each of the two process schemes piloted.

GIARDIA LAMBLIA LITERATURE REVIEW

A search was conducted for available literature regarding the prevention of Giardiasis due to the transmission of this disease by municipal potable water systems. The findings of the literature search are summarized herein.

Cause and Symptoms of Giardiasis

Giardia lamblia is a flagellated protozoan.[1] In its trophozite (or hatched) form, it is not detrimental to man because it cannot survive the early digestion stages in a human. However, in the cyst (or egg) form, Giardia lamblia is potentially infectious to man. Human consumption of approximately ten cysts may result in the disease, "giardiasis." The symptoms of giardiasis are chronic diarrhea; abdominal cramps; bloating; fatigue; weight loss; and frequent loose, pale, greasy, and malodorous stools.[2] The incubation period for the cyst prior to the onset of giardiasis varies from 2 to 35 days.[2,3]

Potential Giardia Sources

The source of Giardia lamblia cysts in public water supplies is typically fecal contamination of the raw water supply by animals such as beavers or humans. It also has been shown that sediment in reservoirs might also contain Giardia lamblia cysts, as the cysts have a higher density than water, and thus with time settle to the bottom of reservoirs.[2] If the reservoir overturns or is drained to a low operating level, the cysts could potentially contaminate the raw water supply.

Giardiasis Outbreaks

Outbreaks of giardiasis attributed to public water supplies have typically occurred in areas where surface water supplies are only disinfected prior to distribution for public use or where direct filtration treatment facilities are operated and/or maintained improperly. In Camas, Washington, an outbreak of giardiasis was attributed to infected beavers near the raw water supply intake and an improperly operated and maintained direct filtration treatment process. In this case, federal drinking water quality standards were being met. In Colorado, outbreaks of giardiasis have been traced to the drinking of untreated surface water or surface water whose only treatment has been chlorination. Once a water distribution system is infected with Giardia lamblia cysts, the system may remain infectious for many months.[2]

Removal of Giardia Lamblia Cysts by Treatment

The treatment of water to remove Giardia lamblia cysts is difficult because the organism is very small, in the range of 7 to 12 microns; is resistant to standard chlorine disinfection practices[2,4]; and is difficult to detect. Although the Giardia lamblia cysts are small, research has indicated they may be effectively removed using diatomaceous earth filters or conventional flocculation, sedimentation, and rapid or slow rate granular media filtration.[3,5,6,7,8,9]

Removal of all Giardia lamblia cysts is not guaranteed when conventional treatment is used; although when properly operated and maintained, water systems using these processes have not had outbreaks of giardiasis among the users due to the raw water being contaminated with the cyst.[5] The proper operation of a conventional treatment plant is essential to preventing the passing of the Giardia lamblia cysts through the processes. It has been demonstrated that meeting the drinking water turbidity guidelines of one turbidity unit and disinfection does not assure that Giardia lamblia cysts have been removed from water treated using coagulation, sedimentation, and granular media filters. Diatomateous earth filters are shown to be effective in removing Giardia lamblia cysts at a treated water turbidity of 1 NTU or greater.[10]

Research has indicated that there is a direct relationship between maintaining a stable treated water turbidity of 0.3 NTU or less from the granular media filter treatment processes and the removal of Giardia lamblia cysts from water.[3]

The treated water turbidity is considered stable as long as the turbidity during a filter run does not increase more than 0.1 to 0.2 NTU when using granular media filtration. If the turbidity rises 0.1 to 0.2 NTU during a filter run, backwashing of the granular media filter is recommended.

Proper operation of the granular media treatment process also includes the use of alum and a polymer for the coagulation and sedimentation processes. A cationic or nonionic polymer has been shown to be an effective coagulant aid for removal of the Giardia lamblia cysts by sedimentation and filtration, since the cysts have a negative zeta potential.[3]

Microbiological guidelines requiring coliform counts on treated water are not indicative of the presence of Giardia lamblia cysts, because the coliform count is intended as an indicator of the effectiveness of the disinfection process on the treated water. The Giardia lamblia cysts are not affected by standard chlorine disinfection; thus the coliform count is not an indicator of the presence of these cysts. Giardia lamblia cysts typically are elongated spheroids ranging from 5 to 7 microns wide and 7 to 12 microns long when fresh and plump. When the cysts age, they become slightly smaller and shrivelled looking. Testing procedures for identification and counting of Giardia lamblia cysts are not yet available for general use. A particle count of one micron or greater sized particles may prove more effective as an indicator of the effectiveness of a system for removing Giardia lamblia cysts by inferring that if a majority of particles greater than 1 micron are removed by the treatment process scheme, then most Giardia lamblia cysts will also be removed.

PILOT TESTING

Two water treatment processes were selected for pilot testing of the removal of Giardia lamblia cyst sized particles from the raw water supply of Palisade, Colorado. The two processes used were modified direct filtration and diatomaceous earth filtration. Henningson, Durham and Richardson, Inc., in conjunction with the Colorado Department of Health, developed the guidelines for pilot testing treatment processes and equipment at Palisade for comparison of the processes to conventional treatment. The guidelines were as follows:

1. Develop a baseline raw water turbidity and particle distribution.

2. Conduct pilot testing with each type of process and equipment for approximately a two-week period.

3. During pilot testing, monitor raw water and treated water quality for turbidity, particle distribution characteristics, and Giardia lamblia cyst removal.

4. Use various chemical dosage rates, and determine their effect on treated water quality.

Pilot Test Equipment

Water treatment equipment suitable for pilot testing was provided by two manufacturers. Neptune Microfloc, Inc. of Corvallis, Oregon, provided a modified direct filtration treatment plant; and the Industrial Filter and Pump Manufacturing Company of Cicero, Illinois, provided a diatomaceous earth filter. Sampling equipment and analyses to determine the presence of Giardia lamblia cysts were obtained from James H. Stewart and Associates, Inc. of Fort Collins, Colorado, with subsequent verification of presence of cysts by Dr. Charles P. Hibler of Colorado State University.

Modified Direct Filtration Treatment Unit

The Neptune Microfloc modified direct filtration treatment equipment consists of a polymer feed tank and pump, an alum feed tank and pump, an eight-foot high, two square-foot combined flocculation and floc trapping section filled with floatable small plastic media chips, and a two square-foot, eight-foot high multi-media filter unit. The trade name of the filter unit is "Trident." The Trident has a treatment capacity of 20 gpm in the flocculation-floc trapping compartment and 10 gpm in the filter section. Because of the difference in design capacities of the two compartments, one-half of the flow out of the flocculation-trapping compartment was wasted prior to entering the filter. Additionally, microelectronics and microprocessor-based instrumentation and controllers adjust and control the unit. The first of the two microprocessor-based control modules used was a box called the "Aquaritrol." This unit has a key pad entry system which displayed all the pertinent parameters of the unit, and in addition, sensed the influent and effluent turbidity and adjusted the alum and polymer feed pumps feed rates based on an operator-adjustable range of effluent turbidity. There was an approximate 20-minute lag period between effluent turbidity sampling and pump adjustment. The second microelectronics-based unit used was a box containing a Gould Corporation microcontroller and electronic control strips which replaced conventional relays, contacts, and mechanical timers. The self-diagnostic microprocessor-based control units have the ability to change the order and duration of each sequence. There was no filter-to-waste connection used on the pilot unit.

A flow schematic of the Trident pilot test unit, adapted from a Neptune Microfloc schematic, is presented as Figure 1. Pictures of the pilot test equipment are included as Figures 2 and 3.

Diatomaceous Earth Filter Unit

A Type 122 diatomaceous earth pressure filter with a capacity of 10 gallons per minute and four filter leaves was provided by the Industrial Filter and Pump Manufacturing Company. The total filter area of the unit is 10 square feet. The filter leaves are stainless steel mesh with a fabric sock made of multifilament Dacron with a 76x67 thread count per inch and a Frazier porosity of 20 to 25 cfm. The unit is equipped to allow dry cake removal of the filter aid from the filter leaves at the end of the filtering cycle. A flow schematic of the diatomaceous earth filter pilot test unit is included as Figure 4. Pictures of the test unit are included as Figures 5 and 6.

Giardia Lamblia and Particle Distribution Sampler

The Town obtained two filtration apparatuses from James H. Stewart and Associates for testing of Giardia lamblia-sized particle removal and particle size distribution. Each filtration apparatus included a feed pump, water meter, cartridge type filter holder, one micron cartridge filter element, and appurtenant valving. Figure 3 shows the filtration apparatus.

Procedure

All pilot testing was done at the Town's treatment site which contains a microstrainer building and a five million gallon treated water reservoir. The Town prepared a level site with electrical power for the test units. A cross pipe fitting with a two-inch tap was installed in the existing raw water supply pipeline to provide raw water to the test units. The line pressure in the raw water pipeline was 20 psig and provided adequate pressure for both the modified direct and diatomaceous earth filter units to function properly.

Both the influent raw water and finished treated water were sampled using the one micron cartridge filter. The sampling procedure involved taking large volume, 1,000 gallon average, samples of both raw water and treated water for every test run. Upon completion of sampling, the filter cartridge was removed from its holder and shipped to James H. Stewart and Associates for particle size distribution analysis and inspection for the presence of Giardia lamblia cysts. The units were shipped in iced coolers in a moisture preservation envelope to prevent dessication. The cartridge filter was used to collect particles for evaluation, rather than shipping large volumes of raw or finished water to a laboratory for testing. A water sample would introduce error by particle agglomeration or dispersal causing erroneous size distribution readings. The particles trapped on a cartridge filter would be less likely to modify their size by agglomeration or dispersal, as the particles would be fixed on the cartridge filter media.

Modified Direct Filtration Treatment Results

Neptune Microfloc conducted the pilot test using Palisade's raw water supply. Raw water turbidity during the test period ranged from 1.1 to 1.3 NTU and final treated water turbidity from 0.06 to 0.13 NTU depending on chemical dosage rates. Table 1 from Neptune Microfloc's Pilot Study Report, dated October 15, 1982, indicates filter runs of approximately 24 hours.

The influent cartridge filter element for each sample clogged completely, its appearance was completely black and muck covered. The effluent cartridge filters had a butterscotch appearance at the end of each sampling period.

Giardia lamblia were not found in either the raw or treated water. Giardia lamblia-sized particles, 12 microns or less, were found in the treated water. Particles in the 1 to 15 micron range accounted for the majority of all size particles present. Based on total particle counts the reduction in particle count from the raw water to treated water averaged 96.2% as shown in Figure 7. Removal of 1 to 15 micron-sized particles averaged 94.9%, thus, although significantly fewer, Giardia lamblia cyst-sized particles were still present in the treated water.

Diatomaceous Earth Filter Results

Pilot testing with the diatomaceous earth filter was conducted by Town personnel with help from a filter manufacturer's sales representative. Four 24-hour test runs were made. The filtering rate was approximately 0.5 gpm/sf. Raw water quality during the testing period ranged from 0.9 to 1.2 NTU; although, during two of the runs, water quality jumped substantially higher due to storm water runoff entering a crack in the raw water gravity pipeline carrying water from the reservoir located 2,500 feet higher on the Grand Mesa. Treated water turbidity during two runs unaffected by stormwater runoff was 0.2 NTU. Turbidity jumped to 0.5 and 0.6 NTU, respectively, during filter runs when the storm water contaminated the raw water supply.

Three types of filter aid materials were used during the testing. They were Dicalite Speedflow (4.8 - 5.8 micron median particle size) and Dicalite Speedplus (8.5 - 10.5 micron median particle size), both diatomaceous earths; and Grade 436, a perlite (6.6 - 8 micron median particle size). Hereafter, references to diatomaceous earth include the perlite filter aid. The perlite filter aid was used during the testing, as the other types rapidly plugged.

Giardia lamblia cysts were not present in the raw and treated water that was sampled. Based on total particle counts, approximately 99.9% of the particles in the raw water were removed by the diatomaceous earth filter as indicated on Figure 8. Removal of 1 to 15 micron-sized particles was also approximately 99.9%. Thus, although significantly fewer, there were still Giardia lamblia-sized particles in the treated water after filtration.

Pilot Test Conclusions

Test results for both modified direct filter and diatomaceous earth filter treatment indicate that Colorado Primary Drinking Water Regulations requiring treated water turbidity less than one turbidity unit can be met by both types of treatment. Also, effluent turbidities less than 0.3 turbidity units, which are reported as effective in removing Giardia lamblia cysts for granular media filters[4], were easily achieved with both test units.

Although Giardia lamblia cyst-sized particles do pass through each type of treatment, the particle count for those particles, 1 to 15 microns, is reduced substantially by approximately 94.9% with modified direct filter treatment and by 99.9% with diatomaceous earth filter treatment. Thus, it is concluded that both types of treatment will remove a large proportion of Giardia lamblia cysts, if any are present.

ACTUAL DIATOMACEOUS EARTH FILTER OPERATING EXPERIENCE

Operators and representatives from five diatomaceous earth filter plants, three filter manufacturers, and one supplier were contacted to determine limitations of diatomaceous earth filters. Generally, high turbidity and algae were identified as potential problems for filter operation. Table 2 summarizes the operating experience of the five diatomaceous earth filter plants which were contacted.

Except for Saratoga, Wyoming, which has poor raw water quality, the four other plants all have long filter run times at low raw water turbidities. At moderate turbidities, all of the plants experience significantly shorter filter run times, although the Golden, Colorado plant had a very reasonable performance. The Golden plant was the only facility which obtained its raw water from a reservoir, as is also proposed for Palisade. To remedy Saratoga's poor raw water quality, one potential solution being considered by their engineering consultant is to construct a reservoir to allow the raw water to partially clarify prior to treatment by the diatomaceous earth filters.

The raw water supply to be treated by diatomaceous earth filters should not have large amounts of algae. With the exception of the plant at El Monte, all of the plants were adversely affected by algae if present in the raw water supply. Algae are removed from the raw water by the filters, but the algae rapidly plugs the filter requiring frequent sluicing and decreased treated water production, thus, algae are not desirable in the raw water supply. Generally, hard discrete particles seem to be more easily removed and cause less operational problems than deformable particles. The problems caused by algae to diatomaceous earth filters are not unique, and similar plugging will occur with algae on granular media filters.

PREDESIGN AND ECONOMIC ANALYSIS

Design criteria were established for modified direct and diatomaceous earth filtration water treatment processes based on the findings of the pilot testing, and by traditional criteria. Flow schematics, building layouts, and capital and operating costs are presented for each process. The estimated capital and operating costs are used in a 15-year present worth analysis to determine the most cost effective treatment process.

Design Criteria

The following design criteria were established as the basis of design for the modified direct and diatomaceous earth filtration water treatment facilities at Palisade:

Maximum day water usage	5 mgd
Average day water usage	2 mgd
Raw water turbidity	0.1 to 5 NTU
Modified direct filtration treatment	
Aluminum sulfate (alum) maximum feed rate	1,600 ppd
Soda ash maximum feed rate	840 ppd
Polymer maximum feed rate	40 ppd
Adsorption clarifier surface loading rate	10 gpm/sf
Filtration surface loading rate	5 gpm/sf
Backwash rate	15 gpm/sf
Surface wash rate	1 gpm/sf
Backwash holding pond	400,000 gal.
Diatomaceous earth filtration	
Filtration rate at maximum day rate	1.0 gpm/sf
Precoat	15 lb./100 sf
Body feed	2 parts DE/1 part solids
Sluice water settling basin	45,000 gal.
Chlorine disinfection	200 ppd

Modified Direct Filtration Treatment

For the purpose of this study, "package" water treatment equipment, as manufactured by Neptune Microfloc, Inc. and pilot tested in Palisade, shall be considered as a form of modified direct filtration water treatment. All equipment layouts, capital costs, and operation and maintenance costs are based on using Neptune's equipment.

Figures 9 and 10 are the proposed flow schematic and building layout for the modified direct filtration water treatment plant, respectively. The major components of the treatment system are:

    chemical mix and feed equipment
.   package treatment modules
    backwash pumps
    surface wash pumps
    automatic control
    chlorinators

As may be seen on Figure 9, soda ash, alum, and polymer are mixed into the raw water just before it enters the treatment modules. Chlorine may also be added at the flash mixer if prechlorination is desired. The package treatment module has two compartments. The first serves to coagulate and remove floc particles and the second is a mixed granular media filter for removing floc and small particles not removed in the first compartment. After passing through the package treatment modules, chlorine is added to the water and it is piped to the existing five million gallon reservoir for distribution to the Town's water system. For backwashing and surface washing the filters, treated water is pumped from the five million gallon treated water reservoir to each of the filters. Backwash wastewater from the filters is disposed of in a holding pond. After the solids in the backwash wastewater have settled in the holding pond, water may be decanted off the pond and recycled through the treatment processes and used as potable water.

The capital cost of the modified direct filtration treatment facilities is estimated to be $1,588,300. Table 3 is an itemized breakdown of the total capital cost.

The estimated annual cost of alum and polymer for the modified direct filtration treatment plant is presented in Table 4.

Electric, chlorine, miscellaneous supplies, and staff salary costs are not presented, as they are anticipated to be approximately equal for both types of treatment. No inflation is considered in the costs.

The 1985 present worth of the alum and polymer annual costs is $1,057,363 based on a fifteen-year economic analysis from the year 1985 to 2000 and a discount rate of 8%.

Diatomaceous Earth Filter Treatment

The major components of a diatomaceous earth filter water treatment plant include the following:

    diatomaceous earth filters   precoat pumps
    fill/precoat recycle tank   recycle tank
    fill/precoat recycle pumps   recycle pumps
    body coat mix tank   air compressor and storage tank
    body coat pumps   automatic control
    precoat mix tank   chlorinators

Figures 11 and 12 indicate the proposed flow schematic and plant layout, respectively, for a diatomaceous earth filter water treatment plant. Prior to each filter run, the filters are precoated with a heavy layer (1/8-inch) of diatomaceous earth. Then, raw water and a low concentration diatomaceous earth slurry, known as a body feed, are continuously fed into the filter. The raw water passes through the diatomaceous earth coated filter elements (leaves) which capture the particles in the raw water, clarifying and filtering the water. The filtered water is discharged from the filter unit, chlorinated, and goes to the five million gallon treated water reservoir for distribution. When the diatomaceous earth filter leaves become dirty enough to restrict flow through the filter, causing a predetermined pressure drop, approximately 40 psig, the filter must be cleaned. The cleaning process removes the diatomaceous earth coating and trapped particles from the filter leaves by sluicing the leaves with clean water. The resulting diatomaceous earth slurry flows to a sluice water settling basin. After a period of settling, relatively clean water from the top of the basin is wasted or decanted for recycling through the treatment process and subsequent use as potable water.

The estimated capital cost of a two filter diatomaceous earth water treatment plant is presented in Table 5. The total capital cost of the five million gallons per day, two filter plant, is $1,171,900.

The estimated annual cost of diatomaceous earth is shown in Table 6.

As with modified direct filtration treatment, no costs are indicated for electricity, chlorine, miscellaneous supplies, and staff salary costs as they are expected to be approximately equal. Inflation of chemical costs are not considered.

The 1985 present worth of the diatomaceous earth treatment plant operating chemicals is $431,421. The present worth was determined based on the period from the year 1985 to 2000 at a discount rate of 8%.

Economic Analysis

An economic analysis of the two types of treatment processes indicates that a diatomaceous earth filter treatment plant would be approximately 40% more economical over a long period of time than a modified direct filtration plant. The economic analysis is based on the present worth of capital and diatomaceous earth cost over the fifteen year period from 1985 to 2000. All capital costs are based on 1982 prices and diatomaceous earth cost is current as of October 1982.

The 1985 present worth of the two types of treatment are indicated in Table 7.

CONCLUSION

The pilot testing program, using both a modified direct filter treatment process and a diatomaceous earth filter, indicates that either type of treatment is suitable for treating Palisade's raw water to drinking water quality. Both types of treatment have the capability, if properly operated, to remove a large proportion of Giardia lamblia cyst-sized particles from the water supply. Both types of treatment allow some Giardia lamblia cyst-sized particles to pass through. The turbidity of the treated water from each type of treatment was well below the drinking water limit of 1 NTU.

An economic evaluation of capital and selected operating costs shows that a two-filter five million gallon per day diatomaceous earth filter plant would be approximately 40% more cost effective than a modified direct filter plant.

BIBLIOGRAPHY

1. "Waterbourne Disease Outbreaks in the U.S. 1971-1974," Graun, G.F., McCabe, Huges, JAWWA, August 1976, pgs. 420-424.

2. "A Waterbourne Outbreak of Giardiasis in Camas, Washington," Kirner, J.C., Littler, Angelo, JAWWA, January 1978, pgs. 35-40.

3. "Removal of Giardia Lamblia Cysts By Flocculation and Filtration," Engeset, J., DeWalle, F.B., AWWA 1979 Annual Conference Proceedings.

4. "A Guide to the Investigation of Waterbourne Outbreaks," Rosenberg, M.L., JAWWA, December 1977, pgs. 648-652.

5. "Alternative Filtration Methods for Removal of Giardia Cysts and Cyst Models," Logsdon, Symons, JAWWA, February 1981, pgs. 111-118.

6. "Removal of Giardia Cysts and Cyst Models by Diatomaceous Earth Filtration," Hoye, R., Logsdon, Symons, University of Cincinnati, Department of Civil and Environmental Engineering.

7. "Outbreaks of Waterbourne Disease in the United States: 1975-1976," Craun, G.F., Gunn, R.A., JAWWA, August 1979, pgs. 422-428.

8. "The Role of Filtration in Preventing Waterborne Disease," Logsdon, Gary S., Lippy, Edwin C., JAWWA, December, 1982, pgs. 649-655.

9. "Getting Your Money's Worth From Filtration," Presentation at Pacific Northwest Section Meeting, Logsdon, Gary S., Fox, Kim R., Drinking Water Research Division, U.S. EPA, Cincinnati, May 8, 1981.

10. "Control of Giardia Cysts By Filtration: The Laboratory's Role," Logsdon, Hendricks, Hibler, et al, paper presented at AWWA Water Quality Technology Conference, Norfolk, Va., December 6, 1983.

Reprinted from Proc. AWWA WQTC, Denver, Colo. (Dec. 1984).

TABLE 1

SUMMARY OF OPERATING CONDITIONS AND RESULTS

(data provided courtesy of Neptune Microfloc of Corvallis, Oregon)

	Rate, gpm/sq ft		Clarifier		Mixed Media Filter			
Run	Clarifier	Filter	Headloss Rate, ft/hr	Run Length, Hours	Headloss Rate, ft/hr	Run Length, Hours	Net Production Percent	Backwash Waste, % Flow
4	10	5	0.51	8(T)	0.26	25	96.6	2.4
5	15	7.5	0.34	8.9(T)	0.27	22	97.4	1.5
6	10	5	0.22	16(T)	0.14	48	98.5	1.1
7	10	5	0.25	8(T)	0.27	24	96.9	2.2
8	10	5	0.21	6(B)	0.24	6(B)	-	-
9	10	5	-	8(T)	0.26	25	96.6	2.4

NOTE: T = Terminated based on time; B = Terminated due to turbidity breakthrough.

Basis for Projections

1. Run lengths computed for terminal headloss of 8 ft. on filter.
2. Total volume is 40 gal./sq. ft. for each clarifier flush; and 100 gal./sq. ft. per Mixed Media backwash.
3. Total down time for each flush is 7 minutes; for each backwash is 10 minutes.

TABLE 2

SUMMARY OF DIATOMACEOUS EARTH FILTER PLANT OPERATING EXPERIENCE

Location	Average Day Plant Capacity	Low Turbidity		Moderate Turbidity	
		Raw Water Turbidity	Filter Run Time	Raw Water Turbidity	Filter Run Time
Eagle, CO	1.0 mgd	3 TU	48 hours	15 TU	2 hours
Golden, CO	2.0 mgd	3-5 TU	1 week	15 TU	72 hours
Mills, WY	3.4 mgd	5 TU	80 hours	20-22 TU	4 hours
Saratoga, WY(a)	3.5 mgd	5 TU	2-3 hours	--	--
El Monte, CA	8.0 mgd	2 TU	2 weeks	10-15 TU	8-12 hours

(a) Plant has poor raw water quality.

TABLE 3

ESTIMATED CAPITAL COST OF MODIFIED DIRECT FILTRATION

Sitework		$ 48,900
Architectural and Structural		306,800
Process Equipment and Piping		768,300
Electrical & HVAC		97,700
	SUBTOTAL	$1,221,700
Engineering (15%)		183,300
Contingencies (15%)		183,300
	TOTAL CAPITAL COST	$1,588,300

NOTE: All costs based on 1982 prices.

TABLE 4

MODIFIED DIRECT FILTRATION TREATMENT ESTIMATED ANNUAL ALUM AND POLYMER COST[a]

Year	ADF (MGD)	Annual Alum Use[b] (lb/yr)	Alum Cost ($/yr)	Annual Polymer Use[c] (lb/yr)	Polymer Cost ($/yr)	Total Cost ($/yr)
1980	0.63	42,191	$ 64,550	384	$1,280	$ 65,830
1985	0.85	56,925	87,100	517	1,730	88,830
1990	1.11	74,337	113,740	676	2,260	116,000
1995	1.46	97,776	149,600	889	2,970	152,570
2000	1.86	124,565	190,580	1,132	3,780	194,360

(a) Based on October 1982 prices of $1.53/lb for alum and $3.34/lb for polymer.
(b) Based on 22 mg/l feed rate.
(c) Based on 0.2 mg/l feed rate.

TABLE 5

ESTIMATED CAPITAL COST OF
DIATOMACEOUS EARTH FILTER PLANT

Sitework		$ 36,000
Architectural and Structural		204,600
Process Equipment and Piping		588,700
Electrical and HVAC		72,200
	SUBTOTAL	$ 901,500
Engineering (15%)		135,200
Contingencies (15%)		135,200
	TOTAL CAPITAL COST	$1,171,900

NOTE: All costs based on 1982 prices.

TABLE 6

ESTIMATED DIATOMACEOUS EARTH ANNUAL COST[a]

Year	ADF (MGD)	Total Operation Time (days/year)	Annual Precoat Use[b] (lb/year)	Precoat Cost ($/yr)	Annual Body Feed Use[c] (lb/yr)	Body Feed Cost ($/yr)	Total Cost ($/yr)
1980	0.63	115	52,555	$ 8,930	105,478	$17,930	$26,860
1985	0.85	155	70,835	12,040	142,312	24,190	36,230
1990	1.11	203	92,771	15,770	185,842	31,590	47,360
1995	1.46	266	121,562	20,670	244,441	41,560	62,230
2000	1.86	339	154,923	26,340	311,411	52,940	79,280

(a) Based on October 1982 price of $0.17/lb for diatomaceous earth and filter runs of 24-hour duration.
(b) Based on 457 lb/precoat and 24-hour filter runs.
(c) Based on 55 mg/l body feed rate.

TABLE 7

1985 PRESENT WORTH OF MODIFIED DIRECT
AND DIATOMACEOUS EARTH FILTER TREATMENT

Type Treatment	Capital Cost	Chemical Cost	Total
Modified Direct	$1,588,300	$1,057,363	$2,645,663
Diatomaceous Earth	1,171,900	431,421	1,603,321

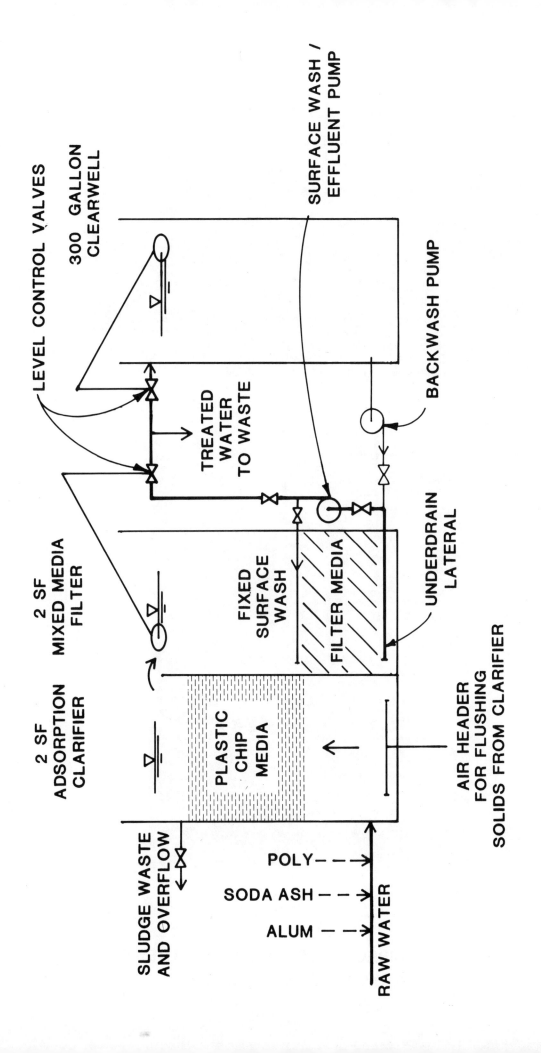

FIGURE 1
NEPTUNE MICROFLOC TRIDENT PILOT UNIT FLOW SCHEMATIC

227

FIGURE 2
MODIFIED DIRECT FILTRATION TREATMENT
PILOT TEST EQUIPMENT

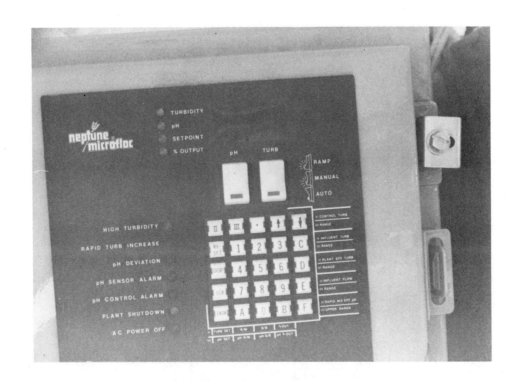

AQUARITROL CONTROL MODULE

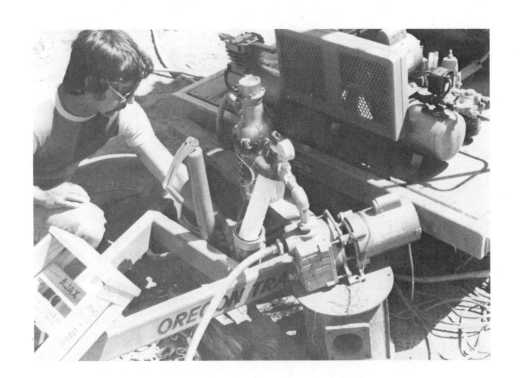

GIARDIA LAMBLIA SAMPLING APPARATUS

FIGURE 3
MODIFIED DIRECT FILTRATION TREATMENT
PILOT TEST EQUIPMENT

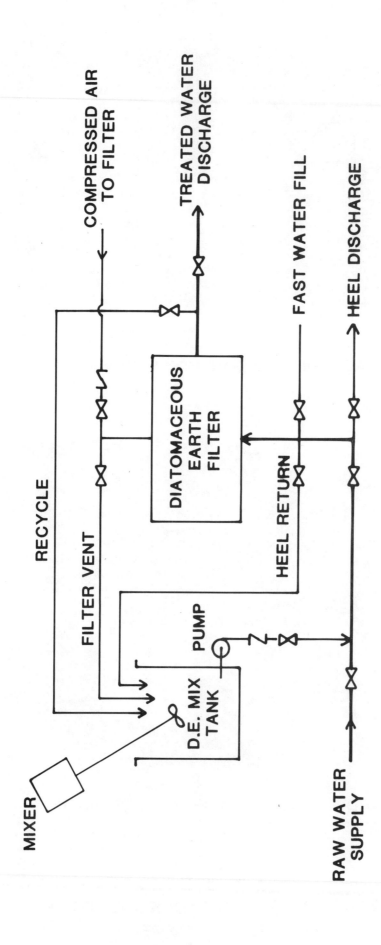

FIGURE 4
DIATOMACEOUS EARTH FILTER PILOT TEST FLOW SCHEMATIC

FIGURE 5
DIATOMACEOUS EARTH FILTER
PILOT TEST EQUIPMENT

D. E. COATED FILTER LEAF

D. E. MIX TANK

FIGURE 6
DIATOMACEOUS EARTH FILTER
PILOT TEST EQUIPMENT

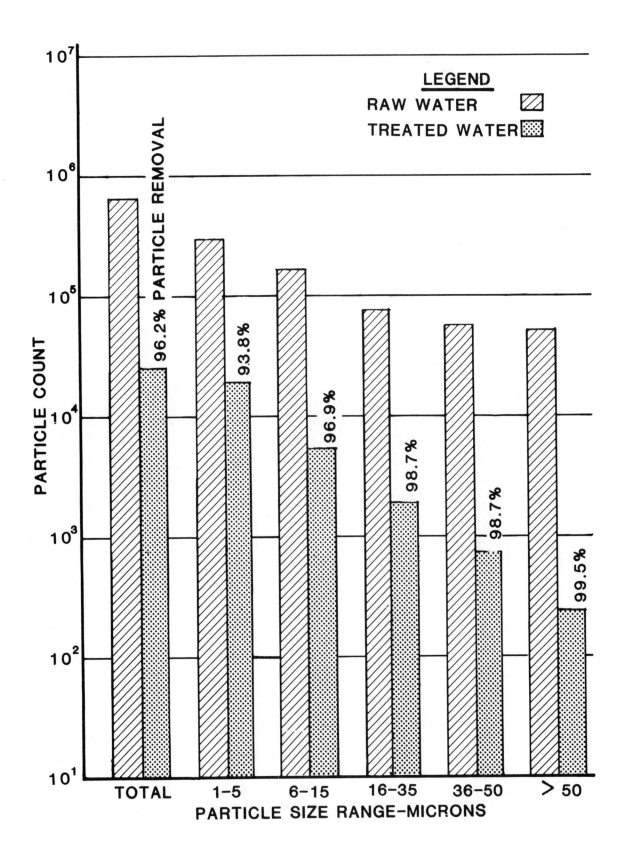

FIGURE 7
PARTICLE COUNT, DISTRIBUTION, AND REMOVAL
FOR NEPTUNE MICROFLOC TRIDENT PILOT UNIT

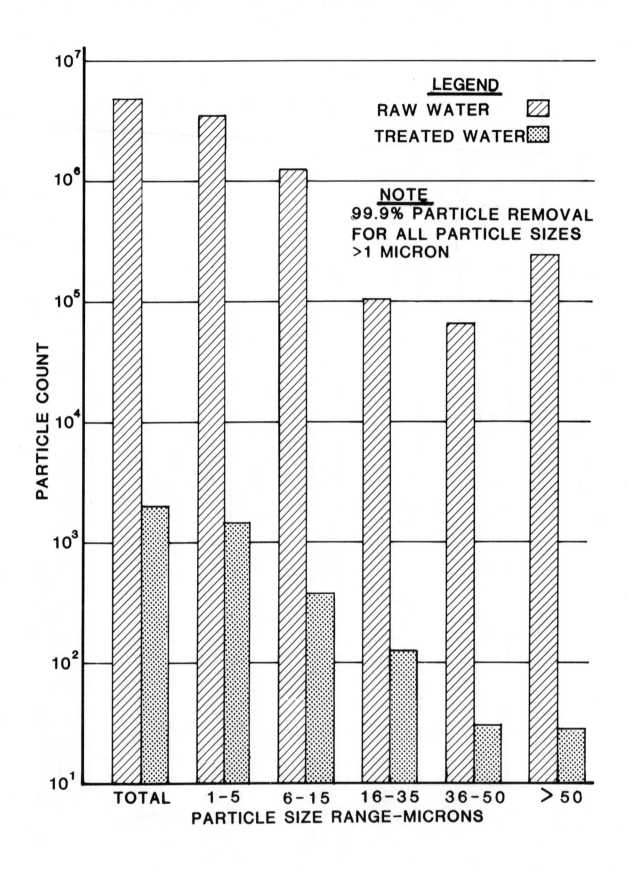

FIGURE 8
PARTICLE COUNT, DISTRIBUTION, AND REMOVAL
FOR DIATOMACEOUS EARTH FILTER PILOT UNIT

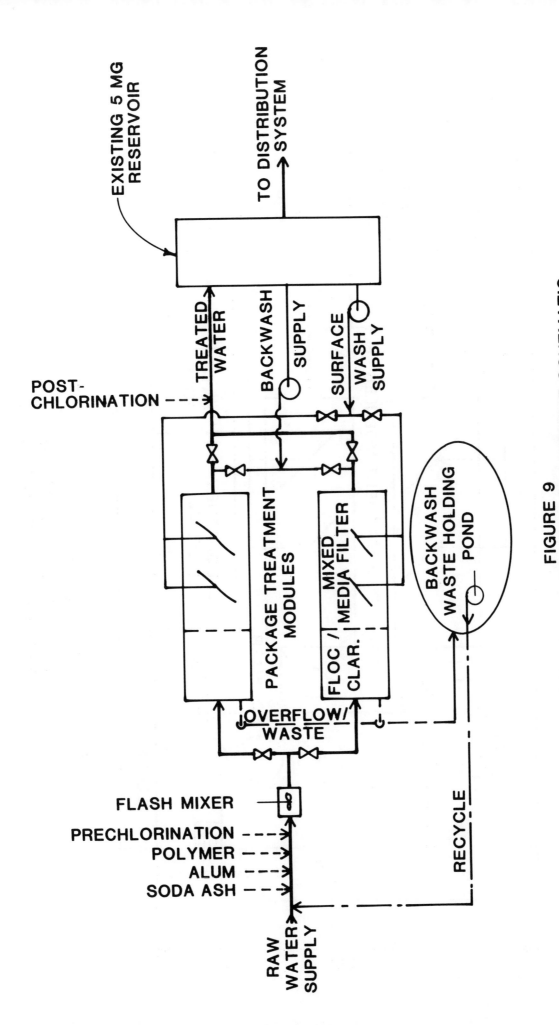

FIGURE 9
MODIFIED DIRECT FILTRATION TREATMENT SCHEMATIC

235

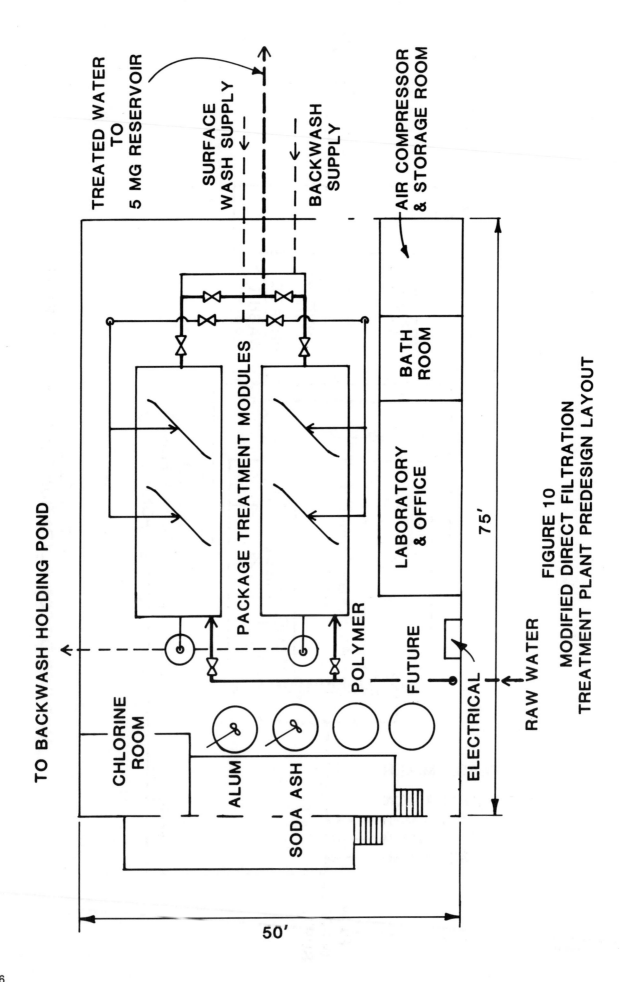

FIGURE 10
MODIFIED DIRECT FILTRATION
TREATMENT PLANT PREDESIGN LAYOUT

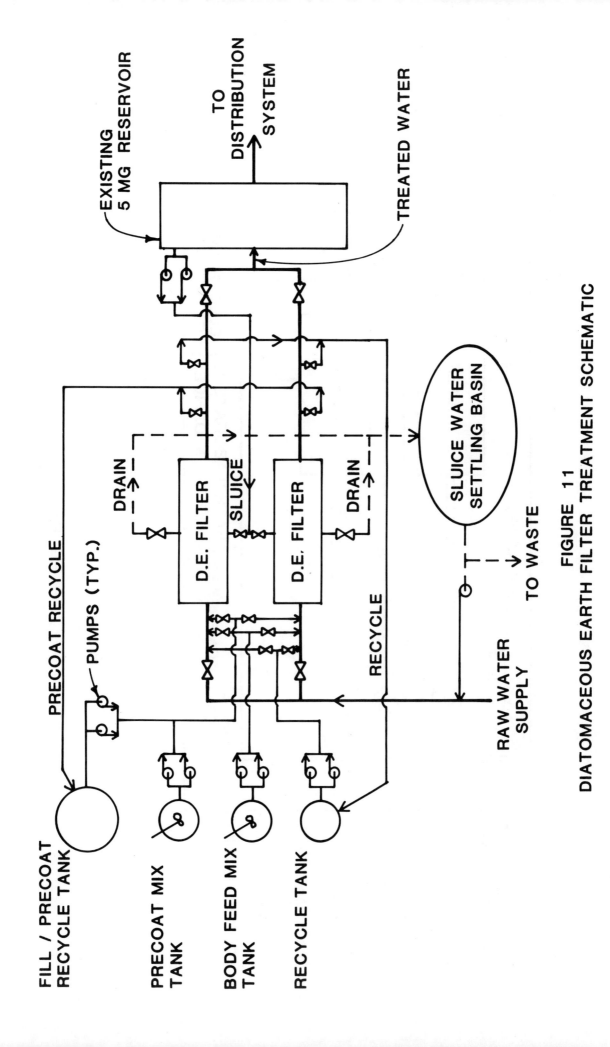

FIGURE 11
DIATOMACEOUS EARTH FILTER TREATMENT SCHEMATIC

237

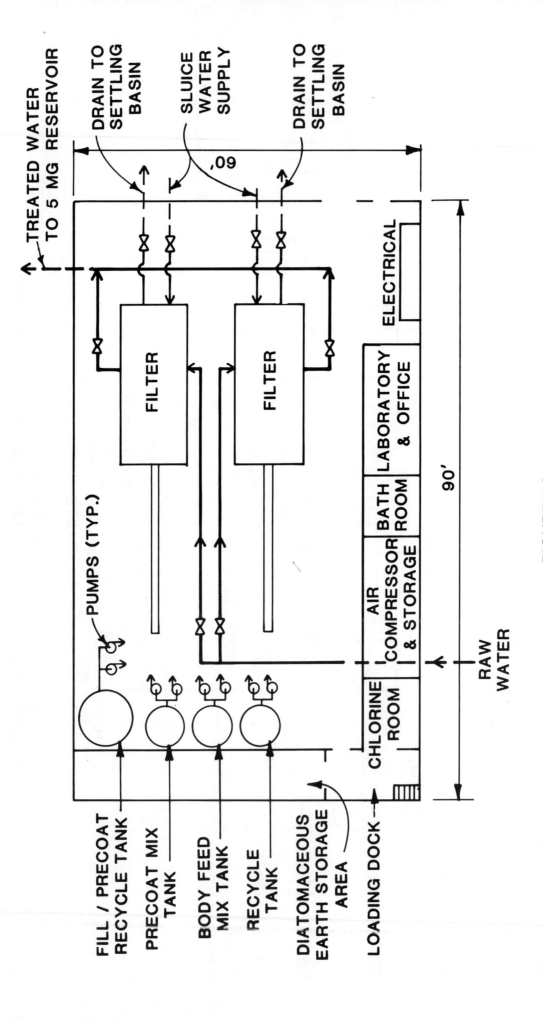

FIGURE 12
DIATOMACEOUS EARTH FILTER PLANT PREDESIGN LAYOUT

EPA Project Summary

Effect of Halogens on *Giardia* Cyst Viability

Ernest A. Meyer

Ernest A. Meyer is with the University of Oregon Health Sciences Center, Portland, OR 97201.

John C. Hoff is the EPA Project Officer (see below).

The complete report, entitled "Effect of Halogens on Giardia Cyst Viability," (Order No. PB 82-102 294) will be available only from:

National Technical Information Service
5285 Port Royal Road
Springfield, VA 22161
Telephone: 703-487-4650

The EPA Project Officer can be contacted at:

Municipal Environmental Research Laboratory
U.S. Environmental Protection Agency
Cincinnati, OH 45268

The report summarized here describes the results of a study in which the effect of halogens on the cysts of *Giardia lamblia* was tested. Halogens were applied under conditions commonly used in drinking water disinfection. The specific effect measured was the ability of the *Giardia* cyst to excyst, under controlled conditions after exposure to halogen; this was compared with the excystation ability of untreated cysts from the same source. Earlier studies of *Giardia* cyst inactivation employed a dye exclusion method; this has been shown to be a less reliable indicator of *Giardia* cyst viability than the excystation procedure.

In one set of experiments, chlorine was tested under a variety of conditions including chlorine concentration, temperature, pH, and chlorine-cyst contact time. Within the range of variables studied, the ability of cysts to excyst after treatment was affected by each of these variables. Percent excystation decreased with (a) increasing chlorine concentration, (b) increasing temperature, (c) decreasing pH, and (d) increasing chlorine-cyst contact time.

In a second set of experiments, six small-quantity water disinfection methods were tested. In every case, directions recommended for the application of the method were strictly followed. Two water qualities (cloudy and clear) and two water temperatures (3° and 20°C) were employed. At 20°C, all of the methods proved effective. At 3°C in cloudy water, however, one method was less than completely effective; and, in clear water, four methods failed to inactivate all of the cysts. The results of these experiments underline the importance of considering water temperature, chlorine demand, contact time, and pH when employing halogens for the disinfection of drinking water.

This Project Summary was developed by EPA's Municipal Environmental Research Laboratory, Cincinnati, OH, to announce key findings of the research project that is fully documented in a separate report of the same title (see Project Report ordering information at back).

Introduction

Parasitic protozoan flagellates in the genus *Giardia* are distributed worldwide and are now the most commonly reported human intestinal parasites in the United States and Great Britain. Host-to-host transmission of *Giardia* occurs when viable cysts, excreted in the feces of an infected host, are ingested directly or in the food or water of another host. The subsequent growth of these organisms in the small intestine frequently results in giardiasis, a disease whose symptoms, including diarrhea, malaise, abdominal cramps and weight loss, may persist for a month or more. Until a decade or so ago, giardiasis was considered to be a disease acquired by Americans outside the United States in parts of the world (particularly in tropical and subtropical areas) where sanitary standards were in need of improvement. Travel to such areas still is the most probable explanation for a significant number of the cases of giardiasis diagnosed in this country.

Since about 1970, evidence has been accumulating that giardiasis can be spread in another way: in epidemic form, in temperate and cold climates. The vehicle for spreading epidemic giardiasis is drinking water. Waterborne giardiasis has now been reported from a number of states in the United States including New York, New Hampshire, Pennsylvania, Colorado, California, Utah, Oregon, and Washington. The disease has been acquired by drinking water from community supplies as well as from untreated sources in recreation areas. Evidence strongly suggests that many giardiasis infections have been acquired in Leningrad in the Soviet Union by drinking the water in that city.

Humans are not the only hosts for *Giardia*, which in nonhuman hosts are morphologically indistinguishable from those that parasitize humans. Although organisms in this genus were long considered to be strictly host-specific, it is now known that this is not the case. Available data suggest that *Giardia* organisms from man are capable of infecting lower animals. There is also evidence to suggest that, conversely, at least some of the *Giardia* parasitic in lower animals can infect humans.

The existence of animal reservoirs of these organisms capable of infecting man simplifies the explanation of how this disease is acquired (a) in areas far from human activity and (b) from water collected from watersheds from which humans have been excluded.

Because *Giardia* cyst survival has proven difficult to study, we don't know how to treat water to ensure that any *Giardia* cysts will be destroyed. The presently used chemical methods of water disinfection are based not on killing *Giardia* but on killing *Entamoeba* cysts. Recently, questions have arisen concerning the ability of these recommended chlorine concentrations to kill *Giardia* cysts.

The development of a method to induce the excystation of *Giardia* cysts has made possible, for the first time, a relatively simple, reliable method of determining *Giardia* cyst viability and, thus, the ability to determine whether a given procedure kills these cysts. The method has recently been used to determine that in cold water, *Giardia* cysts can remain viable for upwards of 2 months.

The report summarized here describes the results of a study to determine the effect on *Giardia* cyst viability.

Effect of Chlorine on *Giardia* Cyst Viability

The variables employed in this study, in addition to chlorine concentration, were pH, contact time, and temperature. By determining the percent of a given *Giardia* cyst suspension capable of excysting after different periods of exposure to chlorine under a variety of experimental conditions, and plotting the resultant data, it was possible to generate a number of curves that describe the rate of *Giardia* cyst inactivation under varying conditions. Within the range of variables studied, the percent of excystation decreased with (a) increasing chlorine concentration, (b) increasing temperature, (c) decreasing pH, and (d) increasing chlorine-cyst contact time. These curves, a more detailed description of these experiments, and a discussion of the significance of the results, were recently published in a journal article to which the interested reader is referred (Jarroll, E.L., A.K. Bingham, and E.A. Meyer. Effect of chlorine on *Giardia lamblia* cyst viability. Applied and Environmental Microbiology *41*:483-487, 1981.).

Effect of Six Small-Quantity Water Disinfection Methods on *Giardia* Cyst Viability

Of the six disinfection methods tested, two (Halazone* and bleach) employed a form of chlorine, and four (Globaline, EDWGT, and elemental iodine in tincture and in saturated form) involved some form of iodine. Because

the recommended amount of halogen to be added, or the recommended contact time, or both varied with some methods according to the water turbidity or temperature, cyst survival using each method was determined using both clear and cloudy water, at 3° and at 20°C.

Two of the methods, one chlorine-based (Halazone) and one iodine-based (EDWGT) inactivated all of the cysts under all test conditions. Three other methods (bleach, Globaline, and tincture of iodine) only failed to inactivate all cysts under one set of experimental conditions: in clear water at 3°C. Finally, one method (saturated iodine) inactivated all cysts in both water samples at 20°C, but failed to inactivate all of the cysts in either clear or cloudy water at 3°C.

These results suggest that *Giardia* cysts can be inactivated by halogen-containing compounds under appropriate conditions; at low water temperatures, however, increased contact time, increased concentrations of halogen, or both may be required.

Details of these experiments have been published in two journal articles to which the reader is referred. (1) Jarroll, E.L., A.K. Bingham, and E.A. Meyer. *Giardia* cyst destruction: effectiveness of six small-quantity water disinfection methods. Am. J. Trop. Med. Hyg., 29, 8-11, 1980. (2) Jarroll, E.L., A.K. Bingham, and E.A. Meyer. Inability of an iodination method to destroy completely *Giardia* cysts in cold water. West. J. Med. 132:567-569, 1980.

The full report was submitted in fulfillent of Grant No. R-806032 by the University of Oregon Health Sciences Center, Portland, OR 97201, under the sponsorship of the U.S. Environmental Protection Agency.

---

*Mention of trade names or commercial products does not constitute endorsement or recommendation for use.

Index

Analytical techniques, 7, 207
  immunofluorescence, 42–46
  microscopic examination, 34
  sampling procedures, 6–7, 32–34,
    50–53, 61–63, 144–45, 207
  sucrose flotation, 33–34
  zinc sulfate flotation, 33
  See also Detection; Methods
Anthracite media, 136–38, 140–41
Atabrine, 5–6

Bacterial spiking, 172, 175
Beaver, 5, 91, 101
Berlin, N.H., giardiasis outbreak
  disease, 83
  epidemiology, 86–87
  Giardia sampler installation, 83–84
  overview, 81–82
  recommendations, 87–88
  system, 82–83
  treatment, 84–86

Camus, Wash., giardiasis outbreak
  epidemiology, 77
  implications, 80
  infection, 76–77
  overview, 75–76
  recommendations, 79
  remedial action, 79–80
  system, 76–78
  water quality, 78–79
Cartridge filters, 6–7, 9
Chlorine
  effect on cysts, 240
Coagulation, 9–11, 110–111
Colorado Department of Health,
  Water Quality Control Division
  Giardia sampling, 7
Connecticut water supplies
  Giardia occurrence, 100–102
  recommendations, 101–102
  watershed animals, 101
Cover slip technique, 7
Cross transmission, 103–105
Curing period, 9–10
Cyst stain
  See Eosin dye exclusion
Cysts, 4, 20–21
  concentration, 189
  counting methods, 145–47, 181
  cross transmission, 103–105
  dosing procedures, 143–44
  effect of chlorine on, 240
  effect of halogens on, 239–40
  measurement, 109–110, 164
  models, 118–19
  overview, 180–81, 239
  recommendations, 185
  recovery, 109–110
  sampling procedures, 6–7, 32–34,
    50–53, 61–63, 144–45, 207

sources, 142–43, 164, 215
  spiking, 137–38, 175
  viability, 5
  See also Filtration; Surrogate
    indicators; Treatment

Detection, 4, 24–25
  reference method, 37–39
  standard methods, 37
  swimming pool filter method, 36
  See also Analytical techniques; Methods
Diatomaceous earth filters, 6, 9, 12, 61–63,
  119–21, 128–29, 147, 150–53, 184–85
  alum coating, 190
  backwash, 13
  bacteria, 175, 189–90
  cyst concentration, 189
  diatomite, 118
  experimental procedures, 187–88
  field testing, 190
  grade of earth, 175, 189
  hydraulic loading rate, 189
  leak testing, 188
  operation, 11–13
  overview, 186–90
  Palisade, Colo., treatment, 217, 219–22
  precoat, 12
  process effectiveness, 188–98
  recommendations, 124, 177–78, 190
  small systems, 173–75
  temperature, 190
  test design, 187
  turbidity, 189–90
Disinfection processes, 13–14

Encystation, 8
Environmental Protection Agency
  Giardia lamblia workshop, 31–32
Eosin dye exclusion, 8
Excystation, 13–14

Filtration, 9–11, 111
  anthracite media, 136–38, 140–41
  diatomite, 118
  direct, 217–18, 220–21
  field tests, 151–52
  granular media, 117–18, 121–24, 149,
    152, 181–84
  overview, 117
  pilot, 148
  rapid rate, 9, 126–28, 147–48, 204–211
  recommendations, 124, 153
  slow sand, 9, 13, 128, 147, 149–50, 152,
    160, 162–65, 167–68, 191–200
  small system, 148–49
  See also Diatomaceous earth filters,
    Surrogate indicators
Flagyl, 5–6
Flocculation, 9–11, 110–111
Fluorescence, 56–57

Fuazolidone
  See Furoxone
Furoxone, 5–6

Giardia lamblia
  Colorado Dept. of Health survey, 94–95
  Connecticut water supplies, 100–102
  culture, 24
  forms, 4
  life cycle, 21
  literature review, 214–16
  mammals susceptible to, 5
  morphology, 20
  overview, 20, 214
  Palisade, Colo., treatment, 214–23
  pathogenesis, 21–22
  recommendations, 26, 105
  transmission, 21
  viability, 8, 215
  workshop, 31–32, 49–54, 56–60
  See also Analytical techniques; Cysts;
    Detection; Methods; Treatment
Giardiasis, 162
  Berlin, N.H., outbreak, 81–88
  Camus, Wash., outbreak, 75–80
  cause, 4, 215
  chronic, 5
  common factors, 95–96
  epidemiology, 22–23, 77, 86–87
  factors associated with outbreaks, 8–9
  Highland (Colo.) Water and Sanitation
    District outbreak, 97–99
  immunity, 6
  McKeesport (Pa.) Municipal·Water
    Authority outbreak, 136–38, 140–41
  outbreaks, 4, 23–24, 109, 215
  prevention, 22
  symptoms, 4–5, 22, 215
  treatment, 5–6, 22
  Washington State outbreak, 90–92
  See also Waterborne disease
Grab samples, 6
Granular media filtration, 117–18, 121–24,
  149, 152, 181–84

Halogens
  effect on cysts, 239–40
Hemocytometer, 7
Highland (Colo.) Water and Sanitation
  District outbreak
  evaluation, 98–99
  findings, 97–98
  overview, 97
  recommendations, 99
Humatin, 5–6

Immunofluorescence
  manifold, 45
  microscope, 45–46
  overview, 42–43

reagents, 43–45
recommendations, 46
Infectivity, 8

Malabsorption, 5
McKeesport (Pa.) Municipal Water
 Authority giardiasis outbreak
   anthracite media, 136–38, 140–41
   cyst spiking, 137–38
   overview, 136
   pilot plant, 137
   plant design, 136
   recommendations, 141
   research, 136–37
   sedimentation results, 138, 140–41
Membrane filters, 6
Methods
   antibody identification, 57–58
   concentrate examination, 53–54
   equipment, 54, 56
   fluorescence, 56–57
   overview, 49
   quality control, 58–60
   recommendations, 60
   reference, 37–39
   results reporting, 53–54
   sample collection, volume, transport,
     and storage, 50–51
   sample processing to concentrate state,
     51–53
   See also Analytical Techniques; Detection
Metronidazole
   See Flagyl
Microscopic observation, 34, 39, 42–43

Palisade, Colo. Giardia treatment
   diatomaceous earth filtration, 217, 219–20
   direct filtration, 217–18, 220–22
   literature review, 214–16
   particle distribution sampler, 218
   pilot testing, 216–19
   predesign and economic analysis, 220–22
   recommendations, 222–23
Paromomycin
   See Humatin
Particle counting, 7
Particle distribution sampler, 218

Quinacrine
   See Atabrine

Rapid rate filtration, 9, 147–48, 149
   bacteria, 127, 207
   coagulant selection and dosage, 209
   experimental design, 206
   field testing, 127
   filtration modes, 209–210
   hydraulic loading rate, 210
   media, 210
   overview, 204–206
   particles, 127, 207
   pilot plants, 206
   recommendations, 211
   removals, 207–209
   surrogate indicators, 210–11
   temperature, 210
   turbidity, 126–27, 207
Reference method
   cultivation, 39
   evaluation, 39

extract concentration, 38–39
   extraction, 38
   microscopic examination, 39
   results reporting, 39
   sampling, 37–38
Research needs, 14–15
Rocky Mountain region giardiasis
   outbreaks, 4
   reasons for high giardiasis risk, 4–5

Sand filters, 6
Sedimentation, 136–38, 140–41
Slow sand filtration, 9, 13, 128, 147,
 149–50, 152, 160, 162
   apparatus, 162, 193
   bacteria, 165, 172, 198–99
   design, 160, 192–93, 197–98
   hydraulic loading rate, 164–65, 195, 199
   microbiological conditions, 165, 167,
     196–97
   operation, 160, 162–63, 192–95
   overview, 160, 191–92
   plant operation, 176
   recommendations, 167–68, 176–77, 200
   recovery of filter after scraping, 171–72
   removals, 164, 195, 197–98
   sand bed depth, 198–200
   sand size, 198, 200
   small systems, 171–73
   temperature, 165, 198–99
   test protocol, 163–64
Small systems
   diatomaceous earth filtration, 173–75
   overview, 169–71
   slow sand filtration, 171–73
Solid particle samplers, 61–63
Sucrose flotation, 33–34
Surrogate indicators
   assessment, 129
   diatomaceous earth filtration, 128–29
   experiments, 125–26
   overview, 125
   rapid rate filtration, 126–28, 210–11
   recommendations, 129–30
   slow sand filtration, 128
Swimming pool filter method, 36, 62

Traveler's disease, 20
Treatment, 215–16
   Berlin, N.H., outbreak, 84–86
   coagulation, 9–11
   curing period, 9
   diatomaceous earth filtration, 11–13
   disinfection processes, 13–14, 25–26
   filtration, 9–11, 26
   filtration plant design, 10–11
   flocculation, 9–11
   particulate removal processes, 9
   plants, 180–85
   research efforts, 8
   sedimentation, 136–38, 140–41
   slow sand filtration, 13
   turbidity levels, 9–10
Trihalomethane precursors, 173, 175
Trophozoite, 4, 20
Turbidity levels, 9–10, 126–27, 173,
 189–90, 207

Washington State giardiasis outbreak
   beaver, 91
   findings, 91–92

human case follow-up, 91
   recommendations, 92
   studies, 90–91
Waterborne disease
   Colorado Dept. of Health survey, 93–94
   community vs. noncommunity systems,
     69–71
   geographic distribution, 69
   illness, 71–72
   outbreaks, 68–71
   overview, 67–68, 142
   recommendations, 74
   temporal distribution, 69
   water system deficiencies, 72–74
   See also Giardiasis
Watershed animals, 5, 101
   See also Beaver

Zeta potential, 110
Zinc flotation, 33